Hydrogen Technologies in Decarbonizing Industries and Energy Systems

**IEEE Press**
445 Hoes Lane
Piscataway, NJ 08854

# Hydrogen Technologies in Decarbonizing Industries and Energy Systems

*Vikas Khare*
*School of Technology, Management and Engineering, NMIMS, Indore, India*

*Sanjeet K. Dwivedi*
*Senior Consultant @ Wind Energy Business RDT Engineers, Denmark*

*Monica Bhatia*
*School of Business Management, NMIMS, Indore, India*

*K. K. Pant*
*Director, IIT Roorkee, India*

IEEE*Press*

WILEY

Published by John Wiley & Sons, Inc., Hoboken, New Jersey.



***Library of Congress Cataloging-in-Publication Data Applied for:***

**Hardback ISBN**: 9781394350063

Cover Design: Wiley
Cover Image: © nongkran_ch/stock.adobe.com

Set in 9.5/12.5pt STIXTwoText by Straive, Pondicherry, India

## Dedication

*This book is humbly dedicated to the spirit of innovation, perseverance, and responsibility that defines humanity's pursuit of a sustainable future. The subject of hydrogen technologies and their role in decarbonizing industries and energy systems is not just a scientific or technical endeavor it is a collective mission to preserve our planet for generations yet to come.*

*We dedicate this work, first and foremost, to the countless researchers, scientists, engineers, and policymakers across the globe who have tirelessly advanced the knowledge and applications of hydrogen. Their relentless efforts in laboratories, industries, and policy arenas have transformed hydrogen from a concept into a reality that can reshape the foundations of global energy systems. Without their groundbreaking discoveries, bold experiments, and enduring commitment, the possibility of a hydrogen-powered, low-carbon world would have remained only a dream.*

*This book is also dedicated to the industrial leaders, entrepreneurs, and innovators who dare to embrace change, challenge the status quo, and adopt cleaner pathways despite economic and operational barriers. Their courage and vision are essential in demonstrating that sustainability is not merely a choice but a necessity. By integrating hydrogen technologies into complex industrial ecosystems, they are creating examples that will inspire others to follow.*

*Equally, this work is dedicated to the educators and students who keep alive the curiosity and passion for knowledge. To the teachers who instill the importance of science, technology, and environmental stewardship, and to the learners who will take these lessons into the future, may this book serve as a resource and an inspiration. It is through education and the transfer of knowledge that society can equip itself to face the global challenge of climate change with resilience and wisdom.*

*We dedicate this book to the communities most vulnerable to the impacts of climate change—those who often bear the brunt of environmental degradation yet contribute the least to its causes. It is for them that sustainable technologies must deliver not only reduced emissions but also equitable growth, improved quality of life, and inclusive opportunities. May hydrogen-based solutions become a bridge toward environmental justice as well as technological progress.*

*On a more personal note, this book is dedicated to our families, friends, and mentors who have stood by us throughout this journey. Their unwavering support, patience, and encouragement provided the strength needed to pursue this work with clarity and determination. Behind every page lies the silent contribution of those who motivated us, believed in us, and reminded us of the purpose behind our efforts.*

*Finally, this book is dedicated to future generations. May they inherit a world where clean energy is not an aspiration but a reality, where industries and energy systems operate in harmony with the environment, and where humanity continues to progress without compromising the health of the planet. Hydrogen, as one of nature's simplest and most abundant elements, has the power to become a symbol of renewal, resilience, and hope. May this work, in some small way, contribute to that brighter tomorrow.*

# Contents

## About the Authors

**DR. VIKAS KHARE**

Dr. Vikas Khare is an Associate Dean, School of Technology, Management and Engineering, NMIMS, Indore, India. He completed his postdoctoral research in 2023. Prior to that, he earned his PhD from the National Institute of Technology Bhopal in 2016 and M. Tech (Honors) in Energy Management from the School of Energy and Environment and Research at DAVV Indore. He also holds an MBA in Information Technology and Management from NGASCE NMIMS Mumbai and a B.E. in Electrical Engineering from Government Engineering College Sagar. He has done Advanced Certification in Data Science and Artificial Intelligence from the Indian Institute of Technology Madras. He is a certified energy manager under the Bureau of Energy Efficiency India and a Fellow Member of the Scholars Academic and Scientific Society, India. His research interests primarily revolve around renewable energy systems, data analysis, artificial intelligence, optimization techniques, game theory, and Big Data. He has authored **14 books**, including 6 published by Elsevier in the USA, 1 by Walnut Publication, and 4 by Dhanpat Rai Publication. Notably, his book titled "Tidal Energy System: Design, Optimization, and Control" served as the basis for designing an M.Tech course in the energy department at IIT Jodhpur. Additionally, he has contributed **15 book chapters** and published **42 (24 SCI/ABDC)** papers in Scopus-indexed international journals, along with presenting 10 papers in international conferences, garnering a Google citation count of over 2800. He has served as a keynote speaker in various Faculty Development Programs (FDP) and international and national conferences. Furthermore, he serves as a reviewer for esteemed journals such as the International Journal of Renewable and Sustainable Energy (Elsevier), Journal of Energy (Elsevier), and International Journal of Sustainable Energy (Taylor and Francis). Recognized for his contributions, he has received several awards, including the Best Faculty Award at STME, NMIMS Indore in 2022, the Best Faculty Award at Medicaps University in 2016, and the Best Paper Award at an Elsevier-affiliated international conference in 2024.

**DR. SANJEET KUMAR DWIVEDI**

Dr. Sanjeet Kumar Dwivedi is a Fellow of the IET (UK) and working as a Senior Consultant @ Wind energy business RDT Engineer, Denmark since 2008. Prior to this, he was an electrical engineer in Larsen and Toubro, an infrastructure company in India (1991–1992), Electrical Engineer in CPWD, Government of India (GoI). He worked as senior faculty member of the Department of Technical Education, MP, India (1993–2006), and Dean R&D, Government Engineering College SAGAR MP, India (2006–2008). He has been awarded two master degrees, first one from IIT Roorkee, ME (Gold Medal) in Power Apparatus and Drives and second one from South Denmark University, M.Sc. Engineering in Innovation and Business. He has completed his Ph.D. degree in Green Technologies from IIT Delhi, India. He also completed his executive leadership education from MIT Boston. He was honored with adjunct professor at Curtin University, Perth, Australia (2016–2018). He is a Member of Faculty Board of South Denmark University in the Department of Innovation and Business. He has authored more than 40 technical papers and holds 11 international patents, three business trade books, and three engineering research books published by Academic Press (UK) and IET Press (UK). He is an Advisory Board Member of International Journal of Power Electronics (IJPE) and Associate Editor of the IEEE Transaction on Industrial Electronics (IEEE TIE). He has given invited presentations, organized, and chaired special sessions in several IEEE and European Power Electronics conferences around the globe. Previously he worked as Technical Editor of IEEE/ASME transaction on Mechatronics, Associate Editor of IET (United Kingdom) Power Electronics Journal, and Associate Editor of Korean Journal of Power Electronics (JPE). He is a recipient of Merit Award from Institution of India IE(I) (2006) for his research publication on permanent magnet machines. He was also awarded with 9th Man on the Moon Global Innovation Award from the CEO and President of Danfoss (2015) and another prestigious recognition as winner of IETE-Bimal Bose Award (2017) for outstanding contribution in power electronics and drive.

**DR. MONICA BHATIA**

Dr. Monica Bhatia, PhD, is a faculty member at the School of Business Management, NMIMS, India. She holds an MBA and a PhD in Marketing and is certified by NET, NCFM, and IRDA. She has extensive professional experience in training and business development within both corporate and educational sectors, covering a wide range of areas such as renewable energy, operational management, data analysis, cognitive science, career development, marketing management, communication skills, interpersonal relations, and workplace professionalism. She has conducted life-skills training for students and trainers from various nationalities and with diverse accents and has contributed to consulting projects on e-commerce, product strategy, marketing strategy, and business processes.

Additionally, she has authored numerous papers in reputable journals and is the author of the book "Cognitive Science, Computational Intelligence, and Data Analytics: Methods and Applications with Python," published by Elsevier.

## DR. KAMAL KISHORE PANT

Prof. Kamal Kishore Pant received his PhD (Chemical Engineering) degree from the Indian Institute of Technology Kanpur in 1997. He took over as the Director of the Indian Institute of Technology Roorkee on October 12, 2022. He previously held the positions of Petrotech Chair and Federation of Indian Petroleum Industries (FIPI) Chair Professor, Chairman GATE & JAM, Chairman, Library, Head of the Department of Chemical Engineering, and Dean Faculty at the Indian Institute of Technology Delhi. He is an Adjunct Faculty at the University of Saskatchewan in Canada. He is also a Fellow of several national and international academies, which include the National Academy of Science India (NASI) and the Indian National Academy of Engineering (INAE). His area of interest is futuristic technologies of national and international importance, e.g., hydrocarbon conversion processes, biomass conversion and biofuel production, $CO_2$ conversion technologies, e-waste, and plastic management.

# Preface

The transition toward sustainable energy has emerged as one of the most pressing imperatives of the 21st century, as industries and societies confront the dual challenges of climate change and resource security. Amid this transformation, hydrogen particularly green hydrogen has gained recognition as a pivotal enabler of decarbonization across industrial sectors and energy systems. Its versatility, scalability, and potential to integrate with renewable sources position hydrogen technologies as essential components of a cleaner, more resilient global energy framework.

This book, *Hydrogen Technologies in Decarbonizing Industries and Energy Systems*, is designed to provide a comprehensive understanding of the technical, managerial, and operational dimensions of hydrogen integration. It blends technological insights with management perspectives, offering readers a holistic approach to navigating the complexities of hydrogen adoption in the evolving energy landscape.

The structure of the book has been thoughtfully developed to guide readers step by step through the multifaceted dimensions of hydrogen technologies:

- **Chapter 1: Introduction to Techno-Management Revolution of Energy** sets the foundation by exploring the convergence of technology and management principles in reshaping the energy ecosystem.
- **Chapter 2: Understanding Green Hydrogen: Production Methods and Technologies** provides an in-depth examination of the processes, tools, and innovations driving the sustainable generation of hydrogen.
- **Chapter 3: The Role of Hydrogen in Energy Transition** situates hydrogen within the broader context of global energy transformation, highlighting its potential to bridge renewable sources and industrial demand.
- **Chapter 4: The Urgency of Decarbonization: Environmental and Climate Imperatives** underscores the critical importance of hydrogen in addressing carbon emissions and mitigating the escalating impacts of climate change.

- **Chapter 5: Operations and Supply Chain Management of Green Hydrogen** delves into the practicalities of building and optimizing supply chains for hydrogen production, storage, and distribution.
- **Chapter 6: Inventory Control and Total Quality Management** concludes the discussion by focusing on the role of inventory strategies, quality assurance, and risk management in ensuring safe, efficient, and sustainable hydrogen systems.

The aim of this book is not merely to present hydrogen as a technological solution but to position it as part of an integrated framework that includes management strategies, policy considerations, and sustainability imperatives. By bringing together these dimensions, this work seeks to serve as a strategic resource for researchers, policymakers, industry professionals, and students who aspire to contribute to the global clean energy revolution.

It is my hope that the insights shared in this volume inspire deeper engagement, critical thinking, and innovative solutions in the pursuit of a decarbonized and sustainable future powered by hydrogen technologies.

# Acknowledgments

The completion of this book, *Hydrogen Technologies in Decarbonizing Industries and Energy Systems*, has been a journey of exploration, collaboration, and inspiration. It would not have been possible without the invaluable support, encouragement, and contributions of many individuals and institutions who guided us along the way.

First and foremost, we express our deepest gratitude to the academic and research communities whose pioneering work laid the foundation for this book. The vast body of knowledge developed by researchers, scientists, and engineers in the fields of hydrogen technologies, renewable energy systems, and industrial decarbonization has been a constant source of learning and insight. Their dedication to advancing sustainable technologies provided both the inspiration and the intellectual framework on which this book is built.

Our sincere thanks go to our colleagues and peers, who offered valuable discussions, feedback, and suggestions at different stages of this project. Their insights helped sharpen our ideas, broaden our perspectives, and refine the content of this book to ensure it remains relevant and impactful. Special appreciation is extended to reviewers and mentors who took the time to critically assess our drafts, ensuring both academic rigor and practical applicability.

We are also grateful to our institutions, whose encouragement and support made this work possible. The infrastructure, academic environment, and resources provided by our universities and research centers have been instrumental in carrying out the research and writing required for this book. Their commitment to promoting research and innovation in sustainable energy has given us the platform to pursue this endeavor with confidence and purpose.

We extend our heartfelt appreciation to the publishers for their professionalism, guidance, and patience throughout the process of bringing this book to publication. Their expertise in shaping manuscripts into polished works has been invaluable, and their belief in the relevance of this subject has motivated us greatly.

On a personal note, we owe immense gratitude to our families and loved ones. Their unwavering patience, encouragement, and understanding gave us the strength to persevere through the challenges of research and writing. The countless hours spent in preparing this manuscript were made possible by their silent sacrifices and continuous support.

Lastly, we would like to thank the future generations of researchers, practitioners, and students who will carry forward the mission of decarbonization. It is our hope that this book serves not only as a source of knowledge but also as an inspiration to innovate and act for the betterment of our planet.

To all who, in one way or another, contributed to the realization of this book we offer our sincere gratitude.

# 1

# Introduction to Techno-Management Revolution of Energy

## 1.1 Energy Landscape Before the Revolution

In various fields such as physics, chemistry, biology, and computer science, an "energy landscape" refers to a conceptual model used to represent the potential energy of a system as it evolves over various configurations or states. It visualizes how the energy of a system changes as it transitions between different states or conformations. In the context of energy landscapes, "states" refer to different configurations or arrangements that a system can occupy. These states could represent different molecular conformations, atomic arrangements, or configurations of a system. Each state in the landscape is associated with a specific energy level. This energy level represents the potential energy of the system when it is in that particular state. The arrangement of states and their energy levels in the energy landscape provides information about the structure and dynamics of the system [1, 2]. The topology of the landscape describes the relationships between different states and the barriers that must be overcome to transition between them. States that are located at the bottom of the energy landscape are considered more stable, as they correspond to lower energy levels. Conversely, states located at higher energy levels are less stable. The energy landscape provides insights into the processes by which a system transitions between different states. Transitions between states often involve overcoming energy barriers, which are represented by higher energy regions in the landscape. Figure 1.1 presents the parameters of the energy landscape.

The "energy landscape" refers to the overall structure, dynamics, and sources of energy within a particular system or society. Before the revolution, which revolution is unspecified, but assuming a general context, such as the industrial revolution or a major technological or societal upheaval, the energy landscape was likely quite different from what it became afterward. Before any major revolution,

*Hydrogen Technologies in Decarbonizing Industries and Energy Systems*, First Edition.
Vikas Khare, Sanjeet K. Dwivedi, Monica Bhatia, and K. K. Pant.

Published 2026 by John Wiley & Sons, Inc.

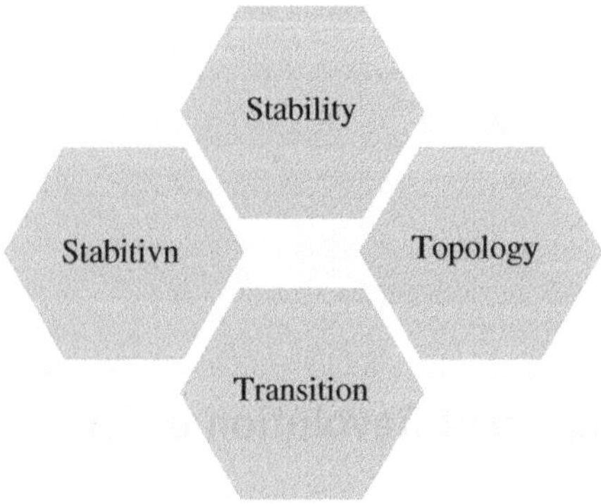


**Figure 1.1** Parameters of the Energy Landscape

societies typically relied heavily on traditional sources of energy such as biomass (wood, dung), animal power (horses, oxen), and human labor. These sources were limited in quantity and often inefficient in their use. Industrialization, if it had begun at all, would have been in its early stages. Energy needs for industrial processes were relatively low compared to later stages when mass production became the norm. Depending on the period and region, fossil fuels like coal may have started to play a role in the energy landscape, particularly in regions where coal deposits were accessible. However, their utilization might not have been widespread or efficient. Infrastructure for energy distribution and utilization would have been rudimentary compared to later periods [3, 4]. Electricity, if available, would have been limited to certain areas and industries. In agrarian societies, the majority of energy consumption would have been in agriculture for tasks such as plowing, planting, harvesting, and processing crops. Energy production would have been largely localized, with communities relying on nearby resources for their energy needs. This would have limited the scale of energy consumption and production. The energy landscape would have been closely tied to social structures, with energy production often controlled by the ruling class or wealthy elites. Access to energy would have been unevenly distributed among different social groups. Before a major revolution, the energy landscape would have been characterized by traditional, localized, and inefficient energy sources and technologies. It would have been a time of transition, with emerging technologies and societal changes laying the groundwork for the transformation that would follow the revolution. Energy landscapes are widely used in various scientific disciplines to study the behavior of complex systems, such as protein folding in biology, phase transitions in physics and materials science, and optimization problems in computer science. By analyzing the energy landscape, researchers can gain valuable insights into the structure, dynamics, and behavior of these systems [5, 6].

### 1.1.1 Renewable Energy Sources

Traditional energy sources refer to the ancient or historical methods that human societies have used to fulfill their energy needs. These sources were typically based on natural resources readily available in the local environment. Figure 1.2 presents the types of renewable energy sources.

#### 1.1.1.1 Biomass Power

A biomass energy system is a renewable energy system that converts organic materials, often derived from plants or animals, into usable energy sources, such as heat, electricity, or liquid fuels. Biomass energy systems utilize the energy stored in organic matter through processes such as combustion, gasification, fermentation, or anaerobic digestion. Biomass energy is derived from organic materials such as wood, crop residues, animal dung, and other plant and animal matter. It was one of the earliest sources of energy used by humans for heating, cooking, and lighting. Biomass remains an important energy source in many parts of the

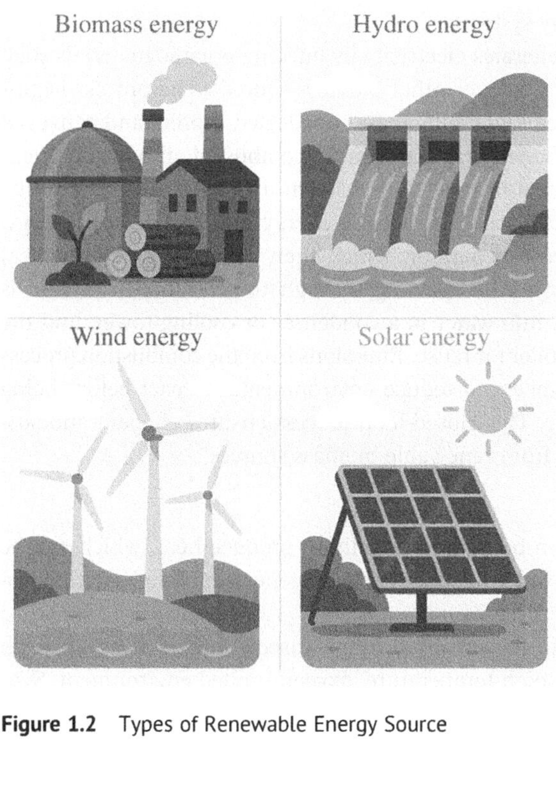


**Figure 1.2** Types of Renewable Energy Source

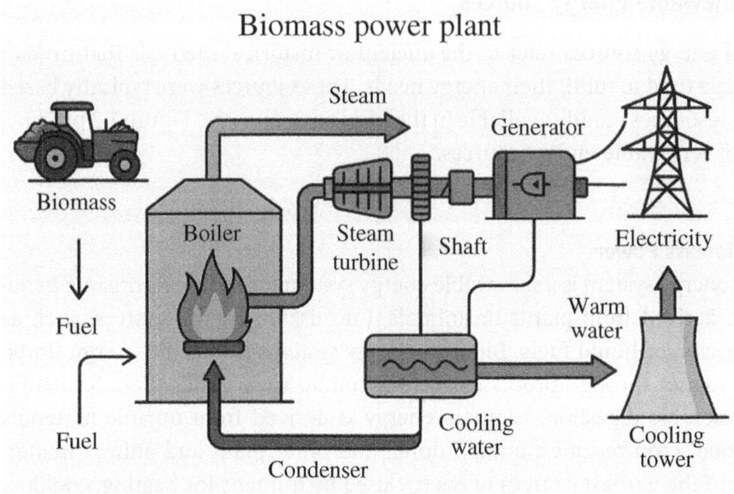


**Figure 1.3** Layout of Biomass Energy System

world, particularly in rural areas of developing countries [7, 8]. Figure 1.3 shows the layout of biomass energy system.

A biomass power plant generates electricity by burning organic materials such as wood chips, agricultural waste, or other biomass sources. The process begins with the fuel-handling system, where biomass is collected, stored, and conveyed to the boiler. Inside the boiler, the biomass is combusted, producing high-temperature steam. This steam is directed to a steam turbine, which rotates due to the force of the steam and converts thermal energy into mechanical energy. The turbine is connected to a generator, which then converts the mechanical energy into electrical energy. After passing through the turbine, the steam is cooled and condensed back into water in a condenser or cooling tower, and the water is cycled back to the boiler for reuse. Emissions from the combustion process pass through filters or scrubbers to reduce environmental impact before being released through a chimney. This closed-loop process ensures efficient and sustainable energy production from renewable biomass sources.

➢ **Conversion Processes**

**Combustion:** Biomass can be burned directly to produce heat, which can be used for space heating, water heating, or steam generation for electricity production or industrial processes.

**Gasification:** Biomass can be converted into a gaseous fuel (syngas) through partial combustion in a high-temperature, oxygen-limited environment. Syngas can be used to generate electricity, produce heat, or as a precursor for liquid fuels and chemicals [9].

**Fermentation:** Biomass can be fermented to produce biofuels such as ethanol and biodiesel, which can be used as transportation fuels or blended with conventional fuels.

**Anaerobic Digestion:** Biomass can undergo anaerobic digestion, a biological process in which microorganisms break down organic matter in the absence of oxygen, producing biogas (a mixture of methane and carbon dioxide) that can be used for electricity generation, heating, or vehicle fuel.

➢ **Energy Applications**

Biomass energy can be utilized in various applications, including:

**Electricity Generation:** Biomass power plants produce electricity through combustion, gasification, or anaerobic digestion.

**Heat Generation:** Biomass boilers and stoves can provide heat for residential, commercial, and industrial applications.

**Transportation:** Biofuels derived from biomass, such as ethanol and biodiesel, can be used as substitutes for gasoline and diesel in vehicles.

**Environmental Considerations:** Biomass energy systems are often considered renewable and carbon-neutral, as the carbon dioxide emitted during combustion or conversion is offset by the carbon dioxide absorbed by the plants during photosynthesis. However, the sustainability of biomass energy depends on factors such as the type of feedstock, land use practices, and the efficiency of conversion processes.

> ***Key: Nonconventional energy sources:*** *Renewable energy sources such as solar, wind, biomass, and hydro energy, which offer sustainable alternatives to traditional fossil fuels, contributing to a more diverse and environmentally friendly energy mix.*

#### 1.1.1.2 Hydro Power

Water power, or hydropower, has been used for thousands of years to drive mechanical processes, such as milling grain and sawing wood. Water wheels and watermills were common in many ancient civilizations, harnessing the energy from flowing water to perform tasks that would otherwise require human or animal labor. Water power, also known as hydro power or hydroelectric power, refers to the generation of electricity from the kinetic energy of moving water. It is one of the oldest and most widely used renewable energy sources, with a long history of human utilization for various purposes, such as mechanical power and electricity generation. Hydro power harnesses the energy of flowing water, typically from rivers, streams, or man-made reservoirs, and converts it into electrical energy through the use of turbines and generators [10]. Figure 1.4 shows the layout of the hydro energy system.

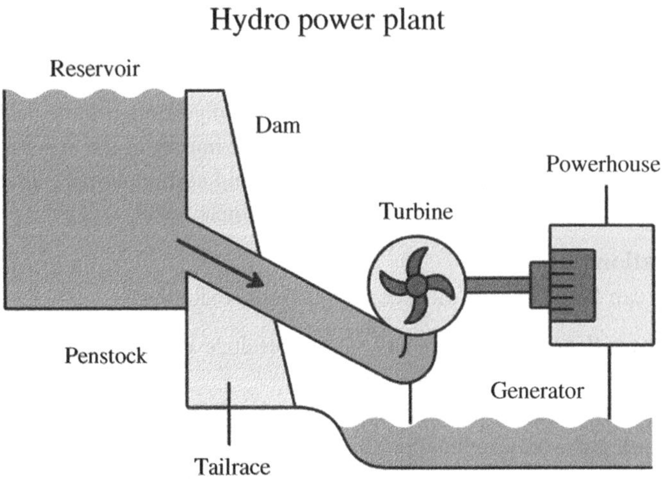


**Figure 1.4** Layout of the Hydro Energy System

Following are some key aspects of water power and hydro power:

**Hydroelectric Power Plants:** Hydroelectric power plants are facilities designed to generate electricity from water flow. These plants typically consist of:

**Dam:** A structure built across a river or stream to create a reservoir or impoundment, which stores water and creates a height difference (head) between the upstream and downstream water levels.

**Generator:** A machine connected to the turbine that converts the mechanical energy into electrical energy through electromagnetic induction.

**Reservoir:** A large storage area that holds water at a high elevation. It serves as the main source of potential energy in a hydro power plant. The height of the water in the reservoir creates pressure, which is later converted into kinetic and mechanical energy as the water flows downward. The reservoir also helps regulate water availability based on electricity demand.

**Intake**: The opening through which water from the reservoir is allowed to enter the penstock. It typically includes gates and screens to control water flow and prevent debris, fish, or unwanted materials from entering the system. The intake ensures a clean and controlled supply of water to the turbines [11].

**Penstock:** A large, high-pressure pipe or tunnel that carries water from the intake to the turbines. As water flows down through the penstock, its potential energy is converted into kinetic energy due to the force of gravity. The design and slope of the penstock play a crucial role in determining the efficiency of energy transfer.

**Turbine:** A rotating mechanical device that is turned by the force of water flowing through it. The kinetic energy of water causes the turbine blades to spin, converting the energy into mechanical energy. There are different types of turbines (e.g., Francis, Kaplan, and Pelton) suited for various water heads and flow rates [12].

**Tailrace:** The channel or river section where the water exits after passing through the turbine and draft tube. It carries the used water back to the downstream side of the river, restoring its natural flow. Maintaining a clean and clear tailrace is essential for environmental balance and proper water circulation.

**Reservoir:** A large storage area that holds water at a high elevation. It serves as the main source of potential energy in a hydro power plant. The height of the water in the reservoir creates pressure, which is later converted into kinetic and mechanical energy as the water flows downward. The reservoir also helps regulate water availability based on electricity demand.

➢ **Types of Hydro Power Plants**

**Impoundment Hydroelectric Plants:** These plants utilize dams to store water in reservoirs, which is released as needed to generate electricity.

**Run-of-River Hydroelectric Plants:** These plants do not require large reservoirs and instead use the natural flow of rivers or streams to generate electricity. They typically have smaller environmental impacts compared to impoundment plants.

**Pumped Storage Hydroelectric Plants:** These plants use surplus electricity during off-peak hours to pump water from a lower reservoir to an upper reservoir. During peak demand periods, the stored water is released to generate electricity [13].

➢ **Advantages**

**Renewable and Sustainable:** Hydro power relies on the natural water cycle and does not produce greenhouse gas (GHG) emissions during operation.

**Reliable:** Hydroelectric power plants can provide a steady and predictable source of electricity, making them suitable for base-load or peaking power generation.

**Long Lifespan:** Hydro power infrastructure, such as dams and turbines, can have a long operational lifespan with proper maintenance.

➢ **Challenges and Considerations**

**Environmental Impact:** Large-scale hydroelectric projects can have significant environmental impacts, including habitat disruption, alteration of water flow regimes, and impacts on fish populations.

**Limited Site Availability:** Not all regions have suitable geography or water resources for hydro power development.

**High Initial Costs:** The construction of dams and hydroelectric facilities can require substantial upfront investment.

#### 1.1.1.3 Wind Power

Wind power has been utilized for centuries to propel sailing ships and grind grains in windmills. Windmills were particularly prevalent in regions with strong and consistent winds, where they provided a renewable source of mechanical energy for various industrial and agricultural purposes. Wind power refers to the generation of electricity using the kinetic energy of wind. It is a renewable energy source that has been harnessed for centuries for various purposes, such as milling grain and pumping water. In modern times, wind power is primarily utilized for electricity generation through wind turbines [14]. Figure 1.5 shows the layout of the wind energy system.

Following are some key aspects of wind power:

**Wind Turbines:** Wind turbines are devices that convert the kinetic energy of wind into mechanical energy, which is then converted into electricity. A typical wind turbine consists of the following components:

**Rotor Blades:** Aerodynamic blades that capture the energy of the wind.

**Nacelle:** The housing that contains the gearbox, generator, and other mechanical components.

**Generator:** Converts the mechanical energy from the rotating blades into electrical energy.

**Tower:** Supports the rotor and nacelle at an elevated height to capture stronger and more consistent wind speeds.

**Onshore and Offshore Wind Farms:** Wind turbines can be installed on land (onshore) or in bodies of water (offshore). Offshore wind farms are often

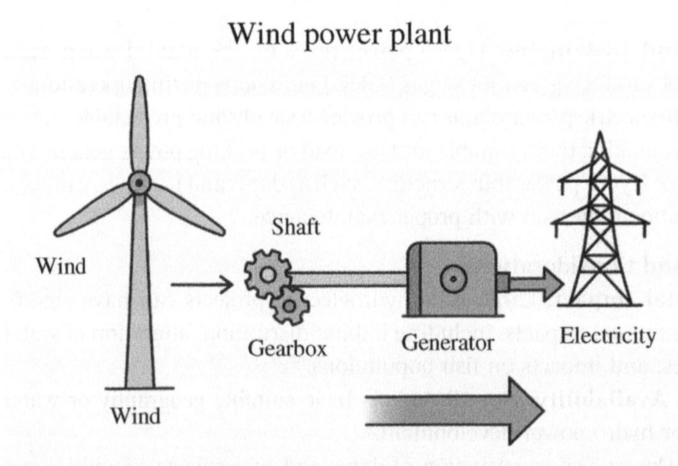


**Figure 1.5** Layout of the Wind Energy System

located in coastal areas or offshore sites with strong and consistent winds. Offshore wind farms typically have larger turbines and higher capacity factors compared to onshore wind farms [15].

➢ **Advantages**

**Renewable and Sustainable:** Wind power relies on the natural wind cycle and does not produce GHG emissions during operation.

**Abundant Resource:** Wind energy is available in abundance in many regions around the world, making it a potentially significant contributor to the global energy mix.

**Low Operating Costs:** Once installed, wind turbines have relatively low operating costs compared to fossil fuel-based power plants.

**Land Multipurpose Use:** Wind turbines can often be installed on agricultural land without significantly impacting farming activities, allowing for dual land use.

➢ **Challenges and Considerations**

**Intermittency:** Wind power generation is variable and dependent on wind speed, which can fluctuate over time. This intermittency can pose challenges for grid stability and reliability.

**Visual and Noise Impact:** Wind turbines can be visually and audibly intrusive, leading to concerns from local communities about aesthetics and noise pollution [16].

**Wildlife Impacts:** Wind turbines can pose risks to birds and bats through collisions with rotor blades and habitat disruption. Proper siting and mitigation measures are necessary to minimize these impacts.

Despite these challenges, wind power has experienced rapid growth in recent decades and is now one of the fastest-growing sources of electricity generation globally. Continued advancements in turbine technology, grid integration, and energy storage are expected to further enhance the role of wind power in the transition to a low-carbon energy future.

> ***Key: Solar energy:*** *Energy derived from harnessing sunlight using photovoltaic panels or solar thermal systems, offering a clean and abundant source of renewable power.*

#### 1.1.1.4 Solar Power

Solar energy has been used for heating and lighting since ancient times. Early civilizations utilized passive solar techniques such as orienting buildings to capture sunlight and constructing buildings with materials that retained heat. Solar energy was also used for drying food and heating water. A solar energy system, also

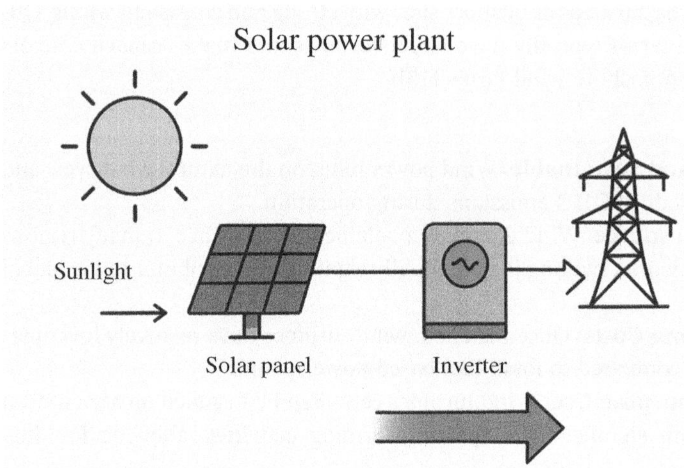


**Figure 1.6** Layout of the Solar Energy System

known as a solar power system, is a renewable energy system that harnesses sunlight to generate electricity or heat for various applications. Solar energy systems utilize photovoltaic (PV) cells or solar thermal collectors to capture and convert sunlight into usable energy. Figure 1.6 shows the layout of the solar energy system.

Following are some key components and aspects of solar energy systems:

➢ **PV System**

**Solar Panels:** PV panels are composed of multiple solar cells made of semiconductor materials, such as silicon, which generate electricity when exposed to sunlight. Solar panels are typically installed on rooftops, ground-mounted structures, or integrated into building facades.

**Inverter:** The electricity generated by solar panels is direct current (DC), but most household appliances and the electricity grid operate on alternating current (AC). An inverter converts the DC electricity from the solar panels into AC electricity suitable for use in homes or businesses.

**Mounting Structure:** Solar panels are mounted on a support structure, which can be fixed, adjustable, or tracking to optimize the angle of sunlight exposure [17].

➢ **Solar Thermal System**

**Solar Collectors:** Solar thermal collectors capture sunlight and convert it into heat energy. There are different types of solar thermal collectors, including flat-plate collectors, evacuated tube collectors, and concentrating collectors.

**Heat Exchanger:** The heat captured by solar collectors is transferred to a fluid (such as water or air) circulating through the system via a heat exchanger. The heated fluid can then be used directly for space heating, water heating, or industrial processes.

**Storage Tank:** Solar thermal systems often include a storage tank to store hot water or heat for use when sunlight is not available, such as during cloudy periods or at night.

➢ **Applications**

**Electricity Generation:** Solar PV systems generate electricity that can be used to power homes, businesses, and utility-scale power plants. Grid-connected PV systems can feed excess electricity back into the grid, while off-grid systems rely on batteries or other storage solutions to store electricity for use when sunlight is not available.

**Water Heating:** Solar thermal systems can provide hot water for domestic or commercial use, including showers, sinks, swimming pools, and industrial processes.

**Space Heating and Cooling:** Solar thermal systems can be used to provide space heating through radiant floor heating systems or to drive absorption chillers for air conditioning.

➢ **Advantages**

**Renewable and Clean:** Solar energy is abundant and does not produce GHG emissions or air pollutants during operation.

**Low Operating Costs:** Once installed, solar energy systems have minimal operating costs compared to conventional fossil fuel-based energy sources.

**Distributed Generation:** Solar PV systems can be installed on rooftops or distributed across a wide area, reducing transmission losses and enhancing grid resilience.

**Scalability:** Solar energy systems can be scaled up or down to meet varying energy needs, from small residential installations to large utility-scale projects.

➢ **Challenges and Considerations**

**Intermittency:** Solar energy is dependent on sunlight, which varies with time of day, weather conditions, and geographic location. Energy storage and grid integration solutions are needed to mitigate intermittency and ensure a reliable electricity supply.

**Upfront Costs:** The initial investment required for solar energy systems, including equipment and installation, can be relatively high. However, costs have been decreasing over time due to technological advancements and economies of scale.

**Land Use and Siting:** Solar energy projects require suitable land or rooftop space for installation, and proper siting considerations are necessary to minimize environmental impacts and land use conflicts.

At the end, we can say that solar energy systems play a crucial role in the transition to a sustainable and low-carbon energy future, offering a clean, abundant, and increasingly cost-effective source of electricity and heat. Continued advancements in technology, policy support, and market development are expected to further accelerate the adoption of solar energy worldwide.

***Key: Wind energy:*** *Energy generated by harnessing the kinetic energy of wind through wind turbines, providing a sustainable and scalable source of electricity production.*

### 1.1.2 Technological Advancements

The global energy landscape has been transformed in recent years due to rapid technological advancements that are paving the way toward cleaner, more efficient, and sustainable energy systems. These innovations are not confined to a single domain but span across renewable energy technologies, energy storage solutions, smart grids, electric vehicles (EVs), and energy efficiency measures. Together, they represent a significant step toward addressing climate change, reducing GHG emissions, and achieving the broader vision of a low-carbon economy.

**Renewable Energy Technologies:** One of the most significant areas of advancement lies in renewable energy technologies, including solar, wind, and hydroelectric power. Solar PV panels have witnessed remarkable improvements in both efficiency and affordability, making solar energy highly competitive with conventional fossil fuel-based sources. Similarly, wind energy has benefitted from better turbine designs, taller towers, and improved aerodynamics, enabling greater electricity generation across diverse locations, including offshore installations. Hydroelectric power has also evolved with innovative approaches such as run-of-river systems and pumped storage plants, both of which enhance energy generation efficiency while reducing environmental impacts. These developments have collectively enabled renewable energy to become a mainstream contributor to the global energy mix.

**Energy Storage:** Complementing the rise of renewables, advancements in energy storage technologies have played a pivotal role in overcoming the intermittency challenges associated with solar and wind energy. Lithium-ion batteries, in particular, have become more affordable, durable, and capable of storing larger amounts of electricity, making them indispensable for both grid-scale storage and electric mobility. Beyond lithium-ion, other emerging storage technologies such as flow batteries, compressed air energy storage, and thermal storage offer promising alternatives that can further enhance grid stability, flexibility, and resilience. Pumped hydro storage remains another reliable option for large-scale energy storage, ensuring a steady power supply during peak demand periods. These innovations collectively strengthen the integration of renewables into national power grids. Figure 1.7 shows the different types of technological advancement.

**Smart Grids and Energy Management Systems (EMSs):** The introduction of smart grids and advanced EMSs has revolutionized the production, distribution, and consumption of electricity. Smart grid technologies allow real-time monitoring and control of electricity flows, optimizing resource use and improving the reliability and resilience of power systems. Key components such as advanced metering infrastructure (AMI), demand response mechanisms, and grid-scale storage solutions enable better integration of

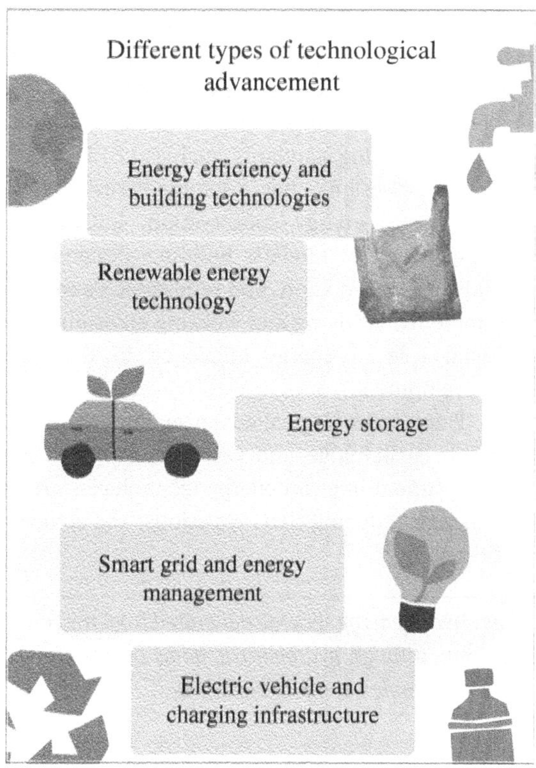


**Figure 1.7** Different Types of Technological Advancement

renewable energy while ensuring stable and efficient electricity distribution. These technologies not only empower utilities but also provide consumers with more control over their energy consumption, thereby fostering a more participatory and efficient energy ecosystem.

**EVs and Charging Infrastructure:** Another area that has witnessed significant growth is the adoption of EVs, which are reshaping the future of transportation. This surge has been driven by breakthroughs in battery technologies, supportive government policies, and increasing public awareness of environmental sustainability. EVs offer a cleaner and more sustainable alternative to traditional internal combustion engine vehicles, reducing reliance on fossil fuels and curbing GHG emissions. Simultaneously, the rapid expansion of EV charging infrastructure, including fast-charging stations and emerging wireless charging technologies, has eased adoption barriers and encouraged greater consumer acceptance. The evolution of EVs is not only transforming mobility but also contributing to the decarbonization of energy systems by creating synergies with renewable energy and smart grids.

**Energy Efficiency and Building Technologies:** In parallel, advances in energy efficiency and building technologies have significantly reduced energy consumption across residential, commercial, and industrial sectors. Energy-efficient appliances, LED lighting, smart thermostats, and building automation systems collectively minimize energy waste while enhancing convenience and lowering utility costs. Beyond individual technologies, the rise of green building standards, such as LEED (Leadership in Energy and Environmental Design), has promoted sustainable construction practices that emphasize energy conservation, indoor air quality, and environmental responsibility. Modern building designs now integrate natural lighting, advanced insulation, and renewable energy systems to create self-sustaining and efficient living and working environments. These efforts underscore the vital role of energy efficiency in the global transition to sustainable development.

These technological advancements represent significant progress toward a more sustainable and resilient energy future. Continued innovation and investment in clean energy technologies are essential for addressing climate change, reducing GHG emissions, and achieving a transition to a low-carbon economy.

***Key: Hydro energy:*** *Energy generated from the gravitational force of flowing water, typically through hydroelectric dams or run-of-river systems, providing a reliable and renewable source of electricity.*

➢ **Impact of Technological Innovations on the Energy Landscape**

Technological innovations have profoundly reshaped the global energy landscape, revolutionizing how energy is produced, distributed, consumed, and managed. These innovations have not only diversified energy sources but also enhanced system resilience, empowered consumers, and accelerated the transition to sustainable models of energy production and use. The impacts are visible across multiple dimensions, including renewable energy adoption, decentralization of generation, modernization of grids, advancements in energy storage, electrification of transportation, and improvements in energy efficiency. Collectively, they represent a paradigm shift toward a more sustainable, resilient, and low-carbon energy future. Following are some key impacts of technological innovations on the energy landscape:

**Diversification of Energy Sources:** Technological advancements have greatly diversified the global energy mix by making renewable energy sources such as solar, wind, and hydroelectric power more accessible, cost-effective, and scalable. Solar PV panels are now far more efficient and affordable than in previous decades, allowing solar power to compete directly with fossil fuels. Similarly, wind turbines have been enhanced in terms of size, aerodynamics, and efficiency, enabling higher energy yields from both onshore and offshore sites. Hydroelectric power generation has also been optimized through innovations such as run-of-river and pumped storage systems, which minimize environmental impacts while maintaining efficiency. This diversification reduces dependency on traditional fossil fuels, lowers GHG emissions, and creates a more resilient and sustainable energy system that is better equipped to withstand global challenges, such as climate change and resource scarcity.

**Decentralization and Distributed Generation:** The rise of renewable energy technologies, energy storage solutions, and smart grid systems has significantly encouraged the decentralization of energy production. Distributed generation models, such as rooftop solar panels, small-scale wind turbines, and community-based renewable energy systems, allow energy to be produced closer to where it is consumed. This reduces transmission and distribution losses, enhances energy security, and creates opportunities for consumers to actively participate as prosumers producing, consuming, and even selling surplus energy. Decentralized systems also offer resilience against large-scale grid disruptions, as localized energy generation can continue to supply power even in cases of grid failure. By empowering individuals and communities, technological innovations have transformed consumers into active participants in the energy ecosystem, promoting both energy independence and sustainability.

**Grid Modernization and Smart Grids:** Another critical impact of technological innovation is the modernization of electricity grids through the development of smart grid technologies. Smart grids integrate advanced sensors, communication networks, and data analytics to allow real-time monitoring, control, and optimization of energy flows. They enhance grid efficiency and reliability by quickly detecting faults, rerouting power, and balancing supply with demand. Moreover, smart grids enable the seamless integration of renewable energy sources, EVs, and demand response programs, which collectively contribute to greater flexibility and resilience of energy systems. By adopting features such as AMI and predictive analytics, smart grids provide utilities with actionable insights while empowering consumers with more control over their energy consumption. This shift represents a cornerstone of modern energy systems that are more intelligent, adaptive, and sustainable.

**Energy Storage Solutions:** The evolution of energy storage technologies has been one of the most transformative innovations in shaping the energy landscape. Solutions such as lithium-ion batteries, pumped hydro storage, and emerging grid-scale storage systems have addressed the intermittency challenge associated with renewable energy sources like solar and wind. By storing excess energy during periods of low demand and releasing it during peak times, these technologies improve grid stability, enhance resilience, and provide backup during outages. Beyond grid integration, energy storage has also supported the electrification of transportation, powering the rapid growth of EVs. Furthermore, it has enabled new business models, including virtual power plants (VPPs) that aggregate distributed storage systems and peer-to-peer (P2P) energy trading platforms that allow consumers to buy and sell energy locally. These innovations illustrate how storage technologies are redefining not only technical operations but also market structures in the energy sector.

**Electrification of Transportation:** Technological innovations in EVs, battery technology, and charging infrastructure have accelerated the global shift toward electrified transportation. EVs present a cleaner and more sustainable alternative to conventional internal combustion engine vehicles, significantly reducing GHG emissions and fossil fuel dependency. Breakthroughs in battery efficiency, energy density, and charging speed have made EVs more practical and appealing for consumers, while the expansion of fast-charging networks and wireless charging systems has reduced range anxiety. The growing adoption of EVs also has far-reaching implications for energy demand, as electricity increasingly replaces petroleum products in the transportation sector. This shift necessitates adaptations in grid infrastructure, urban planning, and energy policy to accommodate new patterns of energy use. As a result, electrified transportation is not only transforming mobility but also creating

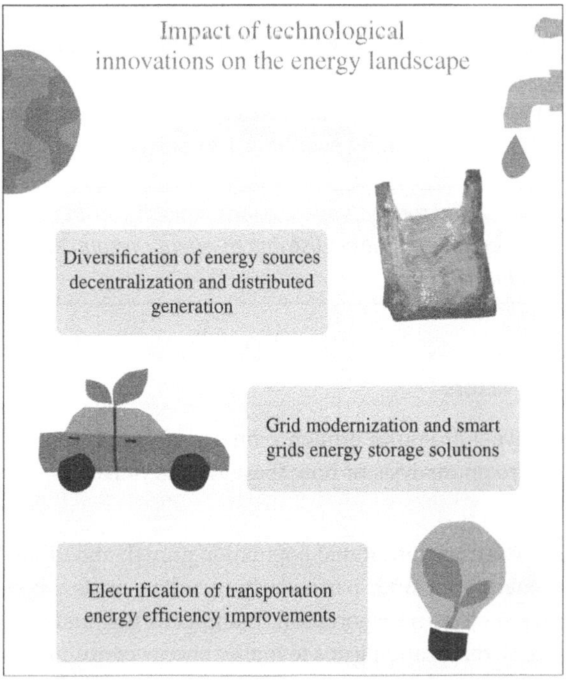


**Figure 1.8** Impact of Technological Advancement

synergies with renewable energy generation and smart grid systems. Figure 1.8 shows the impact of technological advancement.

**Energy Efficiency Improvements:** Energy efficiency remains a fundamental pillar of technological innovation in the energy landscape. Advances in appliances, lighting, high voltage alternating current (HVAC) systems, and building materials have dramatically reduced energy consumption across residential, commercial, and industrial sectors. For instance, LED lighting consumes far less electricity than traditional lighting while offering longer lifespans, while smart thermostats and building automation systems enable precise control of energy use based on occupancy and environmental conditions. Improvements in insulation, windows, and green building standards such as LEED have further minimized energy waste and environmental impacts. These efficiency measures not only reduce energy bills for consumers but also ease pressure on energy infrastructure and decrease GHG emissions. In effect, energy efficiency improvements complement renewable energy adoption by lowering overall energy demand, contributing to a sustainable and balanced energy system.

At the end, technological innovations continue to shape the energy landscape, driving the transition toward a more sustainable, resilient, and low-carbon energy future. Continued research, development, and deployment of clean energy technologies are essential for addressing climate change, enhancing energy security, and promoting economic prosperity.

> ***Key: Microgrid:*** *A localized energy grid that can operate independently or in conjunction with the main grid, integrating various distributed energy resources to enhance reliability, resilience, and efficiency.*

### 1.1.3 Social and Economic Factors

Social and economic factors have a profound influence on energy usage and distribution patterns. Following are an analysis of how these factors impact energy consumption and distribution:

**Population Growth and Urbanization:** Rapid population growth and urbanization drive increased energy demand, particularly in urban areas where energy-intensive activities such as transportation, manufacturing, and construction are concentrated. Urbanization leads to higher energy consumption per capita due to greater access to modern amenities, higher living standards, and increased use of appliances and electronics.

**Income and Economic Development:** Income levels and economic development strongly influence energy consumption patterns. As countries experience economic growth and rising incomes, there is a corresponding increase in energy demand, driven by higher consumption of goods and services, greater industrial activity, and increased mobility. Wealthier societies tend to have higher energy consumption per capita, reflecting higher levels of consumption and energy-intensive lifestyles.

**Industrialization and Manufacturing:** Industrial sectors are major consumers of energy, accounting for a significant portion of total energy demand in many countries. Industrialization leads to increased energy usage for manufacturing processes, heating, cooling, and transportation of goods. Economic factors such as labor costs, energy prices, government policies, and market demand influence industrial energy usage and distribution patterns.

**Energy Prices and Affordability:** Energy prices, including electricity, gasoline, and heating fuels, directly impact energy consumption patterns and distribution. Higher energy prices tend to reduce energy consumption through demand-side management measures, such as energy efficiency improvements, conservation efforts, and fuel switching. Affordability issues can arise when energy costs represent a significant portion of household budgets, disproportionately affecting low-income households.

**Technological Innovation and Efficiency:** Technological advancements play a critical role in shaping energy usage and distribution by improving energy efficiency, increasing access to cleaner energy sources, and enabling new energy services and business models. Innovations in energy-efficient appliances, renewable energy technologies, smart grid systems, and transportation solutions influence consumer behavior and energy consumption patterns.

**Government Policies and Regulations:** Government policies and regulations have a significant impact on energy usage and distribution through incentives, subsidies, taxes, and regulations. Energy policies aimed at promoting renewable energy, energy efficiency, and environmental sustainability can influence investment decisions, market dynamics, and energy distribution patterns. Regulatory frameworks governing energy markets, infrastructure development, and pricing mechanisms also shape energy usage and distribution. Figure 1.9 shows the social and technological factors in the energy transition.

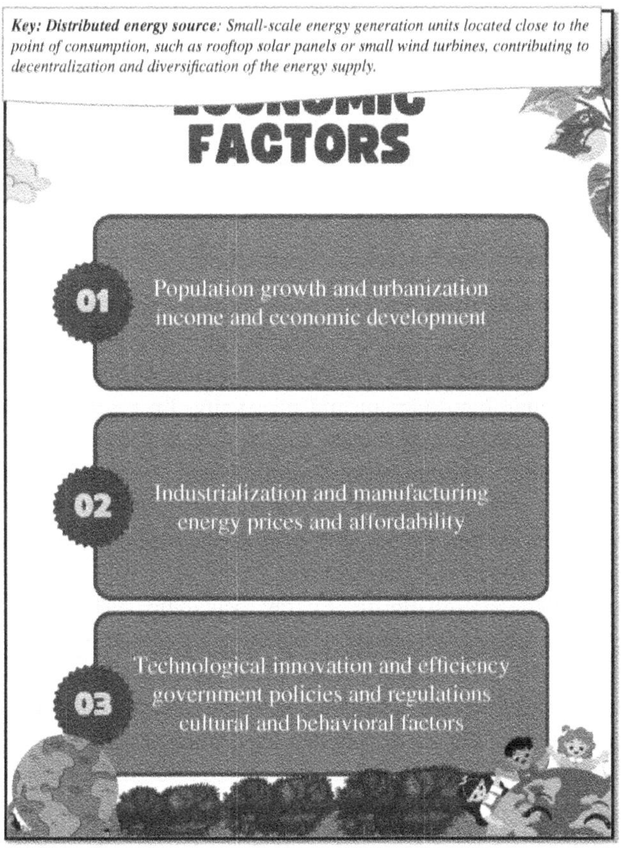


**Figure 1.9** Social and Economic Factors in Energy Transition

**Cultural and Behavioral Factors:** Cultural norms, social practices, and consumer behavior patterns influence energy consumption habits and distribution. Cultural attitudes toward energy conservation, sustainability, and environmental stewardship can drive changes in energy usage patterns and distribution. Behavioral interventions, public awareness campaigns, and social norms can promote energy-saving behaviors and shape energy distribution dynamics.

At the end, social and economic factors interact with technological, regulatory, and cultural influences to shape energy usage and distribution patterns. Understanding these interdependencies is essential for developing effective energy policies, investment strategies, and sustainability initiatives aimed at promoting efficient and equitable energy distribution while addressing societal needs and environmental challenges.

> ***Key: Peer-to-peer energy trading:*** *Direct transactions of electricity between consumers or prosumers (producer–consumers) using digital platforms, enabling more efficient and transparent utilization of renewable energy resources.*

➢ **Relationship Between Energy Availability and Societal Development**

The relationship between energy availability and societal development is multifaceted and complex, with energy playing a fundamental role in shaping various aspects of human civilization. Following are an analysis of the relationship between energy availability and societal development:

**Economic Growth and Productivity:** Energy availability is closely linked to economic growth and productivity. Access to abundant and affordable energy sources enables the efficient operation of industries, businesses, and infrastructure, driving economic development and prosperity. Energy-intensive sectors such as manufacturing, transportation, agriculture, and construction rely heavily on energy inputs to produce goods and services, create jobs, and generate income. Higher levels of energy availability are associated with higher levels of economic output, investment, innovation, and competitiveness, contributing to overall societal development.

**Quality of Life and Human Well-Being:** Energy availability has a direct impact on the quality of life and well-being of individuals and communities. Access to modern energy services such as electricity, clean cooking fuels, heating, cooling, and transportation enhances living standards, health outcomes, and educational opportunities. Reliable energy access enables access to essential services such as healthcare, education, communication, and entertainment, improving overall quality of life and social welfare. Conversely, energy poverty and lack of access to basic energy services can perpetuate poverty, inequality, and social exclusion, hindering societal development and human flourishing.

**Infrastructure Development and Urbanization:** Energy availability drives infrastructure development and urbanization, particularly in rapidly growing cities and urban areas. Energy infrastructure, including power plants, transmission lines, pipelines, and transportation networks, forms the backbone of modern societies, enabling the movement of people, goods, and information. Urbanization and industrialization are fueled by energy-intensive activities such as construction, manufacturing, transportation, and commerce, driving demand for energy and shaping the built environment. Energy availability influences urban planning, land use patterns, housing, transportation systems, and public services, shaping the physical and social fabric of cities and communities.

**Environmental Sustainability and Climate Change:** The availability and use of energy resources have significant environmental implications, including air and water pollution, habitat destruction, and GHG emissions. Fossil fuel-based energy systems contribute to environmental degradation and climate change, posing risks to human health, ecosystems, and biodiversity. Transitioning to cleaner and more sustainable energy sources such as renewable energy, energy efficiency, and low-carbon technologies is essential for mitigating environmental impacts, reducing carbon emissions, and promoting sustainable development. Energy availability and societal development are interconnected with environmental sustainability, requiring a balanced approach that considers social, economic, and environmental objectives.

**Energy Security and Resilience:** Energy availability is essential for ensuring energy security and resilience, particularly in the face of supply disruptions, geopolitical tensions, natural disasters, and other external shocks. Diversification of energy sources, investment in energy infrastructure, and adoption of energy-efficient technologies can enhance energy security by reducing reliance on imported fuels, increasing domestic energy production, and improving energy resilience. Access to reliable and affordable energy services is critical for meeting basic needs, maintaining public safety, and supporting emergency response and disaster recovery efforts.

At the end, energy availability plays a central role in driving societal development, economic growth, and human well-being. Ensuring universal access to affordable, reliable, and sustainable energy services is essential for achieving inclusive and equitable development, addressing global challenges, and building resilient and prosperous societies.

> ***Key: Virtual power plant:*** *A network of decentralized energy resources aggregated and controlled through digital technologies, optimizing their collective operation to provide grid services and support stability.*

## 1.2 Transformation of Energy Distribution

### 1.2.1 Decentralized Energy Generation

Decentralized energy generation refers to the production of electricity or heat on a smaller scale, closer to the point of consumption, rather than relying solely on large centralized power plants. In a decentralized energy generation model, energy is produced locally or regionally using distributed energy resources (DERs), such as solar panels, wind turbines, micro-turbines, fuel cells, and small-scale biomass or hydroelectric systems. Decentralized energy generation involves the production of energy at or near the location where it will be used. This can include rooftop solar panels on homes and businesses, small wind turbines in rural areas, or community-scale solar or biomass facilities. By generating electricity closer to the point of consumption, decentralized energy systems can reduce transmission and distribution losses associated with transporting electricity over long distances from centralized power plants. This can improve overall efficiency and grid reliability. Figure 1.10 shows the types of decentralized energy generation.

Decentralized energy generation can enhance energy resilience and reliability by reducing reliance on centralized infrastructure that may be vulnerable to

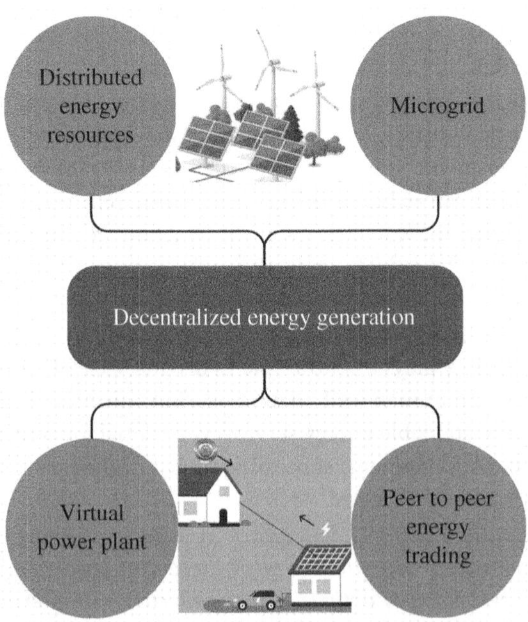


**Figure 1.10** Types of Decentralized Energy Generation

disruptions, such as extreme weather events, cyberattacks, or equipment failures. Localized energy production can provide backup power during outages and support critical facilities such as hospitals, emergency shelters, and telecommunications infrastructure.

Decentralized energy generation is often integrated with smart grid technologies, enabling bidirectional energy flows, real-time monitoring and control, and demand response capabilities. Smart grid systems enable more efficient and flexible management of DERs, improving grid stability and accommodating variable renewable energy sources. Decentralized energy generation can empower communities by giving them greater control over their energy supply, fostering local economic development, and promoting energy independence. Community-owned renewable energy projects, cooperative models, and P2P energy trading platforms enable stakeholders to participate in energy decision-making and benefit from the transition to cleaner and more sustainable energy systems. Decentralized energy generation can contribute to environmental sustainability by reducing GHG emissions, air pollution, and environmental degradation associated with centralized fossil fuel-based power plants. Renewable energy sources such as solar, wind, and biomass offer cleaner alternatives to traditional fossil fuels, helping to mitigate climate change and preserve natural resources.

### 1.2.2 Microgrids

A microgrid is a localized energy system that integrates various DERs and loads within a defined geographical area. Unlike traditional centralized power grids that rely on large-scale power plants and long-distance transmission lines, microgrids operate autonomously or in coordination with the main grid, providing a reliable and resilient energy supply to local communities, facilities, or campuses. Microgrids incorporate a mix of renewable energy sources (such as solar, wind, and biomass), conventional generators (such as diesel generators or natural gas turbines), and energy storage systems (such as batteries or flywheels). These DERs can generate, store, and distribute electricity locally, reducing dependence on centralized infrastructure and enhancing energy resilience. Microgrids leverage advanced control and communication technologies to optimize the operation of DERs, manage energy flows, and balance supply and demand in real-time. Smart grid functionalities enable microgrids to respond to changing conditions, integrate intermittent renewable energy sources, and provide ancillary services to the main grid. Figure 1.11 shows the layout of the microgrid system.

One of the defining characteristics of microgrids is their ability to operate in islanded mode, disconnected from the main grid during emergencies or grid disturbances. Microgrids can automatically isolate themselves from the main grid and continue to provide electricity to critical loads, ensuring continuity of service

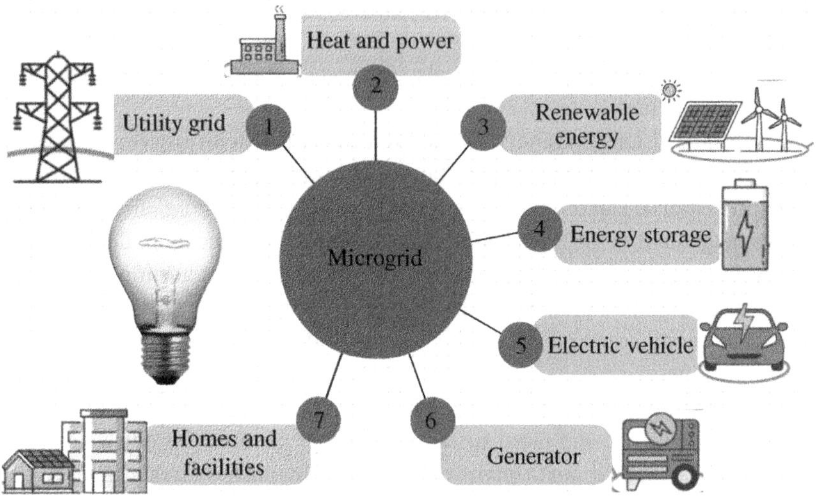


**Figure 1.11** Layout of the Microgrid System

during power outages or disruptions. Microgrids enhance energy resilience and reliability by minimizing the impact of grid failures, reducing downtime, and providing backup power to critical facilities. By incorporating diverse energy resources and redundant components, microgrids can withstand disruptions and recover quickly from outages, supporting critical infrastructure such as hospitals, military bases, data centers, and remote communities. Microgrids employ sophisticated EMSs to monitor, control, and optimize the operation of DERs, energy storage, and loads. EMS software algorithms balance generation and consumption, prioritize critical loads, manage energy storage, and optimize dispatch strategies to minimize costs and maximize efficiency. Microgrids are deployed in various applications and settings, including remote communities, military bases, industrial complexes, university campuses, and commercial developments. They offer a flexible and scalable solution for meeting energy needs, reducing carbon emissions, and enhancing energy security in diverse environments.

### 1.2.3 Distributed Energy Resources (DERs)

DERs refer to a diverse set of small-scale energy sources and storage technologies that are located close to the point of energy consumption. Unlike traditional centralized power plants, which supply electricity from a single location to a wide area through transmission and distribution networks, DERs generate, store, and manage energy locally. DERs play a crucial role in decentralizing the energy system, improving reliability, resilience, and sustainability. Following are the key components of DERs include:

**Renewable Energy Sources:** DERs often incorporate renewable energy technologies such as solar PV panels, wind turbines, biomass generators, and small-scale hydroelectric systems. These renewable energy sources harness naturally occurring energy flows sunlight, wind, water, and biomass to generate electricity or heat locally, reducing dependence on fossil fuels and lowering carbon emissions.

**Energy Storage Systems:** DERs may include energy storage technologies such as batteries, flywheels, compressed air, and pumped hydro storage. Energy storage systems capture excess energy generated during periods of low demand or high renewable energy output and store it for later use when demand is higher or renewable energy production is lower. Energy storage enhances grid stability, improves flexibility, and supports the integration of intermittent renewable energy sources.

**Combined Heat and Power (CHP):** CHP systems, also known as cogeneration, simultaneously generate electricity and useful heat from a single fuel source, such as natural gas, biomass, or waste heat. CHP systems capture and utilize the heat generated during electricity generation for heating, cooling, or industrial processes, increasing overall efficiency and reducing energy waste.

**Demand Response Technologies:** DERs may include demand response technologies and strategies that enable consumers to adjust their electricity usage in response to price signals or grid conditions. Demand response programs encourage consumers to shift energy consumption to off-peak hours, reduce peak demand, and provide grid services such as load balancing and frequency regulation, improving grid reliability and efficiency.

**Microgrids and Smart Grids:** DERs are often integrated into microgrid and smart grid systems, which enable localized energy generation, distribution, and management. Microgrids combine DER, energy storage, and control technologies to create localized energy systems that can operate autonomously or in coordination with the main grid. Smart grid technologies enable real-time monitoring, control, and optimization of DERs, enhancing grid reliability, resilience, and efficiency.

***Key: Net metering:*** *A billing mechanism that credits energy consumers for excess electricity generated from renewable sources and fed back into the grid, promoting the adoption of distributed generation and renewable energy systems.*

### 1.2.4 Peer-to-Peer (P2P) Energy Trading

P2P energy trading represents a transformative approach to energy distribution and consumption, redefining the dynamics of the traditional energy market by enabling direct transactions between energy producers and consumers within a decentralized framework. This emerging concept harnesses advancements in digital technologies, blockchain, and smart grid infrastructure to facilitate the exchange of electricity generated from renewable energy sources, such as solar PV panels, wind turbines, and micro-hydro systems, among individuals, businesses, and communities. At its core, P2P energy trading empowers prosumers—those who both consume and produce electricity—to trade surplus energy with their peers in real-time, thereby promoting energy self-sufficiency, resilience, and sustainability.

The rise of P2P energy trading is propelled by several key factors that have reshaped the energy landscape in recent years. One of the primary drivers is the growing adoption of DERs and renewable energy technologies, which have democratized energy production and decentralization. As the costs of solar panels, wind turbines, and energy storage systems continue to decline, more consumers are investing in on-site generation and seeking opportunities to monetize their excess energy production. P2P energy trading platforms provide an avenue for these energy producers to monetize their surplus electricity by selling it directly to neighboring consumers, thereby unlocking new revenue streams and enhancing the economic viability of renewable energy investments.

Moreover, P2P energy trading addresses the limitations and inefficiencies of the traditional energy market, characterized by centralized generation, long-distance transmission, and monopolistic utility models. By facilitating direct transactions between energy users and producers, P2P trading reduces dependence on centralized utilities, minimizes transmission losses, and fosters local energy autonomy. This localized approach to energy distribution promotes resilience and reliability, as electricity can be generated and consumed within the same community, mitigating the risks of grid disruptions, blackouts, and supply chain vulnerabilities. Furthermore, P2P energy trading incentivizes the deployment of renewable energy infrastructure and accelerates the transition to a low-carbon energy system, aligning with global sustainability goals and climate mitigation efforts.

The advent of digital technologies, particularly blockchain and smart contracts, has been instrumental in enabling the implementation of P2P energy trading platforms. Blockchain technology, a decentralized and tamper-resistant ledger, serves as the backbone of P2P energy markets, recording and verifying transactions securely and transparently. Smart contracts, self-executing contracts with predefined rules encoded in code, automate the exchange of energy between buyers and sellers, ensuring trust, transparency, and efficiency in energy trading processes. Through blockchain-enabled P2P platforms, participants can access real-time data on energy production, consumption, and pricing, enabling informed decision-making and dynamic optimization of energy flows.

In addition to technological innovations, supportive regulatory frameworks and market incentives play a crucial role in driving the adoption of P2P energy trading models. Regulatory reforms that facilitate market access, promote grid interoperability, and enable fair compensation mechanisms for distributed energy producers are essential for fostering a conducive environment for P2P trading. Policies such as net metering, feed-in tariffs (FiTs), and P2P energy trading regulations incentivize investment in renewable energy infrastructure, encourage consumer participation, and level the playing field for decentralized energy providers. Furthermore, financial incentives, subsidies, and tax credits can stimulate market demand for P2P energy trading platforms, making them more accessible and attractive to consumers and investors alike.

***Key: Feed-in tariffs:*** *Government incentives that guarantee a fixed payment for electricity generated from renewable sources, encouraging investment in renewable energy projects and promoting grid integration.*

The benefits of P2P energy trading extend beyond economic considerations to encompass social, environmental, and community-centric outcomes. By empowering individuals and communities to become active participants in the energy transition, P2P trading fosters a sense of ownership, collaboration, and collective action toward sustainability goals. Communities that embrace P2P energy trading models often experience social cohesion, knowledge sharing, and empowerment, as residents collaborate to build resilient and self-sustaining energy ecosystems. Moreover, P2P energy trading promotes energy equity and inclusivity by democratizing access to clean energy resources, particularly in underserved or marginalized communities, where traditional energy infrastructure may be lacking or unreliable.

Furthermore, P2P energy trading contributes to environmental conservation and climate resilience by accelerating the adoption of renewable energy sources and reducing reliance on fossil fuels. By incentivizing local generation and consumption of renewable electricity, P2P trading reduces GHG emissions, air pollution, and environmental degradation associated with conventional energy production. Communities that embrace P2P energy trading models often prioritize sustainability, conservation, and environmental stewardship, leading to enhanced ecological resilience and resource efficiency. Additionally, P2P trading encourages the integration of energy storage technologies, demand response strategies, and energy efficiency measures, further enhancing the environmental benefits of distributed energy systems.

Despite its potential benefits, P2P energy trading faces several challenges and barriers to widespread adoption, including regulatory complexities, technical interoperability, market design issues, and cybersecurity risks. Regulatory frameworks governing energy markets, grid operations, and financial transactions may not be adequately aligned with the decentralized nature of P2P trading, posing legal and compliance challenges for market participants. Technical interoperability and standardization of communication protocols, data formats, and grid interfaces are essential for the seamless integration of DERs and P2P trading platforms into existing energy infrastructure.

Moreover, market design issues such as pricing mechanisms, market transparency, and risk management require careful consideration to ensure fair and efficient energy trading processes. Participants in P2P energy markets must navigate complex pricing structures, tariff regulations, and contractual arrangements, which can impact market liquidity, competitiveness, and consumer confidence. Additionally, cybersecurity threats, data privacy concerns, and vulnerabilities in blockchain-based systems pose risks to the integrity, reliability, and trustworthiness of P2P energy trading platforms, necessitating robust cybersecurity measures and risk mitigation strategies.

In conclusion, P2P energy trading represents a promising paradigm shift in energy distribution and consumption, offering a decentralized, resilient, and sustainable alternative to traditional energy markets. By enabling direct transactions between energy producers and consumers, P2P trading empowers individuals, businesses, and communities to participate in the energy transition, unlock economic opportunities, and promote environmental stewardship. Through technological innovation, supportive policies, and community engagement, P2P energy trading has the potential to revolutionize the way energy is produced, traded, and consumed, paving the way for a more equitable, inclusive, and sustainable energy future.

### 1.2.5 Virtual Power Plants (VPPs)

VPPs represent a cutting-edge approach to energy management and distribution, revolutionizing the traditional centralized model of power generation by aggregating and orchestrating a diverse array of DERs into a unified and dynamic virtual network. At their core, VPPs harness the power of digital technologies, advanced analytics, and real-time communication systems to optimize the operation of DERs, including solar PV panels, wind turbines, energy storage systems, demand response technologies, and flexible loads, across multiple sites and stakeholders. By leveraging the flexibility and scalability of DERs, VPPs enable grid operators, utilities, and energy aggregators to effectively balance supply and demand, enhance grid stability, and unlock new value streams in the energy market. Figure 1.12 shows the layout of the VPP.

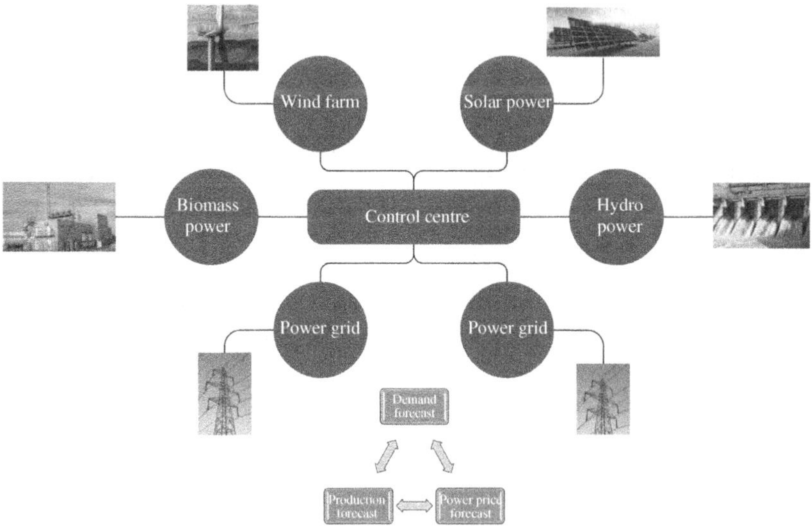


**Figure 1.12** Virtual Power Plant

The rise of VPPs is driven by several key factors that have reshaped the energy landscape in recent years. One of the primary drivers is the proliferation of DERs and renewable energy technologies, which have democratized energy production and decentralization. As the costs of solar panels, batteries, and other DERs continue to decline, more consumers are investing in on-site generation and seeking opportunities to monetize their excess energy production. VPPs provide a pathway for these distributed energy assets to be aggregated and integrated into a VPP, enabling participation in energy markets, grid services, and demand response programs.

Moreover, VPPs address the challenges and limitations of the traditional centralized power grid, characterized by aging infrastructure, increasing demand, and intermittent renewable energy sources. By aggregating and coordinating distributed energy assets, VPPs enhance grid reliability, resilience, and flexibility, enabling efficient integration of renewable energy resources and demand-side management strategies. VPPs serve as a decentralized and adaptive solution to grid optimization, enabling real-time monitoring, control, and optimization of DERs to meet evolving energy needs and market dynamics.

***Key: Energy market reforms:*** *Policy initiatives aimed at restructuring energy markets to promote competition, efficiency, and sustainability, often involving deregulation, market liberalization, and adoption of renewable energy targets.*

The architecture of VPPs is designed to be modular, scalable, and interoperable, allowing for seamless integration of diverse DERs and communication protocols. At the heart of a VPP is a sophisticated EMS that serves as the central intelligence, coordinating the operation of DERs based on grid conditions, market signals, and user preferences. The EMS utilizes advanced algorithms, machine learning techniques, and predictive analytics to forecast energy production, demand patterns, and market prices, optimizing the dispatch and scheduling of DERs to maximize economic value and grid stability.

In addition to technological advancements, supportive regulatory frameworks and market incentives play a crucial role in driving the adoption of VPPs. Regulatory reforms that facilitate market access, promote grid interoperability, and enable fair compensation mechanisms for distributed energy providers are essential for fostering a conducive environment for VPP deployment. Policies such as FiTs, capacity payments, and ancillary service markets incentivize participation in VPP programs, encouraging investment in DERs and grid infrastructure.

Furthermore, financial incentives, subsidies, and tax credits can stimulate market demand for VPP solutions, making them more attractive and accessible to consumers, utilities, and energy aggregators. Public–private partnerships (PPPs), pilot projects, and demonstration initiatives can facilitate collaboration among stakeholders, build trust, and drive innovation in the VPP ecosystem. Moreover, education, outreach, and stakeholder engagement efforts are essential for raising awareness, building capacity, and fostering collaboration among participants in the VPP value chain.

The benefits of VPPs extend beyond economic considerations to encompass social, environmental, and grid-related outcomes. From a societal perspective, VPPs empower individuals, communities, and businesses to actively participate in the energy transition, unlock new revenue streams, and reduce reliance on centralized utilities. By democratizing access to energy markets and grid services, VPPs promote energy equity, inclusivity, and social justice, particularly in underserved or marginalized communities where traditional energy infrastructure may be lacking or unreliable.

Furthermore, VPPs contribute to environmental sustainability and climate resilience by facilitating the integration of renewable energy sources, reducing GHG emissions, and promoting energy efficiency. By optimizing the operation of distributed energy assets, VPPs enhance the reliability and resilience of the grid, mitigating the risks of grid disruptions, blackouts, and supply chain vulnerabilities. Additionally, VPPs enable the deployment of innovative grid services such as frequency regulation, voltage support, and grid congestion management, enhancing overall grid stability and performance.

Despite its potential benefits, the widespread adoption of VPPs faces several challenges and barriers, including technical complexity, regulatory uncertainty, market fragmentation, and cybersecurity risks. Technical interoperability and standardization of communication protocols, data formats, and grid interfaces are essential for seamless integration of DERs and VPP platforms into existing energy infrastructure. Moreover, regulatory frameworks governing energy markets, grid operations, and financial transactions may not be adequately aligned with the decentralized nature of VPPs, posing legal and compliance challenges for market participants.

***Key: Blockchain:*** *A decentralized digital ledger technology used to record and verify transactions securely, offering potential applications in energy trading, peer-to-peer transactions, and grid management.*

Market design issues such as pricing mechanisms, market transparency, and risk management require careful consideration to ensure fair and efficient energy trading processes. Participants in VPP markets must navigate complex pricing structures, tariff regulations, and contractual arrangements, which can impact market liquidity, competitiveness, and consumer confidence. Additionally, cybersecurity threats, data privacy concerns, and vulnerabilities in VPP platforms pose risks to the integrity, reliability, and trustworthiness of VPPs, necessitating robust cybersecurity measures and risk mitigation strategies.

In conclusion, VPPs represent a transformative approach to energy management and distribution, offering a decentralized, resilient, and sustainable alternative to traditional centralized power grids. By aggregating and orchestrating DERs, VPPs enable grid optimization, demand response, and renewable energy integration, unlocking economic, social, and environmental benefits for stakeholders. Through technological innovation, supportive policies, and stakeholder collaboration, VPPs have the potential to revolutionize the way energy is generated, distributed, and consumed, paving the way for a more equitable, inclusive, and sustainable energy future.

## 1.3 Policy and Regulatory Frameworks

### 1.3.1 Renewable Energy Targets

Renewable energy targets serve as pivotal instruments in driving the global transition toward a sustainable energy future, guiding policy development, investment decisions, and technology deployment to accelerate the adoption of renewable energy sources and mitigate climate change. These targets, often established at the national, regional, or municipal level, set ambitious objectives for the share of renewable energy in the overall energy mix, typically measured as a percentage of total electricity generation, heat production, or primary energy consumption.

The establishment of renewable energy targets reflects a growing recognition of the urgent need to reduce GHG emissions, enhance energy security, and promote economic development through the expansion of clean, renewable energy sources, such as solar, wind, hydroelectric, biomass, and geothermal.

The rationale behind renewable energy targets stems from several key drivers and imperatives shaping the global energy landscape. One of the primary drivers is the mounting urgency to address climate change and its adverse impacts on ecosystems, economies, and human well-being. With the scientific consensus on the need to limit global warming to well below 2 °C above preindustrial levels, renewable energy targets serve as critical mechanisms for decarbonizing the energy sector, reducing reliance on fossil fuels, and mitigating the risks of climate-related disasters, extreme weather events, and sea-level rise. By transitioning to renewable energy sources, countries can significantly reduce their carbon footprint, enhance resilience to climate impacts, and contribute to international efforts to achieve the objectives of the Paris Agreement.

Moreover, renewable energy targets are driven by considerations of energy security, geopolitical stability, and economic competitiveness. As the world grapples with finite fossil fuel reserves, geopolitical tensions, and volatile energy markets, diversifying the energy mix with renewable sources is seen as a strategic imperative for enhancing energy security and reducing dependence on imported fuels. Renewable energy targets offer a pathway for countries to harness their indigenous renewable resources, such as sunlight, wind, water, and biomass, thereby reducing vulnerability to supply disruptions, price volatility, and geopolitical risks associated with fossil fuel dependence. Furthermore, the declining costs and technological advancements in renewable energy technologies have made renewables increasingly competitive with conventional energy sources, offering new opportunities for job creation, innovation, and economic growth.

> ***Key: Artificial intelligence:*** *Advanced computer algorithms that enable machines to perform tasks typically requiring human intelligence, including optimization of energy systems, predictive maintenance, and demand forecasting.*

The formulation and implementation of renewable energy targets require a multifaceted approach that involves policy alignment, stakeholder engagement, investment mobilization, and technology innovation. National governments, regional blocs, and local authorities play a pivotal role in setting ambitious and achievable renewable energy targets, backed by supportive policy frameworks, regulatory incentives, and fiscal measures. Renewable energy policies may include FiTs, renewable portfolio standards, tax incentives, grants, and PPPs aimed at promoting renewable energy deployment, attracting investment, and overcoming market barriers.

Furthermore, renewable energy targets rely on the collaboration and participation of various stakeholders, including government agencies, utilities, businesses, civil society organizations, and academia. Stakeholder engagement processes, public consultations, and multi-stakeholder dialogues are essential for building consensus, fostering ownership, and ensuring the inclusivity and legitimacy of renewable energy policies and initiatives. Moreover, international cooperation and knowledge exchange play a crucial role in supporting countries' efforts to achieve their renewable energy targets, facilitating technology transfer, capacity building, and financial assistance to developing countries.

The successful achievement of renewable energy targets hinges on the availability of adequate financial resources, investment mechanisms, and market incentives to scale up renewable energy deployment. Financing mechanisms such as green bonds, carbon pricing, climate funds, and renewable energy auctions can mobilize private sector investment and bridge the funding gap for renewable energy projects. Market-based instruments such as carbon markets, renewable energy certificates (RECs), and tradable green certificates provide economic incentives for renewable energy generation and facilitate compliance with renewable energy targets.

Furthermore, technology innovation and research and development (R&D) play a critical role in advancing renewable energy technologies, improving efficiency, reducing costs, and overcoming technical barriers to deployment. Public investment in renewable energy R&D, technology demonstration projects, and innovation hubs can drive breakthroughs in renewable energy storage, grid integration, and hybrid energy systems, unlocking new opportunities for renewable energy deployment and grid decarbonization. Moreover, capacity-building initiatives, vocational training programs, and knowledge-sharing platforms are essential for developing a skilled workforce and fostering innovation ecosystems that support the transition to renewable energy.

***Key: Internet of Things:*** *A network of interconnected devices embedded with sensors and software, enabling remote monitoring, control, and optimization of energy-consuming assets and systems.*

Renewable energy targets have yielded significant achievements and positive outcomes in many countries and regions worldwide, catalyzing unprecedented growth in renewable energy capacity, investment, and adoption. Countries such as Denmark, Germany, Sweden, and Costa Rica have demonstrated leadership in renewable energy deployment by setting ambitious targets, implementing supportive policies, and leveraging renewable energy resources to achieve high shares of renewable energy in their energy mix. These frontrunner countries have not only reduced their carbon emissions and energy imports but also stimulated economic development, job creation, and innovation in renewable energy sectors.

Moreover, renewable energy targets have contributed to the democratization of energy production, empowering communities, households, and businesses to become active participants in the energy transition. Distributed renewable energy systems, such as rooftop solar panels, community-owned wind farms, and microgrids, enable individuals and local communities to generate their own clean energy, reduce energy costs, and contribute to grid stability. The rise of energy prosumers—those who both consume and produce electricity—has transformed the traditional energy paradigm, fostering energy independence, resilience, and sustainability.

Furthermore, renewable energy targets have catalyzed technological innovation and cost reductions in renewable energy technologies, driving economies of scale, R&D investments, and manufacturing efficiencies. Solar PV panels, wind turbines, and battery storage systems have seen dramatic price declines, making renewable energy increasingly competitive with conventional energy sources in many markets. Breakthroughs in energy storage, grid integration, and digitalization have further enhanced the flexibility, reliability, and efficiency of renewable energy systems, paving the way for a clean energy future.

However, despite the progress made in renewable energy deployment, significant challenges and barriers remain to be addressed to achieve ambitious renewable energy targets and realize the full potential of renewable energy. One of the primary challenges is the intermittency and variability of renewable energy sources, such as solar and wind, which pose challenges to grid stability, reliability, and balancing. Effective integration of renewable energy into the grid requires innovative solutions for energy storage, demand response, grid management, and flexible generation to match supply and demand in real-time.

Moreover, the financing gap for renewable energy projects, particularly in developing countries and emerging markets, poses a significant barrier to achieving renewable energy targets. Limited access to capital, high upfront costs, and perceived risks associated with renewable energy investments deter investors and lenders from financing renewable energy projects in many regions. Addressing barriers to finance through innovative financing mechanisms, risk mitigation instruments, and capacity-building initiatives is essential for unlocking investment in renewable energy and achieving universal access to clean, affordable, and reliable energy.

Furthermore, regulatory and policy uncertainties, market distortions, and political resistance can hinder the implementation of renewable energy targets and slow down the pace of renewable energy deployment. Inconsistent or outdated regulatory frameworks, subsidies, and incentives may create barriers to entry for renewable energy projects, discourage investment, and distort market dynamics. Political opposition from vested interests in the fossil fuel industry, utility monopolies, and conservative policymakers may undermine support for renewable energy targets and impede progress toward a low-carbon energy transition.

> ***Key: Predictive analysis:*** *Data analytics techniques used to forecast future energy consumption patterns, equipment failures, and market trends, supporting proactive decision-making and resource allocation.*

Addressing these challenges and accelerating the pace of renewable energy deployment require coordinated action and collaboration among governments, businesses, civil society organizations, and international institutions. Governments play a central role in setting ambitious renewable energy targets, enacting supportive policies, and creating enabling environments for renewable energy investment and innovation. Businesses and investors can drive market transformation by committing to renewable energy procurement, decarbonization goals, and sustainable business practices that align with renewable energy targets.

Civil society organizations, advocacy groups, and grassroots movements can mobilize public support, raise awareness, and hold policymakers and industry stakeholders accountable for their commitments to renewable energy targets. International institutions, such as the United Nations (UN), World Bank, International Renewable Energy Agency (IRENA), and Green Climate Fund, can provide technical assistance, capacity building, and financial support to countries and regions to achieve their renewable energy goals and enhance climate resilience.

So, renewable energy targets represent a powerful tool for accelerating the global transition to a sustainable, low-carbon energy future, driving investment, innovation, and action to mitigate climate change, enhance energy security, and promote economic development. By setting ambitious and achievable renewable energy targets, countries and regions can unlock the transformative potential of renewable energy, unleash economic opportunities, and build resilient, inclusive, and sustainable energy systems that benefit people and the planet. However, achieving ambitious renewable energy targets requires overcoming significant challenges; addressing barriers to finance, technology, and policy; and mobilizing collective action and collaboration across sectors and stakeholders. With concerted efforts and political will, renewable energy targets can propel the world toward a cleaner, greener, and more prosperous future for generations to come.

### 1.3.2 Net Metering Policies

Net metering policies serve as critical mechanisms for promoting the deployment of distributed renewable energy systems, empowering energy consumers to generate their own clean electricity and offsetting their energy costs through the sale of surplus energy to the grid. Under a net metering arrangement, energy consumers who generate electricity from renewable sources such as solar PV panels, wind turbines, or micro-hydro systems can feed excess electricity back into the grid and receive credit or compensation for the electricity they export. Net metering policies enable energy consumers to offset their electricity bills by netting out the electricity they consume from the grid against the electricity they produce on-site, effectively allowing them to "spin their meters backward" and reduce their overall energy costs.

The rationale behind net metering policies stems from several key objectives and imperatives shaping the energy landscape. One of the primary drivers is the need to promote renewable energy deployment and mitigate climate change by reducing GHG emissions, enhancing energy security, and transitioning to a low-carbon energy system. Net metering policies incentivize investment in distributed renewable energy systems by providing a financial incentive for energy consumers to generate their own clean electricity and contribute to the decarburization of the energy sector. By enabling individuals, businesses, and communities to become active participants in the energy transition, net metering policies foster energy independence, resilience, and sustainability.

Moreover, net metering policies are driven by considerations of energy affordability, consumer empowerment, and economic development. As electricity prices continue to rise, net metering offers a cost-effective solution for energy consumers to reduce their electricity bills, hedge against future price volatility, and lock in long-term savings by investing in renewable energy technologies. Net metering empowers energy consumers to take control of their energy destiny, enabling them to generate, consume, and sell electricity on their own terms, without relying solely on centralized utilities or fossil fuel-based energy sources. Furthermore, net metering policies stimulate local economic activity, create jobs, and spur investment in renewable energy infrastructure, supply chains, and manufacturing, driving innovation and economic growth in the clean energy sector.

The design and implementation of net metering policies vary across jurisdictions and regions, reflecting differences in regulatory frameworks, market structures, and energy market dynamics. At the heart of net metering policies is the concept of "one-for-one" energy exchange, whereby the energy consumed from the grid is offset by the energy exported to the grid at the same retail rate. This ensures that energy consumers receive fair compensation for the electricity they generate, eliminating barriers to entry and providing a level playing field for DERs to compete with centralized power generation.

Furthermore, net metering policies may incorporate additional provisions and incentives to encourage renewable energy adoption, such as net metering caps, interconnection standards, performance-based incentives, and time-of-use (TOU) pricing. Net metering caps, which limit the size or number of eligible net metering installations, are intended to manage grid impacts, prevent cross-subsidization, and ensure equitable access to net metering benefits. Interconnection standards govern the technical requirements, procedures, and costs associated with connecting renewable energy systems to the grid, ensuring safety, reliability, and grid compatibility. Performance-based incentives, such as FiTs, production-based incentives, or tax credits, provide additional financial incentives for renewable energy generation and incentivize higher system performance and output. TOU pricing, which varies electricity rates based on the time of day, season, or grid conditions, can incentivize energy consumers to shift their energy consumption to periods of lower demand or higher renewable energy production, optimizing grid utilization and reducing peak demand.

The implementation of net metering policies requires collaboration and coordination among policymakers, regulators, utilities, energy consumers, and other stakeholders to address technical, economic, and institutional challenges and ensure the effectiveness and sustainability of net metering programs. Policymakers play a central role in designing and enacting net metering legislation, setting eligibility criteria, defining program parameters, and establishing regulatory oversight mechanisms to govern net metering programs. Regulators oversee the implementation and enforcement of net metering policies, ensuring compliance with statutory requirements, consumer protection standards, and grid reliability criteria. Utilities are responsible for administering net metering programs, processing interconnection applications, metering renewable energy production, and crediting energy consumers for surplus electricity exported to the grid. Utilities may also be required to invest in grid upgrades, infrastructure improvements, and smart grid technologies to accommodate the integration of distributed renewable energy resources and ensure grid stability and reliability. Energy consumers, including residential, commercial, and industrial customers, are key participants in net metering programs, investing in renewable energy systems, managing their energy consumption, and engaging in energy trading activities to maximize the benefits of net metering.

Furthermore, public awareness, education, and outreach efforts are essential for promoting understanding, acceptance, and uptake of net metering policies among energy consumers and stakeholders. Outreach campaigns, workshops, and training programs can provide information on the benefits of net metering, eligibility criteria, program requirements, and available incentives, empowering energy consumers to make informed decisions about renewable energy investments and participation in net metering programs. Moreover, transparent communication, stakeholder engagement, and public consultation processes can foster trust, collaboration, and accountability among policymakers, regulators, utilities, and communities, ensuring that net metering policies reflect the needs and priorities of all stakeholders and contribute to the achievement of broader energy and climate objectives.

### 1.3.3 Feed-in Tariffs (FiTs)

FiTs are regulatory mechanisms implemented by governments to incentivize the adoption of renewable energy technologies, such as solar PV, wind power, biomass, and hydropower. Under an FiT system, renewable energy producers are guaranteed a fixed payment for the electricity they generate and feed into the grid, typically over a long-term contract period. These payments are set at a rate higher than the market price for electricity, reflecting the higher costs associated with renewable energy generation and providing a financial incentive for investment in renewable energy projects.

The primary goal of FiTs is to accelerate the deployment of renewable energy and facilitate the transition to a low-carbon energy system by providing stable and predictable revenue streams for renewable energy producers. By offering guaranteed prices for renewable electricity generation, FiTs reduce the financial risks and uncertainties associated with renewable energy investments, making renewable energy projects more attractive to developers, investors, and lenders. Moreover, FiTs promote renewable energy development across a wide range of scales, from small-scale residential installations to large-scale utility projects, fostering diversification and decentralization of the energy supply.

FiTs also help to level the playing field for renewable energy technologies by internalizing the external costs of fossil fuel-based electricity generation, such as air pollution, GHG emissions, and environmental degradation. By ensuring that renewable energy producers receive fair compensation for the environmental and social benefits of their energy generation, FiTs create a more favorable market environment for renewable energy deployment, while also helping to address climate change, improve air quality, and enhance energy security.

Furthermore, FiTs can stimulate local economic development by creating jobs, fostering innovation, and attracting investment in renewable energy industries. By providing a stable and supportive policy framework for renewable energy development, FiTs encourage the growth of domestic renewable energy markets, supply chains, and manufacturing sectors, generating economic opportunities and contributing to sustainable economic growth. Moreover, FiTs can help to revitalize rural communities, promote agricultural diversification, and support rural livelihoods through the development of bioenergy and biomass projects.

### 1.3.4 Energy Market Reforms

Energy market reforms encompass a range of policy measures and regulatory changes aimed at restructuring and modernizing the energy sector to promote competition, efficiency, and sustainability. These reforms seek to address the challenges and opportunities posed by technological advancements, environmental concerns, and shifting market dynamics, while also fostering innovation, investment, and consumer choice in the energy market. Energy market reforms are driven by various factors, including the need to enhance energy security, reduce GHG emissions, and promote economic growth and social welfare. By reshaping the structure and operation of energy markets, reforms aim to create a more resilient, flexible, and sustainable energy system that can meet the evolving needs and preferences of society.

One of the key objectives of energy market reforms is to promote competition and consumer choice by opening up energy markets to new entrants, increasing market transparency, and removing barriers to entry and exit. Traditionally, energy markets have been dominated by vertically integrated monopolies or oligopolies, where a few large utilities control the entire value chain from generation to distribution. However, energy market reforms seek to introduce greater competition and diversity into the market by unbundling energy supply, transmission, and distribution activities, allowing independent power producers, renewable energy developers, and innovative technology providers to enter the market and compete on a level playing field. By fostering competition, energy market reforms can lead to lower prices, higher quality services, and greater innovation, ultimately benefiting consumers and driving economic growth.

***Key: Energy access and equity:*** *Ensuring affordable, reliable, and sustainable access to modern energy services for all individuals and communities, addressing disparities in energy access based on socioeconomic factors.*

Furthermore, energy market reforms aim to enhance market efficiency and optimize resource allocation by introducing market-based mechanisms for price discovery, supply–demand balancing, and investment decision-making. Traditional energy markets have often been characterized by rigid pricing structures, regulated tariffs, and centralized planning, which may not reflect the true costs and values of energy production, consumption, and externalities. However, energy market reforms seek to introduce market-based pricing mechanisms, such as wholesale electricity markets, capacity markets, and emissions trading systems, that allow supply and demand to be matched in real-time and incentivize investments in the most cost-effective and environmentally sustainable energy resources. By aligning prices with market fundamentals, energy market reforms can improve resource allocation, encourage efficient investment decisions, and reduce the need for costly government interventions and subsidies.

Moreover, energy market reforms aim to enhance grid flexibility, resilience, and reliability by modernizing grid infrastructure, improving grid management and coordination, and integrating new technologies and services into the grid. Traditional energy grids were designed for one-way flow of electricity from centralized power plants to end consumers, with limited visibility, control, and flexibility. However, energy market reforms seek to transform the grid into a dynamic and interactive platform that can accommodate two-way energy flows, variable renewable energy generation, DERs, and demand-side management strategies.

By upgrading grid infrastructure, deploying smart grid technologies, and implementing advanced grid management systems, energy market reforms can enhance grid stability, optimize energy flows, and support the integration of diverse energy resources, ultimately improving reliability and resilience in the face of disruptions and emergencies.

Furthermore, energy market reforms aim to promote sustainability and environmental stewardship by incentivizing investments in renewable energy, energy efficiency, and low-carbon technologies and internalizing the external costs of fossil fuel-based energy production through carbon pricing mechanisms and environmental regulations. Traditional energy markets have often failed to account for the negative environmental and social impacts of fossil fuel combustion, such as air pollution, GHG emissions, and ecosystem degradation, leading to market distortions and inefficiencies. However, energy market reforms seek to correct these market failures by introducing policies and regulations that encourage the transition to cleaner and more sustainable energy sources, such as renewable energy FiTs, energy efficiency mandates, and carbon pricing schemes. By internalizing the true costs of energy production and consumption, energy market reforms can create incentives for investments in clean energy technologies and practices, while also reducing environmental pollution and mitigating climate change.

In addition, energy market reforms aim to promote innovation and technological advancement by creating a conducive regulatory environment that fosters experimentation, entrepreneurship, and risk-taking in the energy sector. Traditional energy markets have often been characterized by regulatory barriers, bureaucratic inertia, and risk aversion, which may stifle innovation and impede the adoption of new technologies and business models. However, energy market reforms seek to create a more dynamic and adaptive regulatory framework that encourages experimentation, facilitates technology diffusion, and supports the scaling up of successful innovations. By providing clear and consistent rules, incentives, and support mechanisms, energy market reforms can unleash the creative potential of entrepreneurs, researchers, and investors to develop and deploy innovative solutions to energy challenges, such as advanced energy storage systems, smart grid technologies, and decentralized energy networks.

Moreover, energy market reforms aim to promote social equity and inclusivity by ensuring that the benefits of the energy transition are shared equitably among all segments of society, including low-income households, disadvantaged communities, and vulnerable populations. Traditional energy markets have often been characterized by unequal access to energy services, with marginalized groups facing barriers to energy access, affordability, and quality of service. However, energy market reforms seek to address these disparities by promoting universal access to affordable, reliable, and clean energy services through targeted policies and programs, such as energy subsidies, social tariffs, and community energy projects. By prioritizing social equity and inclusivity in energy policy-making and planning, energy market reforms can help to ensure that no one is left behind in the transition to a sustainable energy future.

Energy market reforms play a crucial role in shaping the future of the energy sector by promoting competition, efficiency, sustainability, and inclusivity. By modernizing energy markets, improving regulatory frameworks, and incentivizing investments in clean energy technologies, energy market reforms can unlock new opportunities for innovation, growth, and development, while also addressing pressing challenges such as climate change, energy poverty, and environmental degradation. However, the success of energy market reforms depends on effective implementation, stakeholder engagement, and ongoing monitoring and evaluation to ensure that policy objectives are achieved and that the benefits of the energy transition are shared equitably among all members of society.

## 1.4 Technological Innovations

### 1.4.1 Blockchain for Energy Transactions

Blockchain technology has emerged as a disruptive force in the energy sector, offering innovative solutions for enhancing transparency, efficiency, and security in energy transactions. Blockchain, often referred to as a decentralized ledger technology, enables the secure recording, verification, and tracking of transactions in a tamper-proof and transparent manner, without the need for intermediaries or centralized control. In the context of energy, blockchain has the potential to revolutionize the way energy is generated, distributed, traded, and consumed, by enabling P2P energy transactions, grid optimization, and decentralized energy management. One of the key applications of blockchain in the energy sector is in facilitating P2P energy transactions, where energy consumers can directly buy, sell, or exchange electricity with each other, bypassing traditional energy utilities and intermediaries. By leveraging blockchain-based smart contracts, energy producers can automatically execute and enforce agreements for energy exchange, ensuring transparency, reliability, and trust in energy transactions. Moreover, blockchain-based energy trading platforms enable energy consumers to monetize their excess energy production from renewable energy sources, such as rooftop solar panels, by selling surplus electricity to other consumers in their local community. This P2P energy trading model not only empowers energy consumers to take control of their energy usage and costs but also promotes the integration of renewable energy into the grid and reduces reliance on centralized fossil fuel-based power generation.

Another application of blockchain in the energy sector is in enabling more efficient and transparent energy markets, by providing a decentralized platform for trading energy products and services, such as electricity, capacity, and RECs. Blockchain-based energy trading platforms allow market participants to access real-time market data, match supply and demand, and settle transactions securely and instantly, without the need for intermediaries or central clearinghouses. By eliminating the need for complex and costly reconciliation processes, blockchain reduces transaction costs, improves market liquidity, and promotes greater market participation and competition. Moreover, blockchain-based energy markets enable the integration of DERs, demand-side management, and grid flexibility services, facilitating the transition to a more decentralized and resilient energy system.

Blockchain technology also holds promise for enhancing grid efficiency and resilience through the implementation of decentralized EMSs and P2P energy sharing networks. By enabling the secure and transparent exchange of energy data and transactions, blockchain can help utilities and grid operators optimize energy flows, balance supply and demand, and manage grid congestion more effectively. Moreover, blockchain-based microgrid platforms enable communities to establish localized energy systems that are resilient to disruptions and disasters, by enabling P2P energy sharing, autonomous energy management, and automatic load balancing. These decentralized EMSs not only enhance grid resilience and reliability but also promote energy autonomy, sustainability, and self-sufficiency at the local level.

> ***Key: Resilience to climate change:*** *Building adaptive capacity and robustness in energy infrastructure and systems to withstand and recover from the impacts of climate change, such as extreme weather events and sea-level rise.*

Furthermore, blockchain technology has the potential to improve energy access and affordability for underserved populations and remote communities, by enabling the deployment of decentralized energy systems and microgrids that are not reliant on centralized infrastructure or grid connections. In regions with limited access to electricity or unreliable grid infrastructure, blockchain-based energy solutions can provide a cost-effective and scalable alternative for meeting basic energy needs, such as lighting, cooking, and communication. Moreover, blockchain-based energy financing platforms enable individuals and communities to access affordable financing for renewable energy projects, such as solar home systems or mini-grids, by leveraging crowdfunding, P2P lending, and tokenization mechanisms. These innovative financing models help overcome traditional barriers to investment in energy access projects, such as a lack of collateral, credit history, or financial infrastructure, and empower communities to take ownership of their energy future.

Despite its potential benefits, blockchain technology faces several challenges and barriers to adoption in the energy sector, including regulatory uncertainty, scalability limitations, interoperability issues, and privacy concerns. Regulatory frameworks governing energy markets and transactions may not be sufficiently adapted to accommodate blockchain-based solutions, leading to legal and compliance risks for market participants. Moreover, the scalability of blockchain networks, particularly public blockchains, may be limited by transaction throughput and processing times, hindering their suitability for large-scale energy transactions and grid operations. Interoperability standards and protocols are needed to ensure compatibility and seamless integration of blockchain with existing energy systems, such as smart grids, EMSs, and market platforms. Additionally, privacy and data security concerns related to the transparency and immutability of blockchain ledgers may raise challenges for protecting sensitive energy data and ensuring consumer privacy and confidentiality.

Blockchain technology holds immense promise for transforming the energy sector by enabling more efficient, transparent, and decentralized energy transactions and management systems. From P2P energy trading to grid optimization and energy access, blockchain has the potential to revolutionize the way energy is generated, distributed, and consumed, paving the way for a more sustainable, resilient, and inclusive energy future. However, realizing the full potential of blockchain in the energy sector requires overcoming technical, regulatory, and institutional challenges, as well as fostering collaboration and innovation among stakeholders across the energy value chain. With continued investment, experimentation, and policy support, blockchain technology can play a key role in accelerating the transition to a cleaner, more efficient, and equitable energy system for all.

### 1.4.2 Artificial Intelligence in Grid Optimization

Artificial intelligence (AI) has emerged as a powerful tool for optimizing the operation and management of electrical grids, offering innovative solutions to complex challenges, such as grid stability, efficiency, and reliability. By leveraging advanced algorithms, machine learning techniques, and real-time data analytics, AI enables grid operators to make informed decisions, optimize energy flows, and anticipate and mitigate potential disruptions or failures in the grid infrastructure. One of the key applications of AI in grid optimization is in the prediction and forecasting of electricity demand, supply, and market prices, which are essential for efficient grid planning, scheduling, and dispatch. AI-based demand forecasting models analyze historical consumption data, weather patterns, economic indicators, and other relevant factors to accurately predict future energy demand at different time scales, from minutes to months ahead. Similarly, AI-powered supply forecasting models analyze data from renewable energy sources, conventional power plants, and energy storage systems to forecast generation output and availability, taking into account factors such as weather conditions, fuel prices, and plant performance. These predictive analytics enable grid operators to optimize resource allocation, balance supply and demand, and optimize energy procurement and dispatch strategies, thereby improving grid efficiency and reducing costs.

> **Key:** ***Energy conservation:*** *The practice of reducing energy consumption through efficiency improvements, behavioral changes, and technology adoption, minimizing waste and environmental impact while preserving resources.*

Another key application of AI in grid optimization is in the optimization of energy distribution and grid operations, where AI-based algorithms and optimization techniques are used to optimize the routing, scheduling, and control of energy flows in the grid infrastructure. AI-based optimization models analyze real-time data from sensors, meters, and other grid devices to identify optimal operating conditions, minimize energy losses, and maximize the utilization of grid assets and resources. For example, AI-based voltage control algorithms adjust voltage levels and reactive power flows in the grid to maintain stability, minimize losses, and optimize power quality. Similarly, AI-based distribution management systems optimize the operation of distribution networks, such as smart grids and microgrids, by coordinating the integration of DERs, managing grid congestion, and optimizing energy storage and demand response programs. These AI-powered optimization techniques enable grid operators to improve grid reliability, resilience, and performance, while also reducing operational costs and environmental impacts.

Moreover, AI plays a crucial role in enhancing grid resilience and reliability by enabling proactive maintenance, fault detection, and outage management strategies. AI-based predictive maintenance models analyze historical performance data, sensor readings, and equipment diagnostics to identify potential equipment failures or degradation trends before they occur, allowing grid operators to schedule preventive maintenance and avoid unplanned downtime. Similarly, AI-based fault detection and diagnostics systems analyze real-time data from grid sensors, meters, and other monitoring devices to detect and localize faults or anomalies in the grid infrastructure, such as equipment failures, line outages, or voltage fluctuations. These AI-powered analytics enable grid operators to quickly diagnose and respond to grid disturbances, restore service, and minimize disruptions to energy consumers, thereby improving grid reliability and resilience.

Furthermore, AI facilitates the integration of renewable energy sources, energy storage systems, and EVs into the grid by optimizing their operation, scheduling, and coordination. AI-based EMSs optimize the operation of renewable energy assets, such as solar PV panels, wind turbines, and hydroelectric plants, by forecasting generation output, optimizing energy capture, and coordinating energy storage and grid interactions. Similarly, AI-powered EV charging and management systems optimize the charging and discharging of EV batteries, taking into account factors such as energy prices, grid constraints, and user preferences. These AI-enabled technologies enable grid operators to integrate variable renewable energy sources and flexible loads into the grid more effectively, balance supply and demand, and optimize grid stability and resilience.

Moreover, AI facilitates the development of intelligent grid control and optimization strategies, such as autonomous grid operation, self-healing grids, and adaptive grid planning. AI-based control systems analyze real-time data from grid sensors, meters, and other devices to autonomously optimize grid operation, respond to changing conditions, and self-adapt to disturbances or failures. For example, AI-based autonomous grid controllers adjust grid parameters, such as voltage levels, power flows, and grid topology, to optimize grid performance, maximize energy efficiency, and ensure grid stability. Similarly, AI-powered self-healing grid systems automatically detect, isolate, and reconfigure grid components in response to faults or disturbances, enabling the grid to continue operating reliably and safely under adverse conditions. These AI-enabled grid control and optimization strategies enhance grid resilience, reliability, and responsiveness, while also reducing the need for human intervention and manual control.

In addition, AI enables grid operators to improve grid cybersecurity and threat detection by analyzing network traffic, monitoring system behavior, and detecting anomalies or suspicious activities that may indicate potential cyber threats or attacks. AI-based cybersecurity systems use machine learning algorithms to identify patterns, detect anomalies, and classify threats in real-time, enabling grid operators to respond quickly and effectively to cyber incidents, protect critical infrastructure, and safeguard grid operations and data. Moreover, AI-powered threat intelligence platforms analyze data from multiple sources, such as security logs, incident reports, and threat feeds, to identify emerging threats, vulnerabilities, and attack vectors, enabling grid operators to proactively mitigate risks and strengthen their cybersecurity defenses. These AI-driven cybersecurity solutions enhance grid resilience, reliability, and security, while also reducing the risk of cyber-attacks and disruptions to energy supply.

Overall, AI holds immense promise for optimizing grid operations, enhancing grid resilience, and enabling the transition to a more sustainable, efficient, and reliable energy system. By leveraging advanced analytics, optimization techniques, and autonomous control systems, AI enables grid operators to improve grid efficiency, reliability, and responsiveness, while also integrating renewable energy sources, energy storage systems, and EVs into the grid more effectively. However, the successful deployment of AI in the grid optimization requires addressing technical, regulatory, and institutional challenges, as well as fostering collaboration and innovation among stakeholders across the energy value chain. With continued investment, research, and development, AI has the potential to revolutionize the energy sector and accelerate the transition to a cleaner, more resilient, and sustainable energy future.

### 1.4.3 Internet of Things (IoT) for Energy Management

The Internet of Things (IoT) has emerged as a transformative technology in the field of energy management, offering innovative solutions for optimizing energy consumption, improving operational efficiency, and enhancing grid reliability and resilience. IoT refers to the network of interconnected devices, sensors, and systems that collect, exchange, and analyze data in real-time, enabling remote monitoring, control, and automation of various energy-related processes and assets. In the context of energy management, IoT enables the integration of smart sensors, meters, actuators, and control systems into energy infrastructure, such as buildings, industrial facilities, and power grids, to enable more efficient, reliable, and sustainable energy use.

One of the key applications of IoT in energy management is in the monitoring and optimization of energy consumption in buildings and facilities. IoT-enabled smart building systems leverage sensors, meters, and connected devices to collect real-time data on energy usage, occupancy patterns, indoor environmental conditions, and equipment performance. This data is then analyzed using advanced analytics and machine learning algorithms to identify energy-saving opportunities, optimize building operations, and reduce energy waste. For example, IoT-enabled building management systems can automatically adjust lighting, heating, and cooling systems based on occupancy levels, weather forecasts, and energy prices, to minimize energy consumption while maintaining comfort and productivity. Similarly, IoT-enabled energy management platforms provide building owners and facility managers with actionable insights and recommendations for improving energy efficiency, reducing costs, and achieving sustainability goals.

> ***Key: Demand response of electricity:*** *The ability of electricity consumers to adjust their energy usage in response to signals from grid operators or price fluctuations, helping to balance supply and demand and optimize grid performance.*

Moreover, IoT facilitates the integration of DERs, such as solar PV panels, energy storage systems, and EVs, into the grid by enabling real-time monitoring, control, and coordination of these assets. IoT-enabled smart grid technologies leverage sensors, communication networks, and control systems to collect data on energy production, consumption, and grid conditions, enabling grid operators to optimize grid operations, balance supply and demand, and integrate renewable energy sources more effectively. For example, IoT-enabled distributed EMSs enable grid operators to aggregate and dispatch DERs, such as rooftop solar panels and battery storage systems, to provide ancillary services, such as frequency regulation and voltage support, to the grid. Similarly, IoT-enabled demand response programs enable energy consumers to adjust their energy consumption in response to grid conditions or price signals, reducing peak demand and alleviating grid congestion.

Furthermore, IoT plays a crucial role in enhancing grid reliability and resilience by enabling proactive maintenance, fault detection, and outage management strategies. IoT-enabled asset monitoring systems collect real-time data on grid equipment, such as transformers, switches, and substations, to assess equipment health, detect anomalies, and predict potential failures or degradation. This data is then analyzed using machine learning algorithms to identify early warning signs of equipment malfunctions or failures, enabling grid operators to schedule preventive maintenance, replace faulty components, and avoid unplanned downtime. Moreover, IoT-enabled fault detection and diagnostics systems analyze real-time data from grid sensors, meters, and other monitoring devices to detect and localize faults or disturbances in the grid infrastructure, such as equipment failures, line outages, or voltage fluctuations. These IoT-powered analytics enable grid operators to quickly diagnose and respond to grid disturbances, restore service, and minimize disruptions to energy consumers, thereby improving grid reliability and resilience.

Additionally, IoT enables the development of intelligent EMSs and optimization algorithms that leverage real-time data and predictive analytics to optimize energy production, distribution, and consumption across the grid. IoT-enabled energy management platforms integrate data from multiple sources, such as sensors, meters, weather forecasts, and market prices, to optimize energy flows, balance supply and demand, and maximize the utilization of grid assets and resources. For example, IoT-enabled energy optimization algorithms can dynamically adjust energy production and consumption schedules based on real-time grid conditions, energy prices, and environmental factors, to minimize costs, reduce emissions, and enhance grid stability. Similarly, IoT-enabled energy forecasting models use historical data and machine learning techniques to predict future energy demand, supply, and market prices, enabling grid operators to optimize energy procurement, dispatch, and trading strategies.

Moreover, IoT enables the development of innovative energy services and business models that empower consumers to actively participate in energy management and contribute to grid reliability and sustainability. IoT-enabled energy monitoring and control devices, such as smart thermostats, smart appliances, and home EMSs, enable consumers to monitor their energy usage in real-time, adjust their energy consumption patterns, and optimize their energy costs. Similarly, IoT-enabled energy management platforms provide consumers with personalized insights and recommendations for reducing energy waste, improving efficiency, and achieving energy savings. Furthermore, IoT-enabled P2P energy trading platforms enable energy consumers to buy, sell, or exchange electricity directly with each other, bypassing traditional energy utilities and intermediaries. These IoT-powered energy services and business models empower consumers to take control of their energy usage, reduce their carbon footprint, and contribute to a more sustainable and resilient energy future.

IoT holds immense promise for transforming energy management and enabling the transition to a more efficient, reliable, and sustainable energy system. By leveraging real-time data, advanced analytics, and interconnected devices, IoT enables grid operators, energy managers, and consumers to optimize energy consumption, integrate renewable energy sources, and enhance grid reliability and resilience. However, the successful deployment of IoT in energy management requires addressing technical, regulatory, and institutional challenges, as well as fostering collaboration and innovation among stakeholders across the energy value chain. With continued investment, research, and development, IoT has the potential to revolutionize energy management and accelerate the transition to a cleaner, more resilient, and sustainable energy future.

### 1.4.4 Advanced Analytics for Predictive Maintenance

Advanced analytics has become increasingly vital in predictive maintenance strategies for energy sources, enabling proactive and data-driven approaches to asset management, optimization, and reliability. Leveraging techniques such as machine learning, AI, and big data analytics, advanced analytics empowers energy operators to predict equipment failures, schedule maintenance activities, and optimize asset performance to ensure uninterrupted energy supply and minimize downtime.

One of the key applications of advanced analytics in predictive maintenance is in forecasting equipment failures and degradation trends. By analyzing historical performance data, sensor readings, and operational parameters, advanced analytics models can identify patterns and anomalies indicative of impending equipment failures. These models can predict the likelihood and timing of failures for critical energy assets such as turbines, generators, transformers, and boilers, enabling operators to proactively schedule maintenance interventions before failures occur. By addressing potential issues before they escalate into costly breakdowns, predictive maintenance helps optimize asset lifespan, reduce repair costs, and enhance overall system reliability.

Moreover, advanced analytics enables condition-based maintenance strategies by continuously monitoring equipment health and performance in real-time. Through the integration of sensor networks, IoT devices, and data analytics platforms, energy operators can collect and analyze vast amounts of data on equipment condition, operational parameters, and environmental factors. By detecting early warning signs of equipment deterioration or abnormalities, advanced analytics systems trigger alerts and notifications to maintenance personnel, enabling timely interventions to prevent failures and optimize asset performance. Condition-based maintenance minimizes downtime, extends equipment lifespan, and maximizes operational efficiency by focusing maintenance efforts on assets that require immediate attention, rather than adhering to fixed schedules.

Furthermore, advanced analytics facilitates predictive maintenance optimization by identifying optimal maintenance strategies and schedules based on data-driven insights and risk assessments. Through the analysis of historical maintenance records, failure data, and asset performance metrics, advanced analytics models can determine the most cost-effective and efficient maintenance practices for different types of energy assets. These models take into account factors such as asset criticality, failure probabilities, maintenance costs, and operational priorities to optimize maintenance schedules, resource allocation, and inventory management. By aligning maintenance activities with business objectives and operational constraints, predictive maintenance optimization maximizes asset availability, minimizes maintenance costs, and enhances overall system performance.

Additionally, advanced analytics enables prognostic maintenance capabilities by forecasting equipment degradation and performance deterioration over time. By analyzing sensor data, operational logs, and historical trends, advanced analytics models can predict the future performance and condition of energy assets, allowing operators to anticipate maintenance needs and plan interventions accordingly. Prognostic maintenance provides valuable insights into the health and reliability of critical assets, enabling operators to make informed decisions about asset lifecycle management, investment prioritization, and risk mitigation strategies. By adopting proactive and forward-looking maintenance approaches, energy operators can minimize downtime, optimize asset utilization, and ensure the long-term reliability and resilience of their energy infrastructure.

Moreover, advanced analytics supports real-time monitoring and optimization of energy generation and distribution systems, enabling operators to optimize energy production, reduce costs, and enhance grid reliability. Through the integration of sensor data, weather forecasts, market prices, and operational constraints, advanced analytics platforms can optimize energy dispatch, scheduling, and trading decisions to maximize revenue and minimize operational costs. These platforms leverage predictive models, optimization algorithms, and scenario analysis techniques to forecast energy demand, anticipate supply fluctuations, and optimize generation and storage resources accordingly. By enabling real-time decision-making and adaptive control, advanced analytics enhances grid flexibility, stability, and resilience, enabling energy operators to respond rapidly to changing market conditions, grid disturbances, and operational challenges.

Advanced analytics plays a crucial role in predictive maintenance strategies for energy sources, enabling proactive and data-driven approaches to asset management, optimization, and reliability. By forecasting equipment failures, optimizing maintenance schedules, and monitoring asset performance in real-time, advanced analytics empowers energy operators to maximize asset uptime, minimize downtime, and ensure the continuous and reliable supply of energy. Moreover, by supporting real-time monitoring and optimization of energy generation and distribution systems, advanced analytics enables operators to optimize energy production, reduce costs, and enhance grid reliability and resilience. With continued advancements in data analytics, machine learning, and IoT technologies, advanced analytics will continue to revolutionize predictive maintenance practices and drive innovation in the energy industry, paving the way for a more efficient, reliable, and sustainable energy future.

## 1.5 Energy Conservation and Demand Response

The implementation of energy-efficient technologies in residential, commercial, and industrial sectors is crucial for reducing energy consumption, lowering GHG emissions, and achieving sustainability goals. Following are the energy-efficient technologies being applied in each sector:

**Residential Sector:** In the residential sector, energy-efficient technologies play a vital role in improving home energy performance, reducing energy bills, and enhancing occupant comfort. One of the most widely adopted measures is energy-efficient lighting, where switching to LED bulbs and fixtures reduces electricity use while providing better illumination. Similarly, energy-efficient appliances, such as refrigerators, washing machines, and air conditioners, help household's lower energy consumption and achieve long-term cost savings. Smart thermostats add further efficiency by optimizing heating and cooling schedules based on occupancy patterns, weather conditions, and energy prices. Additionally, insulation and weatherization measures—such as sealing air leaks, installing high-performance windows, and adding insulation—reduce heating and cooling loads while maintaining indoor comfort. Finally, renewable energy systems like rooftop solar panels, solar water heaters, or small-scale wind turbines allow homeowners to generate clean power on-site, lowering reliance on grid electricity.

**Commercial Sector:** In the commercial sector, energy-efficient technologies are essential for reducing operating costs, improving competitiveness, and meeting sustainability targets. Building automation systems play a central role by integrating HVAC, lighting, and security systems to optimize energy use and enhance operational efficiency. Facility managers also rely on EMSs to monitor and control energy consumption in real-time, identify opportunities for savings, and measure performance against established targets. Upgrading to high-efficiency HVAC systems can drastically reduce energy costs while maintaining comfort and indoor air quality. Furthermore, energy-efficient lighting and controls, such as daylighting systems and occupancy sensors, help lower electricity consumption in commercial spaces. Lastly, building envelope improvements, including retrofits with insulation, advanced windows, and roofing materials, enhance thermal performance, reduce energy demand, and increase efficiency.

**Industrial Sector:** In the industrial sector, adopting energy-efficient technologies is critical for optimizing production processes, reducing energy intensity, and enhancing competitiveness. For example, high-efficiency motors and variable frequency drives (VFDs) improve the performance of pumps, fans, and compressors, resulting in substantial energy savings. Process optimization, supported by automation technologies and advanced EMSs, ensures efficient use of resources and minimizes energy waste. CHP systems further boost efficiency by simultaneously generating electricity and thermal energy from a single fuel source, lowering emissions compared to traditional methods. Industries can also benefit from waste heat recovery, which captures and reuses heat from processes in manufacturing, refining, or chemical production, thereby reducing fuel consumption. Finally, energy-efficient lighting and industrial equipment upgrades contribute to cost reduction, improved productivity, and better energy efficiency.

The implementation of energy-efficient technologies in residential, commercial, and industrial sectors is essential for achieving energy savings, reducing environmental impacts, and promoting sustainable development. By investing in energy efficiency measures and deploying advanced technologies, stakeholders across these sectors can improve energy performance, lower operating costs, and contribute to a more sustainable and resilient energy future. The adoption of energy-efficient technologies across residential, commercial, and industrial sectors is essential for achieving energy savings, reducing environmental impacts, and promoting sustainable development. By investing in advanced technologies and efficiency measures, stakeholders can lower operating costs, improve overall energy performance, and contribute to building a sustainable and resilient energy future.

- **Behavioral Changes and Best Practices for Reducing Energy Consumption**

  Behavioral changes and best practices are essential for reducing energy consumption and promoting energy efficiency in various settings, including homes, workplaces, and communities. Following are some effective strategies for encouraging energy-saving behaviors:

  **Education and Awareness:** Increasing awareness about the importance of energy conservation and providing information about energy-saving practices can motivate individuals to adopt more sustainable behaviors. Educational campaigns, workshops, and outreach programs can help raise awareness about energy efficiency and empower people to take action to reduce their energy consumption.

  **Setting Energy-Saving Goals:** Establishing clear energy-saving goals and targets can help individuals and organizations track their progress and stay motivated to achieve energy efficiency objectives. Setting achievable goals, such as reducing energy usage by a certain percentage or implementing specific energy-saving measures, can provide a sense of direction and purpose.

  **Behavioral Incentives and Rewards:** Providing incentives, rewards, or recognition for energy-saving behaviors can encourage individuals to adopt more sustainable habits. Rewards can take various forms, such as financial incentives, rebates, discounts on energy-efficient products, or nonmonetary rewards like recognition or social acknowledgment.

  **Feedback and Monitoring:** Providing real-time feedback on energy usage and performance can help individuals and organizations understand their energy consumption patterns, identify opportunities for improvement, and adjust their behavior accordingly. Energy monitoring systems, smart meters, and energy dashboards can provide feedback on energy usage, costs, and savings, empowering users to make informed decisions about their energy consumption.

  **Efficient Equipment and Appliances:** Encouraging the use of energy-efficient equipment, appliances, and technologies can significantly reduce energy consumption in homes, workplaces, and businesses. Energy-efficient appliances, HVAC systems, lighting fixtures, and office equipment can help minimize energy waste and lower utility bills without sacrificing performance or comfort.

  **Behavioral Nudges and Prompts:** Employing behavioral science principles such as nudges and prompts can influence people's behavior and encourage energy-saving actions. Simple interventions, such as reminders, prompts, or visual cues, can prompt individuals to turn off lights, unplug electronics, adjust thermostats, or adopt other energy-saving behaviors.

**Energy-Efficient Practices and Habits:** Encouraging energy-efficient practices and habits, such as turning off lights when leaving a room, unplugging idle electronics, using natural light and ventilation, and adjusting thermostat settings, can help individuals reduce energy consumption and save money on utility bills.

**Community Engagement and Collaboration:** Fostering a sense of community and collective responsibility for energy conservation can motivate individuals to adopt energy-saving behaviors and initiatives. Community-based programs, neighborhood competitions, and collaborative projects can encourage collective action and inspire people to work together toward common energy-saving goals.

**Employee Training and Engagement:** Providing training and education for employees on energy-saving practices and workplace policies can empower them to take responsibility for energy conservation in their workplace. Engaging employees in energy-saving initiatives, such as energy audits, green teams, or sustainability committees, can foster a culture of energy efficiency and innovation in the workplace.

**Continuous Improvement and Innovation:** Encouraging a culture of continuous improvement and innovation can drive ongoing energy efficiency efforts and inspire new ideas and solutions for reducing energy consumption. Encouraging experimentation, creativity, and knowledge sharing can help uncover new opportunities for energy savings and promote a culture of sustainability and environmental stewardship.

***Key: Industrial energy conservation:*** *Measures and practices implemented to improve energy efficiency and reduce energy consumption in industrial processes and facilities, lowering operating costs and environmental footprint.*

### 1.5.1 Demand Response Mechanisms

In contemporary energy systems, maintaining a delicate equilibrium between supply and demand is paramount for ensuring grid stability, reliability, and efficiency. However, the growing complexity of energy networks, coupled with the increasing penetration of intermittent renewable energy sources, poses significant challenges to this balance. In this context, demand response mechanisms emerge as pivotal tools for actively managing electricity consumption patterns to align with fluctuating supply conditions. This comprehensive overview explores the fundamentals of demand response mechanisms, their operational frameworks, benefits, challenges, and their indispensable role in balancing supply and demand within modern energy systems.

➢ **Understanding Demand Response Mechanisms**

Demand response mechanisms encompass a diverse array of strategies aimed at incentivizing consumers to adjust their electricity usage in response to signals reflecting grid conditions, price dynamics, or environmental constraints. These mechanisms can be broadly categorized into:

**Price-Based Mechanisms:** Price-based demand response programs leverage dynamic pricing signals to incentivize consumers to reduce or shift their electricity consumption during peak periods when prices are high. TOU tariffs, critical peak pricing (CPP), and real-time pricing (RTP) are common examples of price-based mechanisms.

**Incentive-Based Mechanisms:** Incentive-based demand response programs offer financial or nonfinancial incentives to consumers in exchange for reducing their electricity usage during periods of high demand or supply constraints. These incentives may include rebates, bill credits, or access to value-added services.

**Capacity-Based Mechanisms:** Capacity-based demand response programs focus on reducing overall electricity consumption during times of grid stress to alleviate capacity constraints and prevent system overloads. These programs typically involve curtailment agreements with large industrial and commercial consumers or aggregated demand response from DERs.

➢ **Operational Framework of Demand Response Mechanisms**

The operational framework of demand response mechanisms involves several key components:

**Communication and Signaling:** Effective demand response requires robust communication channels and signaling mechanisms to convey price signals, event notifications, and instructions to participating consumers. AMI, smart grid technologies, and demand response platforms facilitate real-time communication and interaction between utilities and consumers.

**Measurement and Verification:** Accurate measurement and verification of demand response actions are essential for ensuring program integrity, transparency, and accountability. Metering devices, data analytics, and performance evaluation methodologies enable utilities to quantify demand reductions, verify compliance, and allocate incentives accordingly.

**Automation and Control:** Automation and control technologies play a crucial role in enabling seamless and responsive demand response actions. Smart thermostats, programmable appliances, building energy management systems (BEMS), and DERs equipped with automation capabilities empower consumers to adjust their electricity usage automatically in accordance with predefined preferences or external signals.

➤ **Benefits of Demand Response Mechanisms**

Demand response mechanisms offer a myriad of benefits to utilities, consumers, and the broader energy system:

**Grid Stability and Reliability:** By reducing peak demand and alleviating grid congestion, demand response mechanisms enhance grid stability, reliability, and resilience, thereby minimizing the risk of blackouts, brownouts, and voltage fluctuations.

**Optimized Resource Utilization:** Demand response enables more efficient utilization of existing generation, transmission, and distribution infrastructure by flattening demand peaks, reducing the need for costly capacity additions, and optimizing the dispatch of generation resources.

**Cost Savings and Economic Efficiency:** By incentivizing load reductions during periods of high electricity prices or capacity shortages, demand response mechanisms can lower wholesale energy costs, mitigate market price volatility, and reduce overall electricity bills for consumers.

**Environmental Sustainability:** Demand response contributes to environmental sustainability by facilitating the integration of renewable energy sources, reducing reliance on fossil fuel-based peaking units, and lowering GHG emissions associated with electricity generation.

➤ **Challenges and Barriers**

Despite their numerous benefits, demand response mechanisms face several challenges and barriers to widespread adoption:

**Technical and Operational Complexity:** Implementing demand response programs requires sophisticated infrastructure, advanced metering technologies, and robust communication networks, which can be costly and complex to deploy, especially for smaller utilities or consumers with limited resources.

**Behavioral and Institutional Barriers:** Consumer participation in demand response programs may be hindered by inertia, lack of awareness, or perceived inconvenience, particularly in residential and small commercial sectors. Moreover, regulatory barriers, market structures, and contractual arrangements may pose obstacles to the effective implementation of demand response initiatives.

**Privacy and Data Security Concerns:** The proliferation of smart meters and demand response technologies raises concerns about data privacy, security, and potential misuse of consumer information. Safeguarding sensitive data, ensuring consent-based participation, and implementing robust cybersecurity measures are essential to building trust and confidence among consumers.

**Interoperability and Standardization:** Achieving interoperability and standardization across diverse demand response technologies, communication protocols, and grid interfaces remains a challenge, inhibiting seamless integration and scalability of demand response solutions.

- **Future Directions and Emerging Trends**
  Looking ahead, several emerging trends and developments are poised to shape the future landscape of demand response:

  **Advanced Analytics and AI:** The proliferation of data analytics, machine learning, and AI technologies enables utilities to develop more sophisticated demand response algorithms, predictive models, and personalized engagement strategies, enhancing the effectiveness and responsiveness of demand-side management initiatives.

  **Integration of DERs:** The rise of DERs, including rooftop solar, energy storage systems, and EVs, presents new opportunities for demand response integration and aggregation. By harnessing the flexibility of DERs, utilities can enhance grid flexibility, resilience, and renewable energy integration.

  **Demand Flexibility Programs:** Innovative demand flexibility programs, such as demand flexibility auctions, demand-side bidding platforms, and transactive energy systems, enable consumers to monetize their demand response capabilities and participate directly in wholesale energy markets, unlocking new revenue streams and incentives for proactive energy management.

  **Policy Support and Regulatory Reforms:** Policy support, regulatory reforms, and market incentives are essential for fostering an enabling environment conducive to demand response deployment and adoption. Policy measures such as capacity markets, demand response mandates, and performance-based incentives can incentivize utilities and consumers to invest in demand-side resources and infrastructure.

  Demand response mechanisms play a pivotal role in balancing supply and demand within modern energy systems, offering a flexible, cost-effective, and sustainable approach to grid management and optimization. By leveraging dynamic pricing signals, incentives, and automation technologies, demand response enables utilities and consumers to collaboratively manage electricity consumption, enhance grid reliability, and accelerate the transition toward a more resilient, efficient, and low-carbon energy future. While demand response faces challenges and barriers to widespread adoption, ongoing technological innovations, policy support, and market reforms are driving momentum toward broader deployment and integration of demand-side resources. By harnessing the full potential of demand response mechanisms, stakeholders can unlock new opportunities for grid optimization, energy efficiency, and environmental sustainability, ultimately shaping a more resilient and adaptive energy landscape for future generations.

### 1.5.2 Peak Load Management

In modern energy systems, peak load management plays a crucial role in maintaining grid stability, reliability, and efficiency. Peaks in electricity demand, often occurring during periods of extreme weather, industrial activity, or residential consumption, pose significant challenges to grid operators and utilities. Effective peak load management strategies aim to mitigate these challenges by strategically reducing or shifting electricity consumption during peak periods, thereby alleviating strain on the grid, optimizing resource utilization, and enhancing overall system performance. This comprehensive note provides an overview of peak load management, its significance, operational strategies, benefits, challenges, and future directions.

- **Understanding Peak Load**

  Peak load refers to the maximum level of electricity demand experienced within a specific timeframe, typically occurring during periods of high energy consumption. Peaks in electricity demand can result from various factors, including:

  **Weather Conditions:** Extreme temperatures, such as heatwaves or cold snaps, can drive up energy demand for heating, cooling, and ventilation systems, leading to peak load events.

  **Industrial Processes:** Industrial facilities, manufacturing plants, and commercial establishments may experience spikes in electricity usage during peak production hours or operational cycles, contributing to overall system peaks.

  **Residential Consumption:** Peak load can also be influenced by residential consumption patterns, such as increased usage of appliances, lighting, and electronics during evening hours or weekends.

  **Seasonal Variations:** Seasonal factors, such as holidays, vacations, and cultural events, can have an impact on electricity demand patterns and contribute to seasonal peaks throughout the year.

  Understanding the drivers and characteristics of peak load is essential for devising effective peak load management strategies tailored to specific market conditions, geographical regions, and customer segments.

- **Significance of Peak Load Management**

  Peak load management is of paramount importance for ensuring grid stability, reliability, and cost-effectiveness in energy systems:

  **Grid Stability:** Peaks in electricity demand can strain grid infrastructure, leading to voltage fluctuations, transmission congestion, and potential system failures. By managing peak load, utilities can maintain grid stability and prevent disruptions in service delivery.

  **Resource Optimization:** Peak load management enables utilities to optimize the utilization of generation, transmission, and distribution assets, reducing the need for costly investments in additional capacity and infrastructure upgrades.

  **Cost Reduction:** Peaks in electricity demand often coincide with higher wholesale energy prices and increased operating costs for utilities. By reducing peak demand through demand-side management initiatives, utilities can lower their overall energy procurement expenses and pass on cost savings to consumers.

**Environmental Impact:** Peak load management can contribute to environmental sustainability by reducing the reliance on fossil fuel-based peaking power plants, which are often less efficient and more polluting than base load or renewable energy sources.

➢ **Operational Strategies for Peak Load Management**

Peak load management encompasses a variety of operational strategies and demand-side management techniques aimed at mitigating peak demand and optimizing grid performance:

**Demand Response Programs:** Demand response programs incentivize consumers to voluntarily reduce or shift their electricity usage during peak periods in exchange for financial incentives, bill credits, or other rewards. These programs leverage price signals, event notifications, or automated controls to modulate electricity consumption in real-time.

**TOU Pricing:** TOU pricing structures charge consumer's different rates for electricity consumption based on the time of day or day of the week. By incentivizing off-peak usage and discouraging consumption during peak periods, TOU pricing encourages load shifting and demand management.

**Energy Efficiency Measures:** Investing in energy efficiency measures, such as efficient appliances, lighting upgrades, insulation, and building retrofits, can reduce overall electricity demand and mitigate peak load events by lowering the energy intensity of buildings, industrial processes, and transportation.

**Distributed Energy Resources (DERs):** DERs, including rooftop solar PV, energy storage systems, CHP units, and demand-side resources, can play a significant role in peak load management by providing on-site generation, storage, and demand flexibility capabilities.

**Load Shifting and Scheduling:** Load shifting strategies involve rescheduling energy-intensive activities or processes to off-peak hours when electricity demand and prices are lower. For example, scheduling laundry, dishwashing, or HVAC operation during off-peak periods can help flatten demand peaks and reduce overall system stress.

➢ **Benefits of Peak Load Management**

Effective peak load management offers a range of benefits to utilities, consumers, and the broader energy system:

**Grid Reliability:** By reducing peak demand and alleviating grid congestion, peak load management enhances grid reliability, resilience, and stability, thereby minimizing the risk of blackouts, brownouts, and voltage fluctuations.

**Cost Savings:** Peak load management can result in significant cost savings for utilities and consumers by reducing peak demand charges, avoiding capacity investments, and optimizing energy procurement strategies.

**Environmental Sustainability:** By promoting energy efficiency, demand response, and renewable energy integration, peak load management contributes to environmental sustainability by lowering GHG emissions, reducing reliance on fossil fuels, and supporting the transition to a low-carbon energy future.

**Consumer Empowerment:** Peak load management programs empower consumers to actively participate in grid management and energy conservation efforts, fostering a culture of energy awareness, responsibility, and engagement among end-users.

➢ **Challenges and Considerations**

Despite its numerous benefits, peak load management faces several challenges and considerations:

**Behavioral Barriers:** Consumer participation in peak load management programs may be hindered by inertia, lack of awareness, or perceived inconvenience, particularly in residential and small commercial sectors. Overcoming these behavioral barriers requires targeted outreach, education, and incentive structures tailored to specific customer segments.

**Technological Complexity:** Implementing peak load management strategies often requires sophisticated technology infrastructure, including advanced metering systems, data analytics platforms, and automation tools, which can be costly and complex to deploy, especially for smaller utilities or resource-constrained consumers.

**Regulatory and Policy Frameworks:** Regulatory barriers, market structures, and policy frameworks may pose obstacles to the effective implementation of peak load management initiatives, particularly in deregulated or fragmented energy markets where incentives for demand-side management may be limited or unclear.

**Equity and Accessibility:** Ensuring equity and accessibility in peak load management programs is essential to prevent disproportionate impacts on vulnerable or low-income communities. Designing inclusive policies, outreach strategies, and incentive structures can help address equity concerns and ensure equitable distribution of benefits and burdens.

### 1.5.3 Industrial Energy Conservation

Industrial energy conservation is a critical component of global efforts to mitigate climate change, enhance energy security, and promote sustainable development.

The industrial sector accounts for a significant share of global energy consumption and GHG emissions, making it a key focus area for energy efficiency improvements. This short note provides an overview of industrial energy conservation, its importance, challenges, strategies, technologies, and policy frameworks aimed at reducing energy intensity and enhancing competitiveness in manufacturing industries.

- **Importance of Industrial Energy Conservation**

  The industrial sector plays a vital role in the global economy, providing goods and services essential for modern society. However, industrial activities are also major consumers of energy, accounting for a substantial portion of total energy use worldwide. The importance of industrial energy conservation stems from several factors:

  **Energy Intensity:** Industrial processes often require large amounts of energy to produce goods and materials, resulting in high energy intensity levels compared to other sectors. Improving energy efficiency in industrial operations can significantly reduce energy consumption per unit of output, leading to cost savings and environmental benefits.

  **Competitiveness:** Energy costs represent a significant portion of operating expenses for many industrial facilities. By adopting energy conservation measures, manufacturers can improve their competitiveness by reducing production costs, enhancing productivity, and gaining a competitive edge in global markets.

  **Environmental Impact:** Industrial activities are major contributors to GHG emissions, air pollution, and resource depletion. Energy conservation in the industrial sector can help mitigate these environmental impacts by reducing carbon emissions, improving air quality, and conserving natural resources.

  **Resource Efficiency:** Industrial energy conservation is closely linked to resource efficiency, as energy is often derived from finite resources such as fossil fuels. By minimizing energy waste and optimizing resource utilization, industrial facilities can enhance resource efficiency and contribute to sustainable resource management.

- **Challenges in Industrial Energy Conservation**

  Despite its importance, industrial energy conservation faces several challenges and barriers to implementation:

  **Capital Constraints:** Many energy efficiency projects require upfront investments in equipment upgrades, process improvements, and technology adoption. Capital constraints, particularly for small- and medium-sized enterprises

(SMEs), can hinder investment in energy conservation measures, even if the long-term benefits are significant.

**Complexity of Industrial Processes:** Industrial operations are often characterized by complex and diverse processes with unique energy requirements. Identifying and implementing energy conservation opportunities in such environments can be challenging, requiring detailed process knowledge, technical expertise, and customized solutions.

**Behavioral and Organizational Factors:** The success of energy conservation initiatives in industrial settings depends not only on technological solutions but also on the behavior and attitudes of employees, management, and other stakeholders. Resistance to change, lack of awareness, and competing priorities within organizations can impede efforts to adopt energy-efficient practices.

**Split Incentives:** In cases where energy costs are shared between landlords and tenants or between different departments within an organization, split incentives may arise, leading to a lack of motivation to invest in energy conservation measures. Aligning incentives and overcoming barriers to cooperation is essential for effective energy management.

➢ **Strategies for Industrial Energy Conservation**

A variety of strategies and approaches can be employed to achieve energy conservation in industrial settings:

**Energy Audits and Assessments:** Conducting energy audits and assessments is a crucial first step in identifying energy-saving opportunities and optimizing energy use in industrial facilities. These assessments involve analyzing energy consumption patterns, identifying inefficiencies, and recommending tailored solutions to improve energy performance.

**Energy Management Systems (EMSs):** Implementing robust EMSs, such as ISO 50001, provides a systematic framework for establishing energy objectives, targets, and action plans, as well as monitoring, measuring, and continuously improving energy performance.

**Process Optimization:** Optimizing industrial processes to minimize energy waste and maximize efficiency is a fundamental strategy for energy conservation. This may involve upgrading equipment, retrofitting existing systems, implementing process controls, and optimizing production schedules to align with energy demand patterns.

**Technology Upgrades and Retrofits:** Investing in energy-efficient technologies, equipment upgrades, and retrofits can yield substantial energy savings in industrial operations. This may include replacing outdated machinery with high-efficiency models, installing energy-efficient lighting and HVAC systems, and adopting advanced control systems and automation technologies.

**Combined Heat and Power (CHP):** CHP systems, also known as cogeneration, capture waste heat from industrial processes and use it to generate electricity or provide heating and cooling services. CHP systems can significantly improve energy efficiency and reduce GHG emissions in energy-intensive industries.

**Renewable Energy Integration:** Incorporating renewable energy sources, such as solar, wind, biomass, and geothermal, into industrial energy systems can diversify the energy mix, reduce reliance on fossil fuels, and mitigate carbon emissions. On-site renewable energy installations and power purchase agreements (PPAs) with renewable energy providers are common approaches.

➢ **Emerging Technologies for Industrial Energy Conservation**

Advancements in technology continue to drive innovation in industrial energy conservation:

**Industrial Internet of Things (IIoT):** The IIoT enables connectivity, data collection, and real-time monitoring of equipment and processes in industrial settings. IIoT platforms leverage sensors, actuators, and cloud-based analytics to optimize energy use, detect inefficiencies, and enable predictive maintenance.

**Big Data Analytics:** Big data analytics solutions analyze vast amounts of data generated by industrial processes to identify patterns, anomalies, and optimization opportunities. By leveraging machine learning algorithms and predictive analytics, industrial facilities can optimize energy consumption, predict equipment failures, and optimize production schedules.

**Artificial Intelligence (AI):** AI technologies, such as machine learning, neural networks, and optimization algorithms, are increasingly being used to optimize energy-intensive processes, control systems, and supply chain operations in industrial settings. AI-powered EMSs can adapt to changing conditions, learn from historical data, and optimize energy usage in real-time.

**Advanced Materials and Manufacturing Techniques:** Advances in materials science and manufacturing techniques enable the development of lightweight materials, energy-efficient coatings, and high-performance components that reduce energy consumption and improve the efficiency of industrial processes.

➢ **Policy Frameworks and Regulations**

Government policies, regulations, and incentives play a crucial role in driving industrial energy conservation:

**Energy Efficiency Standards:** Establishing energy efficiency standards and regulations for industrial equipment, appliances, and processes incentivizes manufacturers to adopt energy-saving technologies and practices.

**Financial Incentives:** Providing financial incentives, such as grants, rebates, tax credits, and low-interest loans, encourages industrial facilities to invest in energy conservation projects and adopt energy-efficient technologies.

**Voluntary Programs:** Voluntary energy conservation programs, such as Energy Star for Industry and the Better Plants Program, engage industrial stakeholders in collaborative efforts to improve energy efficiency, share best practices, and achieve sustainability goals.

**Carbon Pricing and Emissions Trading:** Carbon pricing mechanisms, such as carbon taxes or emissions trading schemes, create economic incentives for industrial facilities to reduce their GHG emissions and invest in energy conservation measures.

Industrial energy conservation is a multifaceted endeavor that requires concerted efforts from policymakers, industry stakeholders, technology providers, and consumers. By adopting a holistic approach that integrates technological innovation, policy support, financial incentives, and stakeholder engagement, industrial facilities can achieve significant energy savings, reduce GHG emissions, and enhance their competitiveness in a rapidly evolving global marketplace. As the world transitions toward a low-carbon, sustainable energy future, industrial energy conservation will continue to play a central role in driving efficiency, innovation, and prosperity for generations to come.

## 1.6 Conclusion

The techno-management revolution of energy marks a profound shift from traditional, centralized systems toward more decentralized, efficient, and sustainable models. Initially, societies depended heavily on conventional energy sources, which not only fueled economic growth and daily life but also posted significant environmental and social challenges. The transformation has been driven by advancements in technology, evolving policy and regulatory frameworks, and an increasing awareness of environmental impacts.

Decentralized energy generation, coupled with innovative management strategies, has empowered consumers and industries to play an active role in energy production and conservation. Smart technologies such as home automation, smart thermostats, and demand response programs have introduced dynamic methods to optimize energy consumption and manage peak loads effectively. Simultaneously, industrial sectors have embraced energy efficiency measures, contributing to broader sustainability goals.

This chapter sets the stage for understanding how technological innovation, strategic management, and societal shifts are revolutionizing the global energy landscape, paving the way for a future that is smarter, cleaner, and more resilient.

## 1.7 Case Study

### 1.7.1 Case Study 1 The Techno-Management Revolution of Energy at SmartGrid Inc.

➢ **Introduction**

SmartGrid Inc., a leading energy technology company, embarked on a transformative journey to harness the power of technology and strategic management practices to revolutionize the energy sector. This case study explores SmartGrid Inc.'s innovative approach to the techno-management revolution of energy and its impact on reshaping the landscape of energy production, distribution, and consumption.

➢ **Background**

SmartGrid Inc. recognized the pressing need for a paradigm shift in the energy industry to address challenges such as climate change, resource depletion, and growing energy demand. Leveraging its expertise in data analytics, AI, and smart grid technologies, SmartGrid Inc. set out to pioneer a holistic approach that integrates technological innovation with strategic management principles to optimize energy systems and drive sustainability.

➢ **Technological Innovations**

SmartGrid Inc. developed a suite of cutting-edge technologies aimed at enhancing grid efficiency, resilience, and flexibility:

**Advanced Metering Infrastructure (AMI):** SmartGrid Inc. deployed AMI systems equipped with smart meters and communication networks to enable real-time monitoring of energy consumption, enhance billing accuracy, and empower consumers with actionable insights into their usage patterns.

**Grid Optimization Algorithms:** SmartGrid Inc. developed sophisticated algorithms powered by AI and machine learning to optimize grid operations, predict demand patterns, and dynamically adjust energy flows in response to changing conditions, such as weather events or equipment failures.

**Distributed Energy Resources (DERs):** SmartGrid Inc. facilitated the integration of DERs, including rooftop solar, energy storage systems, and EV charging stations, into the grid through innovative control systems and demand response platforms, unlocking new opportunities for grid flexibility and renewable energy integration.

➢ **Strategic Management Initiatives**

In addition to technological innovations, SmartGrid Inc. implemented strategic management initiatives to drive organizational agility, collaboration, and sustainability:

**Cross-functional Collaboration**: SmartGrid Inc. fostered a culture of collaboration and innovation by breaking down silos between engineering, marketing, finance, and operations teams. Cross-functional teams were tasked with developing integrated solutions that address both technical and business challenges, ensuring alignment with corporate goals and market needs.

**Stakeholder Engagement:** SmartGrid Inc. actively engaged with stakeholders, including utilities, regulators, policymakers, and community groups, to build consensus, gather feedback, and co-create solutions that meet diverse needs and priorities. Transparent communication and stakeholder involvement were integral to the success of SmartGrid Inc.'s initiatives.

**Continuous Improvement:** SmartGrid Inc. embraced a culture of continuous improvement and learning, investing in employee training, skills development, and knowledge-sharing initiatives. By fostering a growth mindset and encouraging experimentation, SmartGrid Inc. remained at the forefront of innovation and adaptation in a rapidly evolving energy landscape.

➢ **Results and Impact**

SmartGrid Inc.'s techno-management revolution of energy yielded tangible results and lasting impact:

**Enhanced Grid Resilience:** SmartGrid Inc.'s grid optimization algorithms improved grid resilience and reliability, reducing the frequency and duration of outages and minimizing disruptions to customers' electricity supply.

**Cost Savings:** By optimizing grid operations and reducing energy waste, SmartGrid Inc. helped utilities lower operating costs, avoid expensive infrastructure investments, and pass on cost savings to consumers through more competitive electricity rates.

**Environmental Benefits:** SmartGrid Inc.'s focus on renewable energy integration and demand-side management resulted in significant reductions in GHG emissions, supporting climate goals and promoting environmental sustainability.

**Customer Empowerment:** SmartGrid Inc.'s AMI systems and consumer engagement initiatives empowered customers to take control of their energy usage, make informed decisions, and participate in demand response programs, leading to greater energy efficiency and cost savings.

**Industry Leadership:** SmartGrid Inc. emerged as a leader in the energy technology sector, earning recognition for its innovative solutions, thought leadership, and commitment to sustainability. The company's success inspired industry peers to embrace the techno-management revolution of energy and accelerate the transition to a more resilient, efficient, and sustainable energy future.

➤ **Conclusion**

SmartGrid Inc.'s journey exemplifies the transformative potential of the techno-management revolution of energy in driving innovation, sustainability, and resilience in the energy sector. By leveraging technological advancements and strategic management practices, SmartGrid Inc. not only optimized energy systems but also fostered collaboration, empowered stakeholders, and positioned itself as a catalyst for positive change in the industry. As the energy landscape continues to evolve, SmartGrid Inc. remains committed to shaping a future where energy is clean, affordable, and accessible for all.

### 1.7.2 Case Study 2 Energy Conservation and Demand Response at GreenTech Manufacturing

➤ **Introduction**

GreenTech Manufacturing, a leading producer of industrial machinery, embarked on a comprehensive energy management initiative aimed at reducing energy consumption, enhancing operational efficiency, and promoting sustainability. This case study examines GreenTech Manufacturing's successful implementation of energy conservation and demand response strategies, highlighting the benefits, challenges, and lessons learned from the initiative.

➤ **Background**

Facing rising energy costs and increasing pressure to reduce carbon emissions, GreenTech Manufacturing recognized the importance of adopting proactive measures to manage energy usage effectively. In collaboration with energy consultants and technology partners, GreenTech Manufacturing developed a tailored energy management plan focused on energy conservation and demand response.

➤ **Energy Conservation Measures**

GreenTech Manufacturing implemented a series of energy conservation measures to optimize energy usage and reduce waste:

**Equipment Upgrades:** GreenTech Manufacturing invested in energy-efficient machinery and equipment, replacing outdated models with high-efficiency alternatives equipped with advanced control systems and automation features.

**Process Optimization:** GreenTech Manufacturing conducted thorough audits of its manufacturing processes to identify inefficiencies and implement optimization strategies, such as adjusting production schedules, streamlining workflows, and minimizing idle time.

**Lighting and HVAC Retrofits:** GreenTech Manufacturing upgraded lighting systems to energy-efficient LED fixtures and installed programmable thermostats and HVAC controls to optimize heating and cooling operations based on occupancy and usage patterns.

**Employee Engagement:** GreenTech Manufacturing engaged employees in energy conservation efforts through training programs, awareness campaigns, and incentives for energy-saving behaviors, fostering a culture of sustainability and responsibility across the organization.

➢ **Demand Response Initiatives**

In addition to energy conservation measures, GreenTech Manufacturing implemented demand response initiatives to manage peak energy demand and reduce utility costs:

**Demand Response Programs:** GreenTech Manufacturing enrolled in demand response programs offered by local utility providers, agreeing to curtail energy usage during peak demand periods in exchange for financial incentives or rate discounts.

**Automated Controls**: GreenTech Manufacturing installed automated controls and demand response technologies to remotely adjust energy-intensive processes and equipment in response to signals from utility providers or grid operators.

**Load Shedding Strategies:** GreenTech Manufacturing developed load shedding strategies to prioritize critical operations and temporarily reduce nonessential energy consumption during peak demand events, minimizing disruptions to production while maximizing cost savings.

**Energy Storage Integration:** GreenTech Manufacturing explored the integration of energy storage systems to store excess energy during off-peak periods and discharge it during peak demand events, providing additional flexibility and resilience to the energy management strategy.

➢ **Results and Impact**

GreenTech Manufacturing's energy conservation and demand response initiatives yielded significant results and positive impact:

**Cost Savings:** By optimizing energy usage and participating in demand response programs, GreenTech Manufacturing achieved substantial cost savings on utility bills, reducing operational expenses and improving profitability.

**Energy Efficiency:** Energy conservation measures resulted in significant reductions in energy consumption and GHG emissions, enhancing the company's environmental performance and contributing to its sustainability goals.

**Grid Reliability:** GreenTech Manufacturing's participation in demand response programs improved grid reliability by reducing strain on the electrical grid during peak demand periods, enhancing overall system stability and resilience.

**Employee Engagement:** Employee engagement initiatives fostered a sense of ownership and accountability among staff members, empowering them to identify energy-saving opportunities, implement best practices, and contribute to the company's energy management efforts.

**Industry Recognition:** GreenTech Manufacturing's success in energy conservation and demand response garnered recognition from industry peers, customers, and stakeholders, positioning the company as a leader in sustainable manufacturing practices and responsible corporate citizenship.

➢ **Conclusion**

GreenTech Manufacturing's experience demonstrates the transformative potential of energy conservation and demand response in driving operational efficiency, cost savings, and sustainability in industrial settings. By implementing a holistic approach that integrates technology, process optimization, employee engagement, and stakeholder collaboration, GreenTech Manufacturing successfully reduced energy consumption, lowered costs, and enhanced its competitive position in the market. As businesses worldwide seek to navigate the challenges of energy management and sustainability, GreenTech Manufacturing serves as a model for effective energy conservation and demand response strategies that deliver tangible benefits for both the bottom line and the environment.

## 1.8 Exercise

1. What were the primary traditional energy sources utilized before the revolution, and how were they harnessed?

2. Can you provide an overview of the role of biomass, animal power, and watermills in the pre-revolution energy landscape?

3. How did the availability of traditional energy sources vary regionally before the revolution, and what were the factors contributing to these differences?

4. How significant were traditional energy sources in powering economic activities such as agriculture, manufacturing, and transportation?

5. Can you discuss any challenges or limitations associated with the reliance on traditional energy sources in pre-revolution societies?

6. How did innovations such as improved watermill designs or advancements in harnessing animal power influence energy availability and utilization?

7 Were there any notable innovations in the storage or transportation of traditional energy sources during this time period, and what were their effects?

8 What role did economic factors such as trade, labor availability, and wealth distribution play in shaping the accessibility and distribution of energy resources?

9 Can you discuss any instances where disparities in energy availability contributed to social or economic inequality in pre-revolution societies?

10 What are microgrids, and how do they contribute to decentralized energy generation?

11 Discuss the concept of DERs and their role in modern energy distribution.

12 How does P2P energy trading work, and what are the potential benefits and challenges associated with it?

13 Explain the concept of VPPs and how they integrate various DERs into the grid.

14 Describe the function of net metering policies and their impact on the adoption of DERs.

15 How do FiTs incentivize renewable energy production, and what are their effects on energy markets?

16 Discuss the importance of energy market reforms in facilitating the integration of decentralized energy generation into existing grids.

17 Explain the role of AI in optimizing grid operations and enhancing the efficiency of energy distribution.

18 Discuss how the IoT is utilized in EMSs to monitor and control DERs.

19 What are the benefits of using advanced analytics for predictive maintenance in energy distribution infrastructure?

20 Discuss the importance of ensuring energy access and equity in the context of decentralized energy generation.

**21** How does the shift toward distributed energy generation contribute to job creation and economic development?

**22** Explain the role of decentralized energy systems in enhancing resilience to climate change-related disruptions in energy supply.

**23** How do energy-efficient technologies contribute to reducing energy consumption in residential, commercial, and industrial sectors?

**24** What are some effective behavioral changes and best practices individuals can adopt to decrease energy usage?

**25** Discuss the importance of retrofitting buildings and facilities in improving overall energy efficiency and reducing energy waste.

**26** What are the benefits of implementing automated EMSs in optimizing energy consumption?

**27** How can the integration of smart home technology with demand response programs enhance energy efficiency and grid reliability?

**28** Discuss the incentives offered to consumers and businesses to participate in demand response programs.

**29** Can you provide examples of successful demand response initiatives and their impact on energy conservation and grid stability?

**30** What is the significance of industrial energy audits in identifying opportunities for efficiency improvements?

**31** How can collaborative approaches within industrial sectors promote energy conservation and sustainability?

**32** What types of educational programs are available for consumers and businesses to learn about energy-saving opportunities?

**33** What tools and resources are available for tracking and managing energy consumption at the individual and organizational levels?

## 1.9 Objective Type Questions

1 What is the primary function of microgrids in energy distribution?
   a) To centralize energy production
   b) To distribute energy from a single source
   c) To create smaller, self-sufficient energy networks
   d) To regulate energy consumption in urban areas

2 Distributed energy resources (DERs) include:
   a) Large-scale power plants
   b) Wind and solar farms
   c) Decentralized energy sources like solar panels and small wind turbines
   d) Government-controlled energy distribution centers

3 Peer-to-peer energy trading involves:
   a) Selling energy directly to utility companies
   b) Exchanging energy between neighboring households or businesses
   c) Trading energy on the global market
   d) Bartering energy for goods and services

4 Virtual power plants (VPPs) operate by:
   a) Combining physical power plants into a single centralized location
   b) Utilizing software to aggregate and control distributed energy resources
   c) Creating virtual simulations of power generation facilities
   d) Generating power through virtual reality technology

5 Renewable energy targets are set by governments to:
   a) Encourage the use of fossil fuels
   b) Limit the development of renewable energy sources
   c) Mandate a certain percentage of energy to come from renewable sources by a specific date
   d) Regulate energy consumption in urban areas

6 Net metering policies allow:
   a) Consumers to sell excess energy back to the grid at retail rates
   b) Utility companies to charge consumers for energy used during peak hours
   c) Consumers to receive subsidies for using renewable energy sources
   d) Governments to regulate energy prices based on supply and demand

7 Feed-in Tariffs (FiTs) are designed to:
 **a)** Discourage the use of renewable energy sources
 **b)** Subsidize fossil fuel production
 **c)** Provide financial incentives for generating renewable energy
 **d)** Regulate energy consumption in rural areas

8 Energy market reforms aim to:
 **a)** Increase government control over energy production and distribution
 **b)** Encourage competition and efficiency in the energy market
 **c)** Impose tariffs on imported energy resources
 **d)** Limit consumer choice in energy providers

9 Blockchain technology in energy transactions ensures:
 **a)** Transparency and security in energy trading
 **b)** Anonymity for energy consumers
 **c)** Centralized control over energy distribution
 **d)** Increased energy consumption efficiency

10 Artificial intelligence (AI) in grid optimization helps in:
 **a)** Predicting energy demand and adjusting supply accordingly
 **b)** Creating physical barriers to energy flow
 **c)** Limiting access to energy resources
 **d)** Increasing energy waste

11 Internet of Things (IoT) devices in energy management systems are used for:
 **a)** Monitoring and controlling energy consumption remotely
 **b)** Increasing energy consumption without human intervention
 **c)** Storing energy in virtual servers
 **d)** Decreasing energy efficiency in households

12 Advanced analytics for predictive maintenance help in:
 **a)** Identifying potential failures in energy distribution infrastructure before they occur
 **b)** Increasing energy consumption by detecting energy leaks
 **c)** Delaying maintenance of energy infrastructure
 **d)** Predicting energy prices in the market

13 Carbon emissions reduction is a key environmental impact of:
 **a)** Increasing energy consumption
 **b)** Implementing renewable energy projects
 **c)** Promoting fossil fuel use
 **d)** Encouraging deforestation

**14** Energy access and equity address:
- **a)** Ensuring fair distribution of energy resources among different socioeconomic groups
- **b)** Limiting energy access to specific regions
- **c)** Expanding energy access to urban areas only
- **d)** Privatizing energy distribution networks

**15** Job creation and economic development are potential social impacts of:
- **a)** Decreasing energy consumption
- **b)** Increasing reliance on traditional energy sources
- **c)** Investing in renewable energy projects
- **d)** Implementing energy conservation measures

**16** Resilience to climate change is enhanced by:
- **a)** Ignoring energy conservation efforts
- **b)** Relying solely on fossil fuels for energy generation
- **c)** Implementing measures to reduce carbon emissions and adapt to changing climate conditions
- **d)** Expanding deforestation for energy production.

**17** What were the primary traditional energy sources utilized before the revolution?
- **a)** Fossil fuels
- **b)** Renewable energy
- **c)** Biomass, animal power, and watermills
- **d)** Nuclear energy

**18** How were traditional energy sources like biomass, animal power, and watermills harnessed?
- **a)** Through combustion processes
- **b)** By converting wind energy
- **c)** By utilizing natural resources like wood, animals, and water flow
- **d)** Through nuclear reactions

**19** Which of the following is NOT an example of a traditional energy source used before the revolution?
- **a)** Biomass
- **b)** Coal
- **c)** Solar power
- **d)** Watermills

**20** How did traditional energy sources contribute to meeting the energy demands of daily life activities before the revolution?
   **a)** By providing electricity to households
   **b)** By powering machinery in industries
   **c)** By heating and lighting homes, cooking food, and powering agricultural tools
   **d)** By fueling transportation vehicles

## 1.10 Glossary

1) **Energy:** The capacity to do work, available in various forms such as thermal, mechanical, chemical, and electrical.
2) **Renewable Energy:** Energy derived from natural sources that replenish over time, like solar, wind, hydro, and biomass.
3) **Nonrenewable Energy:** Energy from sources that are finite and deplete with use, such as coal, oil, and natural gas.
4) **Fossil Fuels:** Natural fuels like coal, oil, and gas, formed from the remains of ancient plants and animals, primarily nonrenewable.
5) **Solar Energy:** Power derived from the sun's radiation, harnessed through technologies like photovoltaic cells and solar thermal systems.
6) **Wind Energy:** Energy generated by converting the kinetic energy of wind into electricity using turbines.
7) **Hydropower:** Electricity generated from the movement of water, often through dams or water flow systems.
8) **Biomass Energy:** Energy obtained from organic materials such as plants, wood, and agricultural waste.
9) **Geothermal Energy:** Heat energy extracted from within the Earth, used for power generation and heating.
10) **Energy Conservation:** The practice of reducing energy consumption through efficiency measures and behavioral changes.
11) **Energy Efficiency:** The ratio of useful energy output to the energy input, reflecting how effectively energy is utilized.
12) **Sustainability:** Meeting current energy needs without compromising the ability of future generations to meet their own.
13) **Carbon Footprint:** The total amount of greenhouse gases emitted by human activities, usually measured in $CO_2$ equivalents.
14) **Greenhouse Gases (GHGs):** Gases such as $CO_2$, $CH_4$, and $N_2O$ that trap heat in the Earth's atmosphere, contributing to climate change.

15) **Climate Change:** Long-term changes in temperature, precipitation, and weather patterns caused largely by human-induced emissions.
16) **Global Warming:** The increase in Earth's average temperature due to rising concentrations of greenhouse gases.
17) **Sustainable Development:** A mode of development that balances economic growth, environmental care, and social well-being.
18) **Energy Transition:** The global shift from fossil fuel-based energy systems to renewable and sustainable energy solutions.
19) **Smart Grid:** An advanced electricity network that uses digital technology to monitor, manage, and optimize energy flow.
20) **Energy Security:** The uninterrupted availability of energy sources at an affordable price.
21) **Technological Innovations:** New methods, tools, and systems that enhance energy production, storage, distribution, and efficiency.
22) **Hydrogen Energy:** A clean energy carrier that can be produced from various sources and used for electricity, transport, and industry.
23) **Nuclear Energy:** Power generated through nuclear reactions, usually fission, that release large amounts of energy.
24) **Sustainable Energy Policies:** Government or institutional regulations aimed at promoting renewable energy, efficiency, and conservation.
25) **Energy Storage:** Systems and technologies (like batteries, pumped hydro, or hydrogen storage) that save energy for later use.

## References


**1** Agyekum, E. B., Odoi-Yorke, F., Abbey, A. A., & Ayetor, G. K. (2024). A review of the trends, evolution, and future research prospects of hydrogen fuel cells – A focus on vehicles. *International Journal of Hydrogen Energy*, 72, 918–939 https://doi.org/10.1016/J.IJHYDENE.2024.05.480.

**2** Andrade, T. S., & Thiringer, T. (2024). Low platinum fuel cell as enabler for the hydrogen fuel cell vehicle. *Journal of Power Sources*, 598, 234140 https://doi.org/10.1016/J.JPOWSOUR.2024.234140.

**3** Baker, J., Guler, M., Medonna, A., Li, Z., & Ghosh, A. (2025). Analysis of large-scale (1GW) off-grid agrivoltaic solar farm for hydrogen-powered fuel cell electric vehicle (HFCEV) charging station. *Energy Conversion and Management*, 323, 119184 https://doi.org/10.1016/J.ENCONMAN.2024.119184.

**4** Bernalte Sánchez, P. J., García Márquez, F. P., Papaelias, M., & Lee, D. (2024). Life cycle cost analysis of an autonomous underwater vehicle that employs hydrogen fuel cell. *Ocean Engineering*, 299, 117300 https://doi.org/10.1016/J.OCEANENG.2024.117300.

5 Bi, C., Shen, Y., Wang, C., Zhang, C., & Lv, H. (2024). Experimental and simulation study on hydrogen diffusion and venting inside a container carrying fuel cell vehicles. *International Journal of Hydrogen Energy*, 92, 755–768 https://doi.org/10.1016/J.IJHYDENE.2024.10.210.

6 Chen, L., Wang, Y., Jiang, Y., Zhang, C., Liao, Q., Li, J., Wu, J., & Gao, X. (2024). Life cycle assessment of liquid hydrogen fuel for vehicles with different production routes in China. *Energy*, 299, 131472 https://doi.org/10.1016/J.ENERGY.2024.131472.

7 Filina, O. A., Malozyomov, B. V., & Shchurov, N. I. (2024). Generation of hydrogen fuel on board vehicles with internal combustion engines. *International Journal of Hydrogen Energy*, 93, 320–327 https://doi.org/10.1016/J.IJHYDENE.2024.08.042.

8 Hu, D., Huang, J., Lu, D., & Wang, J. (2024). Hydrogen consumption estimation of fuel cell vehicle based on vehicle energy transfer. *Sustainable Energy Technologies and Assessments*, 68, 103854 https://doi.org/10.1016/J.SETA.2024.103854.

9 Shchurov, N. I., Dedov, S. I., & Shtang, A. A. (2024). Modelling and simulation hybrid electric vehicle with hydrogen fuel cells. *International Journal of Hydrogen Energy*, 95, 481–488 https://doi.org/10.1016/J.IJHYDENE.2024.08.044.

10 Dai, J., Yang, S., Yang, Y., & Fang, Q. (2024). Research on the consequences of hydrogen leakage and explosion accidents of fuel cell vehicles in underground parking garages. *International Journal of Hydrogen Energy*, 77, 1296–1306 https://doi.org/10.1016/J.IJHYDENE.2024.06.278.

11 Khare, V. (2023). Prediction of tidal energy resources at Gulf of Kutch and Gulf of Khambhat by data mining. *Cleaner Energy Systems*, 6, 100090.

12 Khare, V. (2023). Design, optimization, and data analysis of solar-tidal hybrid renewable energy system for Hurawalhi, Maldives. *Cleaner Energy Systems*, 6, 100088.

13 Khare, V. (2023). Assessment of hydro energy potential from rain fall data set in India through data analysis. *E-Prime-Advances in Electrical Engineering, Electronics and Energy*, 6, 100290.

14 Khare, V., & Chaturvedi, P. (2023). Design, control, reliability, economic and energy management of microgrid: A review. *E-Prime-Advances in Electrical Engineering, Electronics and Energy*, 5, 100239.

15 Khare, V., & Bhuiyan, M. A. (2023). Perspective of renewable energy in the BRICS country. *E-Prime-Advances in Electrical Engineering, Electronics and Energy*, 5, 100250.

16 Khare, V., & Chaturvedi, P. (2023). Solar energy system concept change from trending technology: A comprehensive review. *E-Prime-Advances in Electrical Engineering, Electronics and Energy*, 4, 100183.

17 Khare, V., & Chaturvedi, P. (2023). Design and operational management of sustainable multi-networks green energy system for study area at India. *Materials Today: Proceedings* https://doi.org/10.1016/j.matpr.2023.05.575.

# 2

# Understanding Green Hydrogen: Production Methods and Technologies

## 2.1 Introduction

Green hydrogen, often hailed as the fuel of the future, represents a sustainable energy source produced by using renewable electricity such as wind, solar, or hydropower to split water into hydrogen and oxygen through electrolysis. Unlike conventional hydrogen production methods, which rely on natural gas or coal and emit significant amounts of carbon dioxide, green hydrogen is entirely carbon-free, offering a clean alternative that can help mitigate climate change. As a versatile energy carrier, green hydrogen has the potential to decarbonize various sectors, including transportation, industry, and power generation, where direct electrification may be challenging. With the growing global interest and technological advancements, green hydrogen is poised to play a crucial role in transitioning toward a net-zero carbon future, fostering energy security and sustainability [1, 2]. Many countries have set ambitious targets for reducing carbon emissions, with green hydrogen playing a significant role in these strategies. The European Union (EU), Japan, South Korea, Australia, and the United States have announced hydrogen strategies or roadmaps, committing billions of dollars to develop hydrogen infrastructure, research, and market development. The EU has set a target to produce 10 million tons of green hydrogen by 2030 and to import an additional 10 million tons, primarily from regions like North Africa and the Middle East. Countries like Germany, France, and Spain are investing heavily in green hydrogen projects, including electrolyzer manufacturing, hydrogen storage, and distribution networks. Germany has launched the "National Hydrogen Strategy," with €9 billion allocated for developing hydrogen infrastructure and creating partnerships with countries in North Africa and the Middle East. The United States has announced a "Hydrogen Shot" initiative, aiming to reduce the cost of clean hydrogen by 80% to $1 per kilogram within a decade [3, 4]. The Bipartisan Infrastructure

*Hydrogen Technologies in Decarbonizing Industries and Energy Systems*, First Edition.
Vikas Khare, Sanjeet K. Dwivedi, Monica Bhatia, and K. K. Pant.

Published 2026 by John Wiley & Sons, Inc.

Law has earmarked $9.5 billion for developing a national hydrogen economy, focusing on clean hydrogen hubs, manufacturing, and R&D. Countries like Saudi Arabia and the United Arab Emirates are investing heavily in green hydrogen. Saudi Arabia's NEOM project aims to build the world's largest green hydrogen plant powered entirely by renewable energy, while the UAE's Hydrogen Leadership Roadmap plans to capture a 25% share of the global hydrogen market by 2030. The global green hydrogen market is projected to grow from around $1 billion in 2021 to over $56 billion by 2030, with a compound annual growth rate (CAGR) of over 60%. By 2050, green hydrogen could account for up to 24% of global energy demand, reducing $CO_2$ emissions by up to 6 gig tons annually. The scale-up of green hydrogen could drive significant cost reductions, technological innovation, and sustainable economic growth worldwide. The worldwide scenario for green hydrogen is rapidly evolving, driven by strong policy support, technological innovation, and international cooperation. While challenges remain, the potential of green hydrogen to revolutionize the energy landscape and contribute significantly to global decarbonization goals is widely recognized. Before going to the concept of green hydrogen, it is necessary to understand the concept of Hydrogen [5, 6].

### 2.1.1 What Is Hydrogen?

Hydrogen, the lightest and most abundant element in the universe, has garnered significant attention as a potential game-changer in the quest for clean and sustainable energy solutions. Its unique properties and versatile applications make it a promising candidate for addressing the challenges of climate change, energy security, and environmental sustainability. At its core, hydrogen is a simple molecule, consisting of just one proton and one electron. Figure 2.1 shows the structure of hydrogen.

It is denoted by the chemical symbol "H" and holds the distinction of being the most abundant element in the universe, constituting roughly 75% of its elemental mass. One of the hydrogen's most remarkable properties is its high energy density, which makes it an attractive fuel for various applications. When combusted or reacted with oxygen in a fuel cell, hydrogen produces energy, with water vapor being the only by-product [7, 8]. Hydrogen's versatility extends beyond clean energy generation. It serves as a crucial feedstock in industries such as ammonia production, refining, and chemical manufacturing. Additionally, hydrogen can be utilized as a fuel for transportation, with fuel cell vehicles (FCVs) emerging as an alternative to traditional internal combustion engine vehicles.

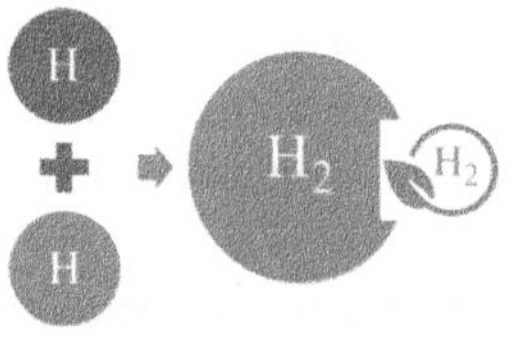

**Figure 2.1** Structure of Hydrogen

Hydrogen can be produced through various methods, each with its own environ mental implications. Steam methane reforming (SMR) and coal gasification are conventional methods that utilize fossil fuels as feedstock's, resulting in emissions of GHGs and other pollutants. However, advancements in carbon capture and storage (CCS) technology offer a pathway to mitigate these environmental impacts, leading to the production of "blue hydrogen." On the other hand, electrolysis offers a cleaner alternative for hydrogen production. By passing an electric current through water, electrolysis splits water molecules into hydrogen and oxygen gases. When powered by renewable energy sources such as wind or solar power, this process yields "green hydrogen," which boasts zero-carbon emissions and holds great potential for decarbonizing various sectors of the economy [9, 10].

Despite its promise, widespread adoption of hydrogen faces several challenges. Cost, infrastructure development, and energy efficiency are key areas that require attention. The production, storage, and distribution of hydrogen infrastructure need significant investment and technological innovation to scale up effectively. Moreover, hydrogen's energy efficiency varies depending on the production method. While electrolysis powered by renewable energy offers a clean solution, it currently faces challenges related to high costs and efficiency limitations. Advances in electrolyzer technology and economies of scale are crucial for driving down costs and improving efficiency. Government policies and market dynamics play a pivotal role in shaping the future of hydrogen. Many countries are recognizing the potential of hydrogen as a clean energy carrier and are implementing policies to support its development and deployment. These include financial incentives, research funding, regulatory frameworks, and international collaborations [11, 12].

Furthermore, private sector investments and partnerships are driving innovation and accelerating the commercialization of hydrogen technologies. Companies across various industries are exploring hydrogen solutions as part of their sustainability strategies, signaling a growing market demand for clean energy alternatives. Hydrogen holds immense promise as a clean, versatile, and sustainable energy carrier. Its potential to decarbonize sectors such as transportation, industry, and power generation offers a pathway to mitigate climate change and reduce reliance on fossil fuels. However, realizing this potential requires concerted efforts from governments, industries, and academia to overcome technical, economic, and policy challenges. With continued innovation, investment, and collaboration, hydrogen could revolutionize the global energy landscape and usher in a cleaner, more sustainable future for generations to come.

> ***Key: Hydrogen gas:*** *The lightest and most abundant chemical element, often used as a clean energy carrier.*

- **Hydrogen as an Element**

Hydrogen, the most abundant element in the universe, is fundamental to the understanding of chemistry and plays a crucial role in various natural and industrial processes. Following are the overview of hydrogen as an element:

**Chemical Symbol:** Hydrogen is denoted by the chemical symbol H and is positioned at the very top of Group 1 in the periodic table. Its atomic number is 1, meaning it contains exactly one proton in its nucleus. The name "Hydrogen" comes from the Greek words *hydro* (water) and *genes* (creator), as it was discovered that hydrogen produces water when burned in oxygen. Unlike other Group 1 elements (alkali metals), hydrogen is not a metal under normal conditions and exhibits both metallic and nonmetallic chemical characteristics depending on its bonding partners.

**Atomic Structure:** The hydrogen atom consists of a single proton in its nucleus and one electron orbiting the nucleus. It is the simplest atom, devoid of neutrons in its most common form. Figure 2.2 shows the atomic structure of hydrogen.

**Isotopes:** While the most common isotope of hydrogen, known as protium, has no neutrons, hydrogen can also exist in two other isotopic forms: deuterium and tritium. Deuterium contains one neutron in addition to the proton, while tritium contains two neutrons. Deuterium and tritium are used in various scientific and industrial applications, such as nuclear fusion research.

**Physical Properties:** Hydrogen is a colorless, odorless, and tasteless gas at standard temperature and pressure. It is the lightest element, with a very low density. Hydrogen gas is highly flammable and burns with a pale blue flame when ignited in air or oxygen.

**Chemical Properties:** Hydrogen is highly reactive, readily forming chemical bonds with other elements. It can form covalent bonds with nonmetals and metal hydrides with metals. Hydrogen is essential in various chemical reactions and is a key component of many important compounds, including water ($H_2O$) and hydrocarbons.

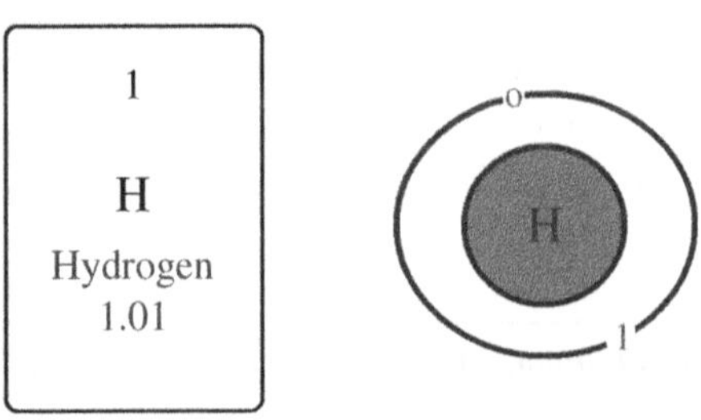


**Figure 2.2** Atomic Structure of Hydrogen

**Occurrence:** While hydrogen is the most abundant element in the universe, it is relatively rare on Earth in its elemental form due to its tendency to combine with other elements. It is primarily found in compounds such as water ($H_2O$), hydrocarbons, and organic matter.

**Uses:** Hydrogen has numerous industrial applications, including in the production of ammonia for fertilizers, petroleum refining processes, and the synthesis of various chemicals. It is also utilized as a fuel for rocket propulsion and in fuel cell technology for generating electricity.

**Energy Carrier:** Hydrogen has gained attention as a clean and renewable energy carrier. It can be produced through electrolysis using renewable energy sources such as wind and solar power, yielding "green hydrogen" with zero-carbon emissions. This green hydrogen holds promise as a sustainable fuel for transportation, energy storage, and power generation, offering potential solutions to address climate change and reduce dependence on fossil fuels.

Hydrogen's unique properties and versatile applications make it a cornerstone of chemistry and a key player in various scientific, industrial, and energy-related fields. Its potential as a clean energy carrier holds promise for a sustainable future, driving innovation and advancements in technology and industry.

➢ **Properties of Hydrogen Gas**

Hydrogen gas, with its simplicity and unique characteristics, exhibits several important properties, which is also illustrated in Figure 2.3:

- **Colorless and Odorless:** Hydrogen gas is colorless and does not have any discernible odor, making it difficult to detect by sight or smell.
- **Highly Flammable:** Hydrogen gas is highly flammable and combustible. When ignited in the presence of oxygen, it burns with a pale blue flame, producing heat and water vapor as by-products.
- **Low Density:** Hydrogen is the lightest element, with a very low density. As a result, hydrogen gas rises rapidly in air and disperses quickly when released.
- **Low Boiling and Freezing Points:** Hydrogen has extremely low boiling (−252.87 °C) and freezing (−259.14 °C) points, making it one of the coldest substances known when in its liquid or solid state.
- **Diatomic Molecule:** In its most common form, hydrogen gas consists of diatomic molecules ($H_2$), where two hydrogen atoms are bonded together covalently. This molecular form is stable and represents the majority of naturally occurring hydrogen gas.
- **Nontoxic:** Hydrogen gas is nontoxic when inhaled in small quantities. However, it can displace oxygen in confined spaces, potentially leading to asphyxiation in high concentrations.

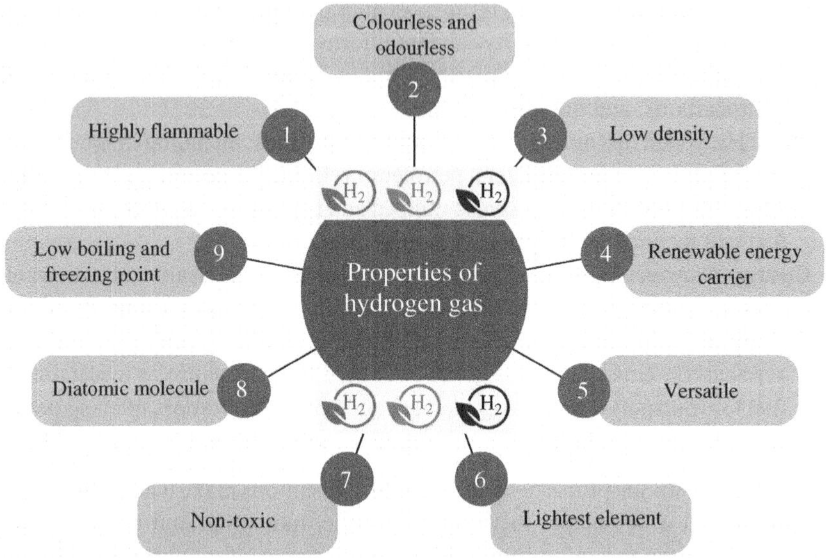


**Figure 2.3** Different Properties of Hydrogen Gas

- **Highly Reactive:** Hydrogen is highly reactive and can readily form chemical bonds with other elements. It reacts with oxygen to form water ($H_2O$) and with various metals to form metal hydrides.
- **Lightest Element:** With an atomic mass of approximately 1 amu, hydrogen is the lightest element on the periodic table. This property contributes to its buoyancy and rapid diffusion in air.
- **Versatile:** Hydrogen gas has versatile applications in various industries. It is used in the production of ammonia for fertilizers, petroleum refining processes, food processing, and as a fuel for rocket propulsion and fuel cell technology.
- **Renewable Energy Carrier:** Hydrogen is gaining attention as a clean and renewable energy carrier. It can be produced through electrolysis using renewable energy sources such as wind and solar power, offering a potential solution for energy storage and decarbonization in sectors like transportation and power generation.

Hydrogen gas possesses unique properties that make it indispensable in a wide range of applications, from industrial processes to clean energy technologies. Understanding these properties is crucial for safe handling and utilization of hydrogen in various fields.

**➢ Importance of Hydrogen in Various Industries**

Hydrogen plays a significant role in various industries due to its unique properties and versatile applications. Following are some key industries where hydrogen is important:

**Chemical Industry:** Hydrogen is an exceptionally versatile element in the chemical industry, serving multiple roles as a raw material, reducing agent, hydrogen donor, and process medium. One of its most important uses is as a feedstock for the production of essential chemicals. The most notable example is ammonia synthesis via the Haber–Bosch process, where hydrogen reacts with nitrogen under high temperature and pressure. Ammonia is a vital ingredient for fertilizers, explosives, and refrigeration systems. Similarly, hydrogen is used in methanol production by hydrogenating carbon monoxide or carbon dioxide. Methanol has wide-ranging applications in the manufacture of plastics, paints, adhesives, and formaldehyde, and it can also serve as a fuel additive. Hydrogen also plays a role in the production of hydrogen peroxide, commonly used in bleaching, disinfection, and chemical oxidation processes, and in manufacturing hydrochloric acid by burning hydrogen in chlorine gas for applications in pH control, metal cleaning, and chemical synthesis [13, 14].

The petroleum refining industry relies heavily on hydrogen to upgrade crude oil into cleaner, more valuable products. Processes like hydrocracking use hydrogen to break down large hydrocarbon molecules into smaller, more useful fractions such as diesel and jet fuel. Hydrotreating, or desulfurization, removes sulfur, nitrogen, and other impurities to meet environmental standards, while hydrogenation of aromatic compounds improves the quality and performance of fuels. In addition to refining, hydrogen is essential in hydrogenation reactions across various industries. In the food sector, it is used to hydrogenate vegetable oils to produce margarine and shortening. In fine chemicals and pharmaceuticals, hydrogenation reactions are employed to synthesize active pharmaceutical ingredients, vitamins, and chemical intermediates. The polymer industry uses hydrogenation to improve the properties of synthetic rubbers, such as enhancing weather resistance in nitrile rubber.

Hydrogen's role as a reducing agent is also significant. In metallurgy, it is used to reduce metal oxides, such as converting tungsten oxide into tungsten metal, and in the production of high-purity metals for electronics. There is growing interest in using hydrogen instead of carbon for direct reduction of iron ore in steel manufacturing, offering a low-carbon alternative to traditional processes. Gas treatment and purification in chemical plants also benefit from hydrogen. It is used in ammonia plant purge gas recovery, syngas composition adjustment for optimal yields, and as a high-purity atmosphere in semiconductor manufacturing. Hydrogen can further serve as an energy and heat source, either through direct combustion or in fuel cells, providing process heat and electricity for chemical production without carbon emissions.

Environmental and pollution control processes incorporate hydrogen as well. It is critical in producing low-sulfur fuels, reducing nitrogen oxides in selective catalytic reduction systems, and even in hydrogen plasma applications for hazardous waste decomposition. Emerging applications for hydrogen in the chemical industry are expanding rapidly with the development of green hydrogen. This includes the conversion of captured $CO_2$ and hydrogen into methanol, synthetic fuels, and polymers. Ammonia is also gaining importance as a hydrogen carrier for global energy trade. Electrochemical synthesis processes are being developed to integrate hydrogen into green chemical manufacturing, while bio-based chemical production is exploring ways to use hydrogen to upgrade biomass into high-value materials. Hydrogen is not only a vital reactant in the production of core industrial chemicals but also a key enabler of cleaner processes and a future driver of low-carbon chemical manufacturing. As green hydrogen technology advances, its role in the chemical industry is expected to expand further, revolutionizing both production methods and supply chains [15, 16]. Table 2.1 illustrates hydrogen application for the chemical industry.

**Petroleum Industry:** Hydrogen is extensively used in oil refineries to improve fuel quality, remove impurities, and ensure compliance with environmental regulations. One of the key processes is hydrocracking, which breaks down heavy petroleum fractions such as vacuum gas oils and residuum into lighter, more valuable products like diesel, kerosene, and jet fuel. In this process, hydrogen prevents coke formation and saturates cracked hydrocarbons, thereby increasing yield and producing cleaner-burning fuels. Another critical process is hydrotreating or hydrodesulfurization (HDS), which removes sulfur, nitrogen, oxygen, and metals from petroleum streams. Here, hydrogen reacts with sulfur compounds to form hydrogen sulfide ($H_2S$), which is further processed, ultimately enabling the production of ultra-low-sulfur diesel (ULSD) and gasoline that meet stringent emission standards. Additionally, hydroisomerization is used to improve the cold-flow properties of fuels, particularly diesel and jet fuel. In the presence of hydrogen and catalysts, hydrocarbon molecules are rearranged without being cracked, thus enhancing fuel quality without sacrificing energy density. Hydrogen in refineries also plays a vital role in producing feedstocks for the chemical industry. In aromatics production, hydrogen assists in reforming naphtha into benzene, toluene, and xylene (BTX), which are essential building blocks for plastics, solvents, and other chemicals. Similarly, olefin hydrogenation is used to remove unsaturated hydrocarbons from feedstocks, preventing polymerization and ensuring stability in petrochemical production.

**Table 2.1** Hydrogen Application for the Chemical Industry

| Application Area | Description | Hydrogen Role | Benefits |
|---|---|---|---|
| Ammonia production | Used in fertilizer manufacturing | Acts as a feedstock in Haber–Bosch process | Supports agriculture and food security |
| Methanol production | Used in plastics, paints, and other chemicals | Hydrogen reacts with $CO/CO_2$ to produce methanol | Versatile industrial chemical with broad applications |
| Refining processes | Desulfurization and hydrocracking in petroleum refining | Hydrogen removes sulfur from fuels and upgrades heavy oils | Produces cleaner fuels and meets environmental regulations |
| Hydrogen peroxide production | Used in bleaching, disinfection, and chemical synthesis | Hydrogen reacts with oxygen to form $H_2O_2$ | Environmentally friendly oxidizing agent |
| Synthetic fuels | Production of synthetic hydrocarbons from hydrogen and $CO_2$ | Hydrogen is combined with captured $CO_2$ to produce synthetic fuels | Reduces carbon footprint and provides low-carbon fuel alternatives |
| Specialty chemicals | Used in production of aniline, cyclohexane, and other intermediates for polymers and fibers | Hydrogenation reactions for specific chemical synthesis | Supports high-value specialty chemical markets |
| Polyurethane production | Feedstock for diisocyanates, used in foams, coatings, and adhesives | Hydrogen is used in the production of ammonia derivatives and related precursors | Enables diverse polyurethane applications |
| Chlor-alkali industry | Produces chlorine, caustic soda, and hydrogen as by-products | Hydrogen can be used internally as fuel or in other chemical processes | Improves energy efficiency and integrates circular hydrogen use |

Many refineries generate hydrogen internally through SMR or gasification. This hydrogen can be used as fuel gas for heaters and furnaces, although this practice results in $CO_2$ emissions. It can also serve as a supplementary fuel in cogeneration units, contributing to the internal energy requirements of the refinery. With the ongoing global energy transition, hydrogen's role in the petroleum sector is evolving toward supporting low-carbon initiatives. One promising area is renewable diesel production, also known as hydrotreated vegetable oil (HVO), where hydrogen is used to remove oxygen from bio-oils and fats to produce drop-in diesel fuels. This enables refineries to pivot toward low-carbon fuel markets. Similarly, hydrogen plays a role in sustainable aviation fuel (SAF) production, where it is utilized in Fischer–Tropsch synthesis or hydroprocessed esters and fatty acids (HEFAs) pathways. In efforts to reduce $CO_2$ emissions from refining, green hydrogen produced via electrolysis using renewable energy can replace conventional gray hydrogen from SMR, while blue hydrogen, produced via SMR with carbon capture, serves as a transitional decarbonization step. Furthermore, hydrogen blending in refinery fuel systems offers another emissions reduction strategy, where natural gas in process heaters is partially substituted with hydrogen to lower $CO_2$ output. Looking ahead, hydrogen could transform petroleum logistics by serving as a clean transport fuel. Refineries may evolve into hydrogen distribution hubs, supplying nearby sectors such as trucking, bus fleets, and shipping. In regions rich in hydrogen resources such as the Middle East and Australia, oil companies are exploring opportunities to produce and export green and blue hydrogen, positioning themselves as key players in the global hydrogen economy [17, 18]. Table 2.2 illustrates utilization of hydrogen in petroleum industry.

**Food Industry:** Hydrogen plays a significant role in various aspects of the food industry, with one of the most prominent uses being the hydrogenation of edible oils and fats. In this process, hydrogen is used to saturate unsaturated fats and oils by reacting them with hydrogen gas in the presence of a catalyst, commonly nickel. This reaction is applied to produce margarine, shortening, and certain spreads, as well as to improve the texture, stability, and shelf life of oils. Additionally, hydrogenation helps reduce rancidity and oxidation in food products. A common example is the conversion of soybean oil into semi-solid fat for baked goods.

Hydrogen also contributes to food preservation and packaging. In modified atmosphere packaging (MAP), it can be blended with inert gases such as nitrogen to reduce oxidation and bacterial growth in packaged foods. Furthermore, molecular hydrogen acts as an antioxidant agent, scavenging free radicals in food storage environments and thereby extending shelf life.

Food safety and sterilization also benefit from hydrogen-based processes. Hydrogen peroxide ($H_2O_2$), which is produced from hydrogen, is widely used to sterilize food packaging such as aseptic cartons for milk and juice, as well as for surface sterilization of food processing equipment. Another advanced application is hydrogen plasma technology, which can be used for microbial decontamination of food surfaces without causing heat damage. In the beverage industry, hydrogen is being introduced in innovative ways.

**Table 2.2** Hydrogen in Petroleum Industry

| Application Area | Hydrogen Role | Current/ Emerging | Benefits |
|---|---|---|---|
| Hydrocracking | Saturates cracked hydrocarbons, prevents coke | Current | Higher yield of diesel, jet fuel |
| Hydrotreating/HDS | Removes sulfur, nitrogen, metals | Current | Compliance with ultra-low-sulfur fuel standards |
| Hydroisomerization | Rearranges hydrocarbons for better fuel properties | Current | Improves cold flow properties |
| Aromatics production | Assists catalytic reforming | Current | Produces BTX chemicals |
| Olefin hydrogenation | Stabilizes feedstock | Current | Prevents polymerization in storage |
| Renewable diesel (HVO) | Deoxygenation of bio-oils | Emerging | Low-carbon diesel |
| SAF production | Feedstock processing | Emerging | Decarbonizes aviation |
| Green/blue hydrogen replacement | Substitutes gray hydrogen | Emerging | Cuts refinery $CO_2$ emissions |
| Hydrogen fuel blending | Partially replaces natural gas in heaters | Emerging | Lower $CO_2$ emissions |
| Hydrogen export/ logistics | Energy carrier for transport | Emerging | New revenue stream |

Hydrogen-enriched water is marketed as a functional beverage due to its antioxidant properties and potential health benefits. Additionally, hydrogen can be infused into drinks as an alternative to carbonation, creating novel beverage products similar to sparkling water but containing hydrogen instead of carbon dioxide [19].

Hydrogen also plays an important role in sustainable food production. In agriculture, green ammonia derived from hydrogen serves as a fertilizer for crops used in the food supply chain. In protein production, certain microbes can use hydrogen as an energy source to grow edible biomass, such as in Solar Foods' Solein. In aquaculture, hydrogen–oxygen generators are used to oxygenate water efficiently in fish farming systems. The sugar industry also utilizes hydrogen in specific bleaching processes to remove color and impurities, improving product quality. Hydrogen is increasingly being explored as a clean energy source in food processing plants. It can be used to power boilers, dryers, and ovens for baking, roasting, or pasteurizing food, significantly reducing the carbon footprint in energy-intensive operations like milk powder production, beer brewing, and chocolate manufacturing. Finally, hydrogen supports the production of specialty food ingredients. In chemical synthesis, it is used to produce food-safe chemicals, sweeteners, and flavor enhancers such as sorbitol and xylitol. Hydrogenation also plays a role in cocoa processing by stabilizing cocoa butter and fats, which is essential for chocolate manufacturing [20, 21]. Table 2.3 illustrates utilization of hydrogen in the food industry.

**Electronics and Semiconductor Manufacturing:** Hydrogen plays a crucial role in the field of electronics and semiconductor manufacturing, serving as both a chemical agent and a process enhancer. The electronics industry demands materials and processes that offer extreme precision, purity, and control, as even minor impurities can drastically affect the performance of semiconductor devices. Hydrogen, due to its unique chemical and physical properties such as its small molecular size, reducing nature, and ability to form high-purity atmospheres, has become indispensable in multiple stages of semiconductor fabrication. From wafer cleaning to deposition, etching, annealing, and cooling, hydrogen contributes to ensuring that the final devices achieve the required standards of efficiency, miniaturization, and reliability demanded by modern applications like smartphones, computers, and high-performance computing systems.

One of the most important uses of hydrogen in semiconductor manufacturing is in epitaxial growth, where thin crystalline layers are deposited onto a wafer to form the base structure for electronic devices. In this process, hydrogen is often employed as a carrier gas, transporting precursor molecules such as silane ($SiH_4$), dichlorosilane ($SiH_2Cl_2$), or gallium precursors in compound semiconductor growth. The inert nature of hydrogen in these contexts ensures that unwanted chemical reactions are minimized while maintaining a high degree of control over the layer's thickness, uniformity, and purity. Additionally, hydrogen helps in reducing surface oxides and contaminants from silicon wafers, thereby creating a clean substrate for epitaxial layer deposition. This role is critical in the development of high-speed integrated circuits, power devices, and optoelectronics components. Table 2.4 illustrates utilization of hydrogen in electronics and semiconductor manufacturing.

Hydrogen is also integral to plasma-enhanced chemical vapor deposition (PECVD) and low-pressure chemical vapor deposition (LPCVD) processes, which are widely used in fabricating thin films, insulating layers, and dielectric coatings in semiconductor devices. In these processes, hydrogen helps

**Table 2.3** Hydrogen in the Food Industry

| Application Area | Hydrogen Role | Details and Benefits | Example Use Cases |
|---|---|---|---|
| Hydrogenation of oils and fats | Reactant | Converts unsaturated fats to saturated fats, improves texture, stability, and shelf life | Production of margarine, shortening, and bakery fats |
| Food preservation | Inert atmosphere | Used in modified atmosphere packaging (MAP) to displace oxygen, preventing oxidation and spoilage | Packaged meats, dairy products, fresh produce |
| pH control and processing aid | Processing gas | Used with hydrogen chloride to adjust pH during certain processing steps | Sugar refining, certain beverage preparations |
| Food additive processing | Chemical reagent | Involved in producing additives, sweeteners, and preservatives | Sorbitol, xylitol, and certain food-grade chemicals |
| Clean energy source | Fuel | Powers food processing plants via hydrogen fuel cells or combustion, reducing carbon footprint | Beverage bottling plants, large-scale bakeries |
| Ammonia for fertilizers (indirect) | Feedstock | Hydrogen used in ammonia production supports agriculture and hence raw food material supply | Fertilizers for crop production |
| Ice cream and frozen food manufacturing | Cryogenics | Hydrogen used in cooling systems (blended with other gases) for ultra-fast freezing | Ice cream production lines, frozen meals |
| Hydrogen peroxide production | Oxidizing agent | Hydrogen as a raw material in making hydrogen peroxide for food-grade sterilization | Equipment cleaning, packaging sterilization |
| Alcohol and beverage industry | Process aid | Used in hydrogenation of certain beverage ingredients to alter flavor or reduce bitterness | Beer flavor adjustment, wine production |
| Biotechnological food production | Feed gas | Hydrogen supports certain microbial and algal processes in novel food production | Single-cell protein, cultured meat |

control the chemical environment, enhances reaction rates, and enables the precise tailoring of film properties such as density, stress, and refractive index. Moreover, hydrogen contributes to the reduction of defects in deposited layers by passivating dangling bonds, particularly in amorphous silicon used in

**Table 2.4** Hydrogen in Electronics and Semiconductor Manufacturing

| Application Area | How Hydrogen Is Used | Benefits | Challenges/ Considerations |
|---|---|---|---|
| Wafer cleaning and surface treatment | Hydrogen gas or hydrogen plasma is used to clean silicon wafers by removing surface oxides and contaminants before further processing. | Produces ultra-clean surfaces, improves adhesion for subsequent layers, and prevents contamination | Requires ultra-high purity hydrogen; handling must be extremely controlled |
| Epitaxial growth processes | Hydrogen acts as a carrier gas for transporting silicon precursors during epitaxial growth of semiconductor layers. | Ensures uniform crystal growth, reduces defects, and improves electrical properties of semiconductors | High operational costs for maintaining purity; safety concerns in high-temperature reactors |
| Annealing and passivation | Hydrogen is used during annealing to passivate dangling bonds in silicon, improving semiconductor performance. | Enhances device lifetime, reduces leakage current, improves yield. | Requires precise control of temperature and hydrogen concentration |
| Chemical vapor deposition (CVD) | Hydrogen is used as a reducing agent and carrier gas in processes like low-pressure CVD (LPCVD) for depositing thin films. | Enables high-quality film deposition, reduces unwanted oxidation, and improves film adhesion | Hydrogen flammability risk in CVD chambers; high energy cost |
| Hydrogen plasma etching | Hydrogen plasma is used to selectively etch certain layers without damaging underlying structures. | High precision, reduced damage compared to chemical etching | Requires specialized plasma control systems |
| Reduction of metal oxides | Hydrogen is used to reduce oxides of metals (e.g., copper, tungsten) during chip manufacturing. | Produces pure metal surfaces essential for microelectronics | Needs extremely controlled atmosphere to avoid re-oxidation |
| Cooling and heat transfer | Hydrogen's high thermal conductivity is used in cooling systems for certain semiconductor fabrication equipment. | Improves heat dissipation, ensuring process stability and equipment longevity | Requires leak-proof systems due to flammability |

**Table 2.4** (Continued)

| Application Area | How Hydrogen Is Used | Benefits | Challenges/ Considerations |
|---|---|---|---|
| Fuel for on-site power generation | Hydrogen fuel cells can provide uninterrupted clean power for fabs, avoiding downtime. | Reduces carbon footprint, provides stable backup power | High infrastructure cost for fuel cell systems |
| Purging and inerting | Hydrogen is used to purge process chambers, replacing other inert gases in some applications. | Removes impurities and residual gases more effectively than some inert gases | Hydrogen is reactive; must ensure it doesn't cause unwanted reactions in certain steps. |

thin-film transistors for displays. This passivation improves electronic properties, enhances charge carrier mobility, and ensures longer device lifetimes—qualities that are essential in advanced electronics like OLED displays, PV cells, and microelectromechanical systems (MEMS).

Another vital application is in annealing and surface treatment processes, where hydrogen is used in high-temperature furnaces or in forming gas mixtures (commonly 5–10% hydrogen in nitrogen). Annealing in hydrogen-rich environments helps repair crystal defects, activate dopants, and smooth out the silicon surface after ion implantation or etching steps. The reducing nature of hydrogen also removes residual oxides and metallic impurities, which, if left untreated, could lead to leakage currents or performance degradation in microchips. The passivation effect of hydrogen is particularly notable in reducing interface states at the silicon–silicon dioxide boundary, which is crucial for improving the electrical characteristics of MOSFETs (metal-oxide-semiconductor field-effect transistors). Without hydrogen, it would be nearly impossible to achieve the extremely low defect densities required for modern semiconductor nodes.

In addition to process applications, hydrogen is employed in cooling and purging operations in semiconductor fabrication facilities. Due to its high thermal conductivity, hydrogen serves as an excellent coolant in certain reactor designs, allowing for uniform temperature distribution and preventing localized overheating during high-temperature processes. It is also used for purging chambers and equipment to maintain ultra-clean environments, ensuring that no moisture, oxygen, or other contaminants interfere with sensitive deposition and etching steps. The purity of hydrogen used in semiconductor manufacturing is exceptionally high—often at the "six-nines"

(99.9999%) level—underscoring the precision required in the electronics industry.

Looking at broader trends, the use of hydrogen in electronics and semiconductor manufacturing is expected to expand with the push toward miniaturization, advanced materials, and sustainability. As device geometries shrink to the nanometer scale, the demand for cleaner processes, better defect control, and more efficient materials will grow, making hydrogen's role even more critical. Additionally, hydrogen's potential as a clean energy source aligns with the industry's push to reduce its carbon footprint, as semiconductor fabrication is traditionally energy-intensive. Future innovations may include hydrogen-based plasma treatments for next-generation transistors, hydrogen-enabled etching for 3D device architectures, and integration of hydrogen in emerging materials like gallium nitride (GaN) and silicon carbide (SiC) for high-power and high-frequency electronics.

Hydrogen is not just a utility gas in semiconductor manufacturing; it is a strategic enabler of technological progress. Its applications span from carrier gas in epitaxial growth and deposition, to defect passivation, oxide reduction, annealing, purging, and cooling. Each of these roles contributes to the precision, efficiency, and miniaturization of modern electronic devices. As the semiconductor industry continues to drive innovations in computing, communication, renewable energy, and artificial intelligence, hydrogen will remain at the core of ensuring that the devices of tomorrow are faster, more reliable, and more sustainable.

**Clean Energy Sectors:** One of the most significant roles of hydrogen in the clean energy sector is as an energy storage medium. Renewable sources such as solar and wind often generate electricity intermittently, leading to surplus energy during low-demand periods and shortages during peak demand. Hydrogen enables large-scale, long-term storage using surplus renewable electricity for water electrolysis, producing hydrogen that can be stored indefinitely. This stored hydrogen can later be converted back to electricity via fuel cells or turbines, helping to balance the grid and ensure a stable energy supply even when renewables are not actively generating power.

Hydrogen produced through electrolysis powered by renewable energy, commonly referred to as "green hydrogen," is entirely carbon-free. It has the potential to replace fossil fuels in multiple sectors without contributing to GHG emissions. The clean energy sector sees green hydrogen as a critical link for decarbonizing industries that are difficult to electrify directly, such as heavy manufacturing, steelmaking, cement production, and high-temperature industrial processes. By scaling green hydrogen production, countries can accelerate their transition to a net-zero energy system. In clean transportation, hydrogen plays a crucial role through fuel cell electric vehicles

(FCEVs) and hydrogen-powered buses, trains, and trucks. Hydrogen fuel cells generate electricity through an electrochemical reaction between hydrogen and oxygen, emitting only water vapor. This makes them an environmentally friendly alternative to fossil fuels for sectors such as heavy-duty freight and public transport, where battery-electric vehicles (BEVs) may be limited by range, refueling time, or payload capacity.

Hydrogen can be blended with natural gas in existing pipeline networks to gradually reduce the carbon footprint of energy used for heating and power generation. This "hydrogen-enriched natural gas" approach allows a smooth transition toward a low-carbon economy without requiring an immediate and complete overhaul of existing infrastructure. Over time, pipelines could shift entirely to carrying pure hydrogen as the clean energy infrastructure matures.

Hydrogen can also be a major clean energy export commodity for countries with abundant renewable resources. Nations with surplus renewable electricity can produce green hydrogen and ship it overseas in the form of liquid hydrogen, ammonia, or other hydrogen carriers. This could create a global clean energy trade market, replacing fossil fuel exports and enabling countries with limited renewable resources to access clean energy for industrial and transportation needs. Beyond storage, hydrogen can be directly used in power plants as a clean fuel. Turbines modified to run on hydrogen, or hydrogen–natural gas blends, can generate electricity without carbon dioxide emissions. This capability allows hydrogen to serve as a reliable backup to renewables, replacing coal and natural gas plants in providing firm, dispatchable power while still maintaining a low-carbon footprint. Hydrogen can combine with captured carbon dioxide to produce synthetic fuels such as methanol, synthetic diesel, or jet fuel. These fuels can be used in existing engines and aircraft without modification, making them an attractive solution for decarbonizing aviation and shipping two of the hardest-to-electrify sectors. This integration of hydrogen into clean energy solutions supports a circular carbon economy. Table 2.5 illustrates hydrogen in the clean energy sector.

**Aerospace and Rocket Propulsion:** Hydrogen plays a critical role as a high-efficiency rocket fuel. In liquid form, when combined with liquid oxygen (LOX), it produces one of the highest specific impulses among chemical propellants, meaning rockets can achieve greater thrust per unit of fuel. NASA's Space Launch System (SLS) and earlier space shuttles relied on liquid hydrogen for main engine propulsion. This capability makes it ideal for deep space exploration missions where high performance is essential. Hydrogen's high energy density per unit mass and clean combustion make it a potential fuel for hypersonic and supersonic aircraft. Using hydrogen in air-breathing engines such as scramjets can enable sustained high-speed travel (Mach 5+), opening possibilities for rapid global transportation and future spaceplane designs.

**Table 2.5** Hydrogen in the Clean Energy Sector

| Application Area | Description | Potential Benefits | Current Challenges | Future Outlook |
|---|---|---|---|---|
| Green hydrogen production | Producing hydrogen via electrolysis using renewable energy sources like solar and wind. | Zero-carbon emissions, scalable, enables seasonal energy storage | High cost of electrolysis, efficiency limitations, need for renewable power availability | Expected cost reduction with technological advancements and large-scale deployment |
| Grid energy storage | Hydrogen as a medium for long-duration energy storage to balance intermittent renewable generation. | Enables stable grids, reduces curtailment of renewables, supports energy security | Conversion losses in storage and retrieval, infrastructure requirements | Key role in future renewable-dominant grids with hybrid storage solutions |
| Fuel cells for power generation | Using hydrogen fuel cells for clean electricity production in stationary applications | High-efficiency, zero emissions at point of use, quiet operation | High cost of fuel cells, hydrogen supply chain limitations | Widespread adoption in backup power, distributed generation, and microgrids |
| Industrial decarbonization | Using hydrogen to replace coal or natural gas in industrial processes (e.g., steel, cement, chemicals) | Significant reduction in industrial $CO_2$ emissions, enables green manufacturing | High hydrogen demand and production costs, retrofitting challenges | Large-scale industrial hydrogen hubs expected by 2035 |
| Blending with natural gas | Mixing hydrogen into existing natural gas pipelines to reduce carbon intensity | Utilizes existing infrastructure, gradual decarbonization of heating and power | Material compatibility issues, limited blending ratios (~20%) | Transitional solution until full hydrogen networks are developed |

**Table 2.5** (Continued)

| Application Area | Description | Potential Benefits | Current Challenges | Future Outlook |
|---|---|---|---|---|
| Hydrogen-based ammonia for energy | Producing ammonia from green hydrogen for easier storage, transport, and energy use | High energy density, established transport network, multi-sector use | Conversion inefficiencies, ammonia toxicity | Emerging as a global hydrogen carrier and clean fuel for shipping |
| Off-grid and remote power | Hydrogen fuel cells for powering isolated communities or facilities without grid access | Reduces reliance on diesel generators, improves energy resilience | High logistics cost for hydrogen delivery, storage constraints | Increased adoption in remote islands, mining, and disaster-relief power |

Beyond propulsion, hydrogen is used in fuel cells aboard spacecraft to generate electricity and water. Fuel cells combine hydrogen with oxygen to produce power for onboard systems while generating pure water as a by-product, which can be consumed by astronauts. This dual-use capability reduces the need to carry separate power and water supplies, improving mission efficiency. Liquid hydrogen's extremely low boiling point (−253 °C) makes it useful for cooling sensitive instruments, detectors, and superconducting components in satellites and telescopes. For example, hydrogen cooling is used in infrared telescopes to minimize heat interference and improve observation accuracy. Hydrogen can be central to future in-space refueling stations, where spacecraft can top up their fuel tanks during interplanetary missions. Technologies for producing hydrogen on the Moon or Mars (via electrolysis of water ice) could enable a sustainable fuel supply chain for deep space travel, reducing the dependency on Earth-based fuel launches. In aerospace beyond rockets, hydrogen fuel cells are being explored for powering smaller electric aircraft. These systems can achieve zero-emission flight while offering longer ranges than battery-only aircraft, making them suitable for regional air travel and urban air mobility solutions. Hydrogen is used in aerospace material testing to replicate extreme conditions in propulsion systems. For instance, it is involved in testing turbine blades, thermal protection systems, and combustion chambers to ensure durability against high heat and reactive environments. Table 2.6 illustrates hydrogen in the aerospace and rocket propulsion.

**Table 2.6** Hydrogen in the Aerospace and Rocket Propulsion

| Application Area | Description | Benefits | Challenges |
|---|---|---|---|
| Rocket fuel (liquid hydrogen) | Used as a primary propellant in combination with liquid oxygen (LOX) in space launch vehicles and spacecraft | Extremely high specific impulse, lightweight fuel, proven performance in NASA and ISRO missions | Requires cryogenic storage at −253 °C, high insulation cost, and boil-off losses |
| Fuel cells for spacecraft power | Hydrogen fuel cells generate onboard electricity for spacecraft, satellites, and space stations. | Provides clean, reliablé, and continuous power without combustion, emits only water | Limited operational life in long missions, storage complexity |
| Hydrogen-based air-breathing engines | Hydrogen-powered scramjets and turbojets for hypersonic and high-altitude flight | Higher energy density than conventional jet fuels, potential for faster and longer missions | Requires advanced materials to handle extreme temperatures and stresses |
| Auxiliary power units (APUs) | Hydrogen fuel cells or combustion systems used for powering auxiliary systems in aircraft | Reduces $CO_2$ emissions, provides silent and efficient power | Infrastructure for hydrogen refueling at airports is still underdeveloped. |
| Emission reduction in commercial aviation | Hydrogen combustion or fuel cell propulsion for passenger and cargo aircraft | Zero-carbon emissions, potential compliance with future environmental regulations | Aircraft redesign needed, long-term certification and safety approvals required |
| High-efficiency energy storage | Hydrogen used to store excess renewable energy for aerospace ground operations and launch systems | Enhances sustainability of launch facilities, supports grid stability | Energy losses during hydrogen production and conversion |

### 2.1.2 Hydrogen Embrittlement

Hydrogen embrittlement (HE) refers to the process by which metals, particularly high-strength steels and certain alloys, lose their ductility and toughness when exposed to hydrogen. This phenomenon makes materials brittle and prone to sudden fracture under stresses much lower than their normal design limits. It is especially critical in industries dealing with hydrogen storage, transport, and energy applications, where material failure could have catastrophic consequences. HE does not have a single universal explanation because the way hydrogen interacts with metals depends strongly on the metal's microstructure, chemical composition, and stress conditions. Instead, several mechanisms have been proposed. These mechanisms often act together in a synergistic manner, where hydrogen simultaneously affects bond strength, plasticity, microstructure, and internal stress. Below are the four most widely discussed mechanisms.

**Hydrogen-Enhanced Decohesion (HEDE):** The HEDE mechanism suggests that hydrogen reduces the strength of atomic bonds within the metal lattice or at grain boundaries. Hydrogen atoms accumulate at highly stressed regions, such as crack tips or grain boundaries, where atomic bonds are already under tension. The presence of hydrogen weakens these interatomic bonds, lowering the cohesive energy needed to hold atoms together. Microcracks form more easily and propagate rapidly because less stress is required to break atomic bonds. The fracture in this mechanism is often brittle and intergranular (along grain boundaries), though transgranular cleavage can also occur. High-strength steels exposed to hydrogen gas pipelines or electroplating environments often fail through decohesion at grain boundaries.

**Hydrogen-Enhanced Localized Plasticity (HELP):** The HELP mechanism argues that hydrogen actually makes metals *more* locally deformable, not less. Hydrogen atoms interact with dislocations (defects that allow metals to plastically deform). Normally, dislocations are impeded by obstacles such as grain boundaries or precipitates. Hydrogen reduces the stress required for dislocations to move, essentially "lubricating" them. Instead of uniform deformation across the material, plastic strain becomes concentrated in small regions near crack tips. This leads to localized slip bands and stress concentrations that promote crack initiation. Unlike HEDE, HELP is associated with localized ductility that paradoxically makes the material appear more brittle on the macroscopic scale. Austenitic stainless steels and nickel-based alloys show slip localization due to hydrogen, consistent with the HELP model.

**Hydride Formation Mechanism:** In some alloys (notably titanium, zirconium, and some rare-earth metals), hydrogen does not just dissolve; it chemically reacts to form hydride phases. When hydrogen concentration exceeds solubility limits, it precipitates as brittle hydride compounds (e.g., $TiH_2$ or $ZrH_2$). These hydrides are harder and more brittle than the parent metal and often have mismatched volume and lattice structure. This introduces residual stresses and acts as crack initiation sites. Hydride cracking is often delayed because hydrogen must accumulate over time before precipitation occurs. This explains why some embrittlement failures appear hours or days after hydrogen exposure. Titanium alloys used in aerospace are vulnerable to hydride embrittlement during hydrogen exposure, while zirconium alloys in nuclear fuel cladding suffer hydride-related cracking in reactor environments.

**Pressure Build-Up from Molecular Hydrogen:** This mechanism focuses on hydrogen recombination into $H_2$ molecules inside the metal. Atomic hydrogen is highly mobile and can diffuse deep into metals. When it encounters voids, inclusions, or microstructural traps, hydrogen atoms can recombine into $H_2$ molecules. Since $H_2$ molecules are much larger and immobile in the lattice, they create high internal pressure. The pressure buildup leads to blistering, void growth, and microcrack formation, especially under cyclic or tensile stress. This mechanism is more common in metals exposed to high hydrogen pressures or cathodic hydrogen charging. The cracks often propagate transgranularly (through grains) due to internal bursting. Pipeline steels exposed to high-pressure hydrogen gas sometimes exhibit hydrogen-induced cracking (HIC) linked to this mechanism.

**Synergistic Action of Mechanisms:** In real materials, multiple mechanisms often operate simultaneously. In high-strength steels, HELP and HEDE may act together, hydrogen enhances slip near the crack tip (HELP), while also weakening grain boundary cohesion (HEDE). In titanium alloys, hydride formation may occur along with pressure buildup, accelerating crack propagation. The operating mechanism depends on temperature, pressure, hydrogen concentration, alloy composition, and stress state. Table 2.7 illustrates a comparative analysis of HE mechanisms.

HE is a multifaceted phenomenon that cannot be explained by one theory alone. HEDE emphasizes weakened atomic cohesion, HELP highlights dislocation-mediated slip localization, hydride formation shows chemical phase transformation effects, and pressure buildup reflects internal stress from trapped hydrogen. Together, these mechanisms explain why HE can appear in different forms, sometimes intergranular, sometimes transgranular, sometimes delayed, and sometimes immediate. Understanding these mechanisms is critical for designing hydrogen-compatible materials and ensuring the long-term safety of hydrogen infrastructure.

**Factor Affecting HE: Material Type:** The susceptibility of a material to HE depends strongly on its intrinsic composition and mechanical properties. High-strength steels, nickel-based alloys, titanium, and even some aluminum alloys are particularly vulnerable. This is because their crystal lattice is tightly packed and under high internal stresses, which increases the driving force for hydrogen atoms to occupy interstitial sites. In contrast, softer or low-strength steels, with more ductile microstructures, can often tolerate hydrogen without catastrophic cracking. For instance, martensitic steels exhibit much higher sensitivity compared to ferritic steels, where dislocation mobility can help accommodate hydrogen.

**Hydrogen Source:** Hydrogen can enter metals from various sources during processing and service. Industrial processes like electroplating, acid pickling,

**Table 2.7** Comparative Table of Hydrogen Embrittlement Mechanisms

| Mechanism | Description | Commonly Affected Materials | Failure Characteristics |
|---|---|---|---|
| Hydrogen-enhanced decohesion (HEDE) | Hydrogen reduces atomic bond strength at grain boundaries, lowering cohesive energy. | High-strength steels, nickel alloys, some stainless steels | Intergranular brittle fracture, sudden crack propagation at relatively low stresses |
| Hydrogen-enhanced localized plasticity (HELP) | Hydrogen increases dislocation mobility, causing localized slip and strain concentration. | Austenitic stainless steels, nickel-based alloys, some high-strength steels | Transgranular cracking, localized plastic deformation near crack tips, reduced fatigue life |
| Hydride formation | Hydrogen precipitates as brittle hydride phases (e.g., $TiH_2$, $ZrH_2$) once solubility limit is exceeded. | Titanium alloys, zirconium alloys, some rare-earth metals | Delayed cracking, brittle hydride-induced fracture, crack initiation at hydride interfaces |
| Pressure buildup (hydrogen recombination) | Hydrogen atoms recombine into $H_2$ in voids/inclusions, creating internal pressure. | Pipeline steels, cast irons, weldments exposed to high-pressure hydrogen | Blistering, microvoid coalescence, transgranular cracking, hydrogen-induced cracking (HIC) |

or welding introduce atomic hydrogen directly into the metal lattice. In service environments, hydrogen may come from gaseous exposure (e.g., pipelines or storage tanks), corrosion reactions (such as cathodic protection in seawater), or from hydrocarbon cracking in petrochemical industries. The more severe and sustained the hydrogen exposure, the greater the risk of embrittlement. Importantly, environmentally induced cracking (EIC) often combines corrosion and hydrogen uptake, making it a synergistic threat.

**Stress State:** HE is highly sensitive to the mechanical stress field in a component. Applied tensile stresses accelerate hydrogen diffusion toward crack tips (a phenomenon known as stress-directed diffusion). Residual stresses from machining, cold forming, or welding also act as stress concentrators where hydrogen accumulates. This stress-hydrogen synergy lowers the fracture toughness significantly. Even when hydrogen concentration is relatively low, a highly stressed region such as a sharp notch or weld heat-affected zone may experience premature cracking.

**Temperature and Pressure:** Environmental conditions strongly affect hydrogen–metal interactions. At low temperatures, hydrogen mobility is reduced, leading to local accumulation of hydrogen at traps and interfaces, which increases embrittlement severity. On the other hand, higher temperatures enhance hydrogen diffusion but may allow hydrogen to escape, reducing embrittlement. Similarly, high pressures increase hydrogen solubility in metals, raising the likelihood of crack initiation. For example, hydrogen pipelines operating under extreme pressures are much more prone to HE compared to ambient environments. Thus, both thermodynamics (solubility) and kinetics (diffusion) dictate hydrogen behavior under different temperature pressure regimes.

**Microstructure:** The microstructural features of a metal play a decisive role in HE. Grain boundaries, dislocations, voids, and inclusions act as hydrogen traps, where hydrogen can concentrate and destabilize atomic bonding. Finer grain sizes may increase resistance by distributing stresses more uniformly, while coarse grains and brittle phases provide easy crack pathways. For instance, carbides and sulfide inclusions can serve as initiation sites for cracks. The presence of martensitic or tempered phases in steels strongly correlates with HE severity, while annealed structures typically resist hydrogen uptake better.

The graph illustrates the relative influence of different factors on HE, providing a clear comparison of how strongly each parameter contributes to the degradation of materials when exposed to hydrogen (Figure 2.4). The vertical axis represents the severity scale (1–10), where higher values indicate stronger influence, while the horizontal axis lists the major influencing factors. Starting with material type, it ranks the highest (close to 10) because the susceptibility of a metal to HE is fundamentally determined by its composition and strength level. High-strength steels, nickel-based alloys, and titanium are far more prone to embrittlement compared to low-strength steels or softer metals. Thus, the intrinsic properties of the material dominate the extent of damage.

The next strong contributor is the stress state, which is given a high severity score (around nine). Both external tensile stresses and residual stresses from welding, machining, or forming significantly accelerate hydrogen-induced crack growth. Stress provides the driving force for cracks to initiate and propagate once hydrogen has weakened atomic bonds. Without stress, even susceptible materials may not experience catastrophic embrittlement.

Hydrogen source is shown with a moderately high impact (around eight). The severity depends on whether hydrogen exposure occurs through electroplating, pickling, welding, or direct hydrogen gas environments. Prolonged or repeated exposure increases absorption, making this factor critical in industrial

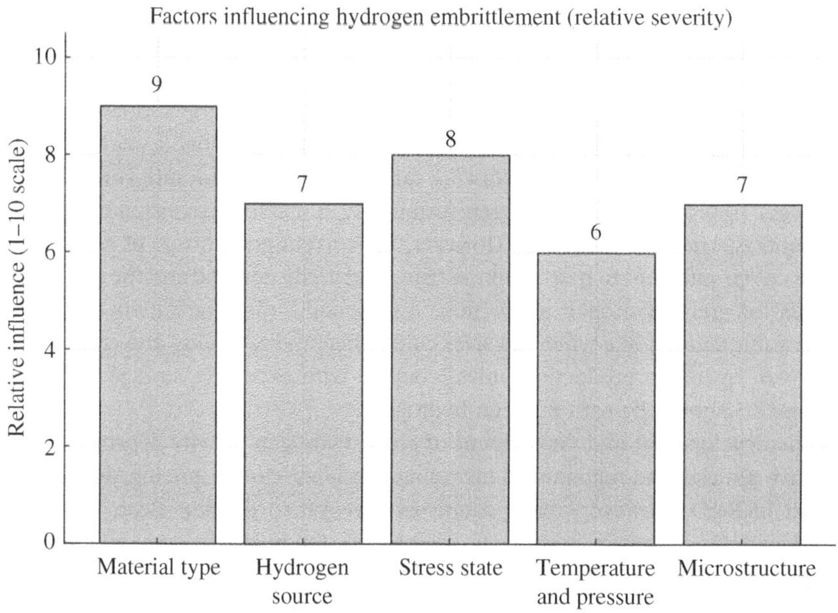


**Figure 2.4** Factor Influencing Hydrogen Embrittlement

applications. Microstructure is plotted with a mid-to-high influence (about seven). Characteristics such as grain size, inclusions, and dislocation density dictate how hydrogen is trapped and how cracks spread within the material. Fine-grained or defect-free structures generally resist embrittlement better, whereas materials with inclusions or voids act as hydrogen traps, making them more vulnerable. Finally, temperature and pressure have a relatively moderate influence (score around six). At low temperatures, hydrogen mobility decreases, allowing local concentrations of hydrogen to build up at sensitive sites, worsening embrittlement. At high pressures, hydrogen solubility increases, raising the amount absorbed by the metal. However, compared to material type and stress state, the effect of temperature and pressure is more situational and depends heavily on the operating environment. The graph emphasizes that while all five factors contribute to HE, material selection and stress management are the dominant variables, followed by hydrogen exposure conditions, microstructural features, and environmental factors such as temperature and pressure.

### 2.1.3 Overview of Green Hydrogen

Green hydrogen refers to hydrogen gas that is produced through a process called electrolysis, utilizing renewable energy sources such as wind or solar power.

During electrolysis, water ($H_2O$) is split into hydrogen ($H_2$) and oxygen ($O_2$), with the hydrogen being the desired product. Unlike conventional hydrogen production methods, which often rely on fossil fuels and emit GHGs, green hydrogen production generates zero-carbon emissions when renewable energy sources are used. As a result, green hydrogen is considered a clean and sustainable energy carrier that can play a crucial role in reducing carbon emissions and mitigating climate change. To produce green hydrogen sustainably, a massive expansion of renewable energy capacity is required. However, the current growth rate of renewables may not be sufficient to meet both the rising electricity demand and the additional needs for green hydrogen production. Additionally, the intermittent nature of renewable sources like wind and solar could affect the reliability and consistency of green hydrogen production unless coupled with expensive storage solutions. Figure 2.5 shows the path of green hydrogen.

The development and deployment of green hydrogen heavily depend on supportive policies and regulations, including subsidies, carbon pricing, and investment in R&D. However, many countries have yet to develop clear, long-term strategies for hydrogen, leading to uncertainty for investors and stakeholders. Without a stable regulatory environment, the growth of green hydrogen markets may face significant obstacles. While green hydrogen is considered clean, its production and use are not without risks. The process of electrolysis involves handling large amounts of electricity, and hydrogen itself is a highly flammable gas.

> ***Key: Green hydrogen:*** *Hydrogen produced using renewable energy sources, contributing to sustainable energy transitions.*

Leakage of hydrogen into the atmosphere could have indirect climate effects, as it is a potent indirect GHG that can contribute to ozone formation. Safety protocols, leak detection, and mitigation measures will need to be developed and implemented rigorously. While green hydrogen presents a significant opportunity for decarbonizing hard-to-abate sectors, its future hinges on overcoming various technical, economic, and infrastructural challenges. Substantial investments in research, technological innovation, infrastructure development, and supportive

**Figure 2.5** Path of Green Hydrogen

policy frameworks are crucial for addressing these issues. Until these hurdles are effectively managed, green hydrogen's role in the energy transition will likely remain limited compared to other clean energy alternatives.

Green hydrogen, gray hydrogen, and blue hydrogen are terms used to differentiate between different methods of hydrogen production based on their environmental impact and the source of energy used in their production:

**Green Hydrogen:** Green hydrogen is produced using renewable energy sources such as solar, wind, or hydropower to split water ($H_2O$) into hydrogen and oxygen through the process of electrolysis. Since the electricity used comes entirely from clean energy, there are no carbon dioxide ($CO_2$) emissions during production, making it the most environmentally friendly form of hydrogen. Green hydrogen plays a crucial role in decarbonizing industries, transportation, and energy storage, offering a sustainable alternative to fossil fuels while supporting climate change mitigation goals. However, it is currently more expensive to produce due to the high cost of renewable infrastructure and electrolysis technology. Electrolysis uses electricity (from renewable sources) to split water into hydrogen and oxygen.

$$2H_2O(l) \rightarrow \text{electricity}2H_2(g) + O_2(g)$$

No $CO_2$ is produced in this process.

**Gray Hydrogen:** Gray hydrogen is the most common and widely produced form of hydrogen today, primarily generated from natural gas through a process called SMR or from coal through gasification. In this method, methane reacts with steam at high temperatures to produce hydrogen and carbon dioxide. The $CO_2$ generated is released directly into the atmosphere, making this process carbon-intensive and environmentally harmful. Gray hydrogen is cheaper to produce than green hydrogen because the technology is mature and fossil fuels are readily available, but it contributes significantly to GHG emissions, which is a major drawback. Methane reacts with steam at high temperature to produce hydrogen and carbon monoxide, followed by the water–gas shift (WGS) reaction to produce more hydrogen and carbon dioxide.

**Step 1: SMR**

$$CH_4 + H_2O \rightarrow CO + 3H_2$$

**Step 2: WGS Reaction**

$$CO + H_2O \rightarrow CO_2 + H_2$$

**Overall Reaction**

$$CH_4 + 2H_2O \rightarrow CO_2 + 4H_2$$

$CO_2$ is released into the atmosphere, causing emissions.

**Blue Hydrogen:** Blue hydrogen is produced in a similar way to gray hydrogen mainly from natural gas via SMR but with a crucial difference: the carbon dioxide emissions are captured and stored using carbon capture, utilization, and storage (CCUS) technologies instead of being released into the atmosphere. This makes blue hydrogen a cleaner alternative to gray hydrogen, as it significantly reduces $CO_2$ emissions. While it is not entirely emission-free due to the potential for leakage during extraction and storage, blue hydrogen is considered a transitional solution, helping industries move toward a low-carbon economy, while green hydrogen technology becomes more cost-effective and scalable. The chemical reactions are the same as in gray hydrogen:

$$CH_4 + 2H_2O \rightarrow CO_2 + 4H_2$$

However, the $CO_2$ produced is captured and stored (not released into the atmosphere) using CCUS technology:

$$CO_2 \rightarrow \text{capture and storage Sequestration Underground}$$

Figure 2.6 shows the depiction of green, gray, and blue Hydrogen. Figure 2.7 and Table 2.8 show the comparison between green, gray, and blue hydrogen.

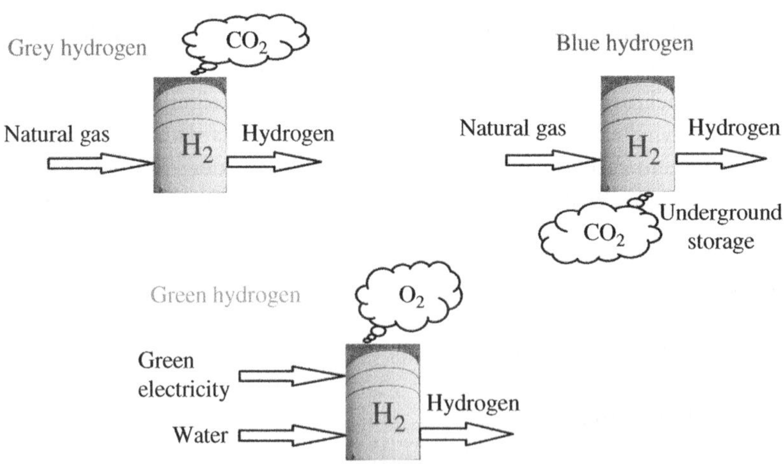


**Figure 2.6** Depiction of Gray, Blue, and Green Hydrogen

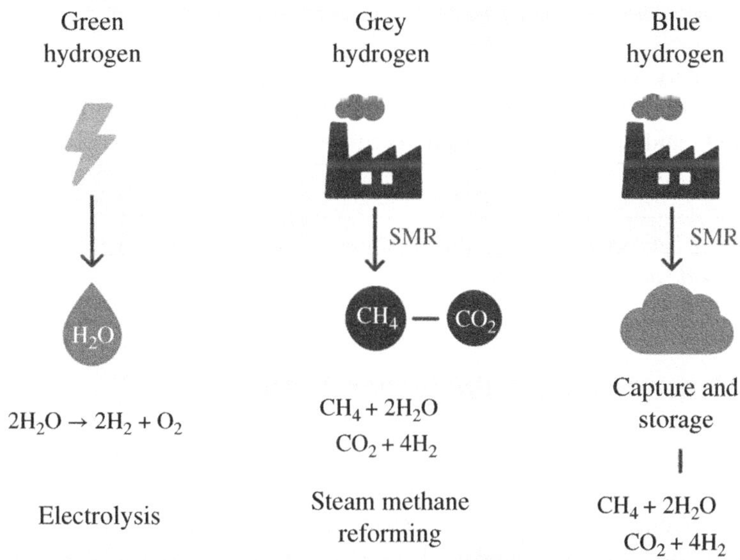


**Figure 2.7** Comparison of Green, Gray, and Blue Hydrogen

**Table 2.8** Comparison Between Green, Gray, and Blue Hydrogen

| Aspect | Green Hydrogen | Gray Hydrogen | Blue Hydrogen |
|---|---|---|---|
| Production method | Electrolysis using renewable energy sources (e.g., wind, solar, hydro) | Steam methane reforming (SMR) using natural gas | SMR using natural gas with carbon capture and storage (CCS) |
| Environmental impact | Zero-carbon emissions during production | Significant carbon emissions during production | Carbon emissions reduced through CCS during production |
| Source of energy | Renewable energy sources | Fossil fuel (natural gas) | Fossil fuel (natural gas) with CCS |
| Carbon emissions | None | Significant emissions of $CO_2$ | Reduced emissions of $CO_2$ due to CCS |
| Sustainability | Sustainable and environmentally friendly | Contributes to greenhouse gas emissions | Offers a lower carbon footprint compared to gray hydrogen |
| Cost | Typically higher due to reliance on renewable energy | Economically competitive due to low cost of natural gas | May have higher costs due to CCS implementation |
| Role in transition to sustainable energy | Integral part of a renewable energy future | Transitional solution toward cleaner energy | Transitional solution toward cleaner energy |

In summary, green hydrogen is produced using renewable energy sources and emits no carbon emissions, while gray hydrogen is produced from natural gas and emits $CO_2$, and blue hydrogen is produced from natural gas but employs CCS to reduce its carbon footprint.

> ***Key: Gray hydrogen:*** *Hydrogen produced from natural gas, typically emitting carbon dioxide during production.*

## 2.2 Green Hydrogen Production Methods

Green hydrogen production through electrolysis, powered by renewable energy sources, is heralded as a key solution for achieving a carbon-neutral future. However, this method faces several critical challenges that question its current viability and scalability. The most pressing issue is the high cost associated with green hydrogen production, which remains significantly higher than that of "gray" or "blue" hydrogen produced from fossil fuels. This cost premium stems from both the expense of electrolyzers and the intermittent nature of renewable energy sources, which leads to lower utilization rates and fluctuating output. Additionally, the efficiency of the electrolysis process itself is limited; a significant portion of the electrical energy used is lost as heat, resulting in lower overall energy conversion efficiency. Moreover, the infrastructure needed to support green hydrogen, including storage, transport, and refueling networks, is underdeveloped, posing a major barrier to widespread adoption. There are also environmental concerns related to the land use and material requirements for large-scale renewable energy installations, such as wind farms and solar panels, which are necessary to produce green hydrogen in substantial quantities. While green hydrogen offers a promising pathway for decarbonizing hard-to-abate sectors, its production method currently faces significant economic, technical, and infrastructural hurdles that must be addressed to realize its full potential as a cornerstone of sustainable energy systems.

> ***Key: Electrolysis:*** *The process of splitting water into hydrogen and oxygen using electricity.*

1) **Steam Methane Reforming (SMR):** SMR is the most common and cost-effective industrial method for hydrogen production, used predominantly in gray hydrogen (without carbon capture) and blue hydrogen (with carbon capture). The feedstock is typically methane ($CH_4$), obtained from natural gas,

biogas, or other hydrocarbon sources. In SMR, methane reacts with high-temperature steam (typically 700–1,000 °C) in the presence of a nickel-based catalyst. The catalyst helps lower the energy barrier for the reaction, allowing it to proceed efficiently at industrial scale. The process occurs in a reformer unit, where methane and steam are introduced and passed over the catalyst inside heated tubes.

SMR is an endothermic reaction, meaning it requires a continuous supply of heat, usually provided by burning a portion of the methane feed. The process is divided into two main reaction stages:

**Step 1: SMR**

$$CH_4 + H_2O \rightarrow CO + 3H_2$$

- Methane ($CH_4$) reacts with steam ($H_2O$) to produce carbon monoxide (CO) and hydrogen gas ($H_2$).
- This reaction takes place at high temperatures in the reformer tubes.
- Being highly endothermic, it consumes a large amount of heat.

**Step 2: WGS Reaction**

$$CO + H_2O \rightarrow CO_2 + H_2$$

- The carbon monoxide (CO) produced in Step 1 is reacted with additional steam to yield carbon dioxide ($CO_2$) and more hydrogen gas ($H_2$).
- This reaction is mildly exothermic (releases a small amount of heat).
- It is usually carried out in a separate shift converter unit, which has its own catalyst optimized for the WGS reaction.

By combining both steps, the net reaction becomes:

$$CH_4 + 2H_2O \rightarrow CO_2 + 4H_2$$

- One molecule of methane and two molecules of steam produce one molecule of carbon dioxide and four molecules of hydrogen.
- The hydrogen is then purified (often through pressure swing adsorption—PSA) to remove any remaining $CO_2$, CO, or methane before use. Figure 2.8 illustrates a schematic diagram of SMR.

The carbon capture phase can be carried out using several industrial methods. One common approach is amine-based absorption, where $CO_2$ binds chemically to an amine solvent, which is later regenerated to release the concentrated $CO_2$. Another method is membrane separation, in which special polymer or ceramic membranes selectively allow $CO_2$ molecules to pass through while retaining

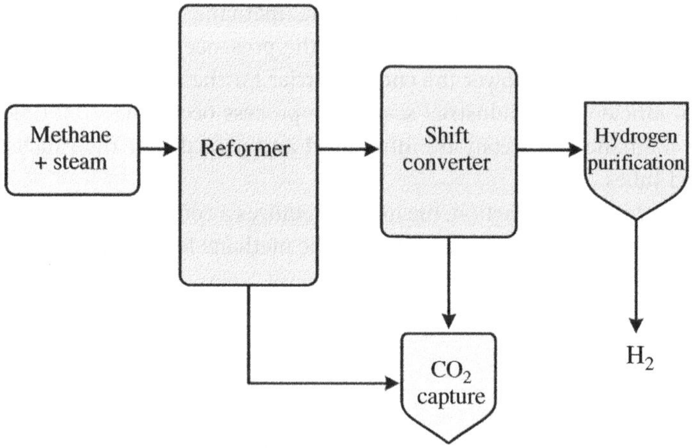


**Figure 2.8** Schematic Diagram of Steam Methane Reforming

hydrogen and other gases. Cryogenic processes can also be used, cooling the gas mixture to extremely low temperatures so that $CO_2$ condenses and can be separated. These technologies can capture a significant portion often over 90% of the $CO_2$ produced during SMR. Once captured, the $CO_2$ must be compressed into a dense form for transport. This is typically achieved using high-pressure pumps, after which the compressed $CO_2$ can be transported via pipelines for short-to-medium distances or via specialized ships for long-distance transfers. The goal is to deliver the $CO_2$ to secure storage sites or facilities where it can be put to beneficial use. Storage is usually achieved in geological formations deep underground. Common options include depleted oil and gas reservoirs, where the $CO_2$ can also be used in enhanced oil recovery (EOR) to extract remaining hydrocarbons, and deep saline aquifers, which are porous rock formations filled with brine that can permanently trap $CO_2$. Proper monitoring and sealing methods are essential to ensure that the stored carbon remains securely underground for thousands of years.

While this integration of CCUS technology with SMR drastically reduces $CO_2$ emissions, it is not entirely emission-free. A significant concern is methane leakage from the natural gas supply chain, from extraction to transport. Methane ($CH_4$) is a potent GHG with a much higher global warming potential (GWP) than $CO_2$, meaning even small leaks can undermine climate benefits. Consequently, for blue hydrogen to be a truly sustainable transitional fuel, both carbon capture efficiency and methane management must be prioritized.

SMR has become the dominant hydrogen production technology globally, responsible for producing more than 70% of the world's hydrogen supply.

Its widespread use is largely due to its maturity, cost-effectiveness, and scalability compared to other hydrogen production methods. SMR plays a vital role in several key industrial sectors. In oil refineries, hydrogen is used for hydrocracking and desulfurization processes to produce cleaner fuels. In ammonia production, hydrogen serves as a key feedstock for the Haber–Bosch process, which is essential for manufacturing fertilizers. Methanol synthesis also depends heavily on hydrogen from SMR, as methanol is widely used in plastics, adhesives, and as a fuel additive. The steel industry utilizes hydrogen in direct reduction processes and other metallurgical applications.

In recent years, blue hydrogen produced via SMR combined with CCUS has gained traction, especially in countries with strong carbon reduction policies but an existing natural gas infrastructure. This makes it an attractive transitional solution for decarbonizing heavy industries and energy systems, as it can leverage existing SMR plants while significantly lowering emissions compared to gray hydrogen.

Despite its dominance, SMR has several notable drawbacks. In its gray hydrogen form, the process emits large amounts of carbon dioxide, contributing significantly to global GHG emissions. Even with carbon capture in blue hydrogen, the process remains energy-intensive, requiring extremely high temperatures (700–1,000 °C), which increases operational costs and often relies on burning additional fossil fuels for heat. Another major concern is methane leakage during natural gas extraction, processing, and transportation. Methane ($CH_4$) is over 25 times more potent as a GHG than $CO_2$ over a 100-year period, meaning even small leaks can erode the climate benefits of blue hydrogen. Additionally, CCUS technology, while effective, requires substantial capital investment for installation and ongoing operational expenses for compression, transport, and monitoring of stored $CO_2$. Finally, SMR fundamentally depends on fossil fuels mainly natural gas, which means it cannot be considered fully renewable. While blue hydrogen reduces emissions compared to gray hydrogen, it does not offer the complete decarbonization potential of green hydrogen, which is produced using renewable energy sources and generates no direct GHG emissions.

2) **Biomass Gasification:** Biomass gasification is a thermochemical process that converts organic material such as agricultural residues (corn stalks, rice husks, sugarcane bagasse), forestry waste (wood chips, sawdust), energy crops (switchgrass, miscanthus), and even organic municipal solid waste into hydrogen-rich syngas. This process occurs in a gasifier at high temperatures (700–1,200°C) in the presence of a controlled amount of oxygen, air, or steam. Unlike combustion, which completely oxidizes biomass to $CO_2$ and water, gasification partially oxidizes it, producing a mixture of gases: hydrogen ($H_2$), carbon monoxide (CO), carbon dioxide ($CO_2$), methane ($CH_4$), and other trace compounds.

Feedstock variability is an important factor because different biomass sources have different moisture content, ash content, carbon-to-hydrogen ratio, and energy density, all of which affect the efficiency and composition of the syngas produced. Dry, low-ash biomass generally yields higher hydrogen concentrations. Conversion efficiency can range between 35% and 55%, depending on the feedstock type, gasifier design (fixed bed, fluidized bed, or entrained flow), and operating conditions such as temperature, pressure, and gasifying agent. The raw syngas produced during gasification contains not only useful gases but also impurities such as tars, particulates, sulfur compounds ($H_2S$, COS), nitrogen compounds ($NH_3$), and trace metals. These must be removed to ensure the hydrogen is pure enough for industrial applications, fuel cells, or chemical synthesis. Figure 2.9 illustrates the layout of biomass gasification.

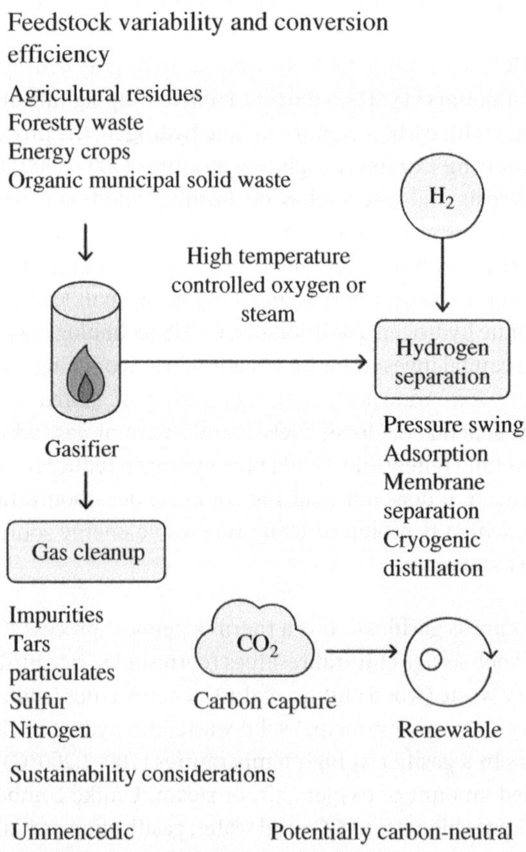


**Figure 2.9** Layout of Biomass Gasification

The gas cleanup process usually involves multiple steps:

- Cyclones and filters to remove particulates.
- Scrubbers or catalytic reformers to crack and remove tars.
- Acid gas removal systems (amine absorption or PSA) to remove sulfur compounds and $CO_2$.

After cleanup, the hydrogen is separated from other gases using PSA, membrane separation, or cryogenic distillation. In many cases, the WGS reaction is applied first to convert carbon monoxide into additional hydrogen:

$$CO + H_2O \rightarrow CO_2 + H_2$$

This step boosts hydrogen yield before final purification.

Biomass gasification is considered renewable and potentially carbon-neutral because the $CO_2$ released during hydrogen production is roughly equal to the $CO_2$ absorbed by the plants during growth. If biomass is sourced from waste streams or sustainably managed crops, the process can significantly reduce dependence on fossil fuels and lower net GHG emissions.

However, sustainability depends on several factors:

- **Land Use and Biodiversity:** Growing dedicated energy crops can compete with food production and affect ecosystems.
- **Supply Chain Emissions:** Collection, transport, and preprocessing of biomass can add carbon costs if not managed efficiently.
- **Residue Management:** Using agricultural and forestry residues avoids competition with food crops and reduces waste.
- **Water Consumption:** Some gasification processes require steam, which must be sourced sustainably.

When coupled with CCUS, biomass gasification can even achieve negative emissions, as the $CO_2$ stored comes from atmospheric carbon absorbed by plants. This makes bioenergy with carbon capture and storage (BECCS) a promising pathway for producing low-carbon or even carbon-negative hydrogen.

## 2.3 Electrolysis Technologies

**Principles of Electrolysis:** Electrolysis is a key process in the production of green hydrogen, leveraging principles of chemistry and electricity to split water molecules into hydrogen ($H_2$) and oxygen ($O_2$). This process occurs within an electrolyzer, a device equipped with electrodes and an electrolyte solution. The principles underlying electrolysis involve the application of an electric current to initiate a chemical reaction within the electrolyte, resulting in the separation of water into its constituent elements. The electrolysis process begins with the placement of electrodes, typically made of conductive materials such as platinum or graphite, into a container filled with an electrolyte solution. The electrolyte, often composed of water mixed with substances like potassium hydroxide (KOH) or sulfuric acid ($H_2SO_4$), facilitates the movement of ions and enhances the conductivity of the solution. When an electric current is applied to the electrodes, it induces an electrochemical reaction at the electrode–electrolyte interface. Figure 2.10 shows the hydrogen production structure by electrolysis. At the anode (positive electrode), water molecules undergo oxidation, releasing oxygen gas and positively charged hydrogen ions (protons):

$$2H_2O(l) \rightarrow O_2(g) + 4H + (aq) + 4e^-$$

Simultaneously, at the cathode (negative electrode), hydrogen ions from the electrolyte gain electrons and are reduced to form hydrogen gas:

$$4H + (aq) + 4e^- \rightarrow 2H_2(g)$$

The electrolysis of water yields hydrogen gas at the cathode and oxygen gas at the anode, with the electrical energy input driving the separation of the water molecules.

Green hydrogen production through electrolysis differs from conventional hydrogen production methods in its reliance on renewable energy sources to power the electrolysis process. Solar, wind, or hydropower-generated electricity

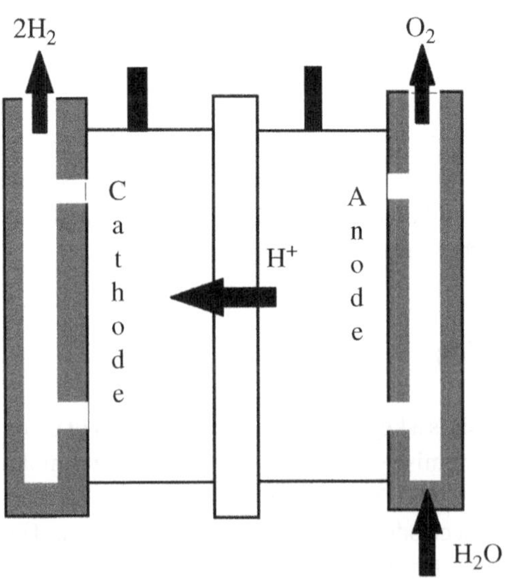


**Figure 2.10** Hydrogen Production: Electrolysis

is utilized to supply the electric current needed for electrolysis, ensuring that the hydrogen produced is carbon-free and environmentally sustainable. This contrasts with conventional methods such as SMR, which rely on fossil fuels and emit GHGs during hydrogen production.

Furthermore, electrolysis offers flexibility in scaling hydrogen production to match demand and integrate with renewable energy systems. Electrolyzer technologies vary, with options including proton exchange membrane (PEM), alkaline electrolysis, and solid oxide electrolysis cells (SOECs), each with its advantages and applications. PEM electrolysis, for instance, operates at relatively low temperatures and is suitable for smaller-scale applications and intermittent operation, while alkaline electrolysis is well-established and suitable for large-scale industrial production. The principles and processes of green hydrogen production through electrolysis involve the application of electricity to split water molecules into hydrogen and oxygen gases. By utilizing renewable energy sources to power the electrolysis process, green hydrogen offers a sustainable alternative to conventional hydrogen production methods, contributing to efforts to reduce GHG emissions and combat climate change.

> ***Key: Fuel cell:*** *A device that converts chemical energy from hydrogen into electricity and heat, with water as the only by-product.*

Following are the key components and steps of the processes of electrolysis for the green hydrogen production:

**Electrolyzer:** An electrolyzer is the device where electrolysis occurs. It consists of an electrolyte solution and electrodes. The electrolyte allows for the flow of ions and enhances the conductivity of the solution. There are different types of electrolyzers used in green hydrogen production, including PEM electrolyzers, alkaline electrolyzers (AELs), and solid oxide electrolyzers.

**Electric Current:** An electric current is applied to the electrolyzer, typically generated from renewable energy sources such as solar, wind, or hydropower. This electricity provides the energy needed to drive the electrolysis reaction.

**Anode:** At the anode (positive electrode), water molecules undergo oxidation, releasing oxygen gas ($O_2$) and positively charged hydrogen ions (protons). The oxidation reaction at the anode is represented as follows:

$$2H_2O(l) \rightarrow O_2(g) + 4H + (aq) + 4e^-$$

**Cathode:** At the cathode (negative electrode), hydrogen ions from the electrolyte gain electrons and are reduced to form hydrogen gas ($H_2$). The reduction reaction at the cathode is represented as follows:

$$4H + (aq) + 4e^- \rightarrow 2H_2(g)$$

**Overall Reaction:** The overall electrolysis reaction combines the oxidation and reduction reactions occurring at the anode and cathode, respectively. This results in the production of hydrogen gas at the cathode and oxygen gas at the anode:

$$2H_2O(l) \rightarrow 2H_2(g) + O_2(g)$$

**Hydrogen Purification:** After electrolysis, the produced hydrogen gas may require purification to remove impurities and moisture, ensuring its quality for various applications.

**Storage and Transportation:** The produced green hydrogen can be stored and transported in various forms, including compressed gas, liquid hydrogen, or converted into chemical compounds such as ammonia or methanol for easier storage and transportation.

Electrolysis is a fundamental process in green hydrogen production, leveraging renewable energy sources to split water molecules into hydrogen and oxygen gases without emitting GHGs. By harnessing the principles of electrolysis, green hydrogen offers a sustainable and low-carbon alternative to conventional fossil fuel-based hydrogen production methods, contributing to efforts to mitigate climate change and transition toward a cleaner energy future.

### 2.3.1 Types of Electrolysis

#### 2.3.1.1 Alkaline Electrolysis: Overview and Operation

Alkaline electrolysis is a process that involves using an electrolytic cell to split water ($H_2O$) into its constituent elements, hydrogen ($H_2$) and oxygen ($O_2$), using electricity. It operates on the principle of passing an electric current through a water-based electrolyte solution, which facilitates the decomposition of water molecules into hydrogen and oxygen gas. Figure 2.11 shows the structure of alkaline electrolysis process.

Following are the detailed overview and explanation of the operation of alkaline electrolysis:

**Electrolysis Cell Configuration:** The alkaline electrolysis cell typically consists of two electrodes, an anode (positive electrode) and a cathode (negative electrode), immersed in an alkaline electrolyte solution. The most commonly used electrolyte is a solution of potassium hydroxide (KOH) or sodium hydroxide (NaOH), which provides the necessary ions for conduction and facilitates the electrolysis process.

**Electrode Materials:** The choice of electrode materials is crucial for the efficiency and durability of the electrolysis cell. Typically, the anode is made of a corrosion-resistant material like titanium coated with a catalyst such as iridium oxide ($IrO_2$) or platinum (Pt) to facilitate the oxidation of water

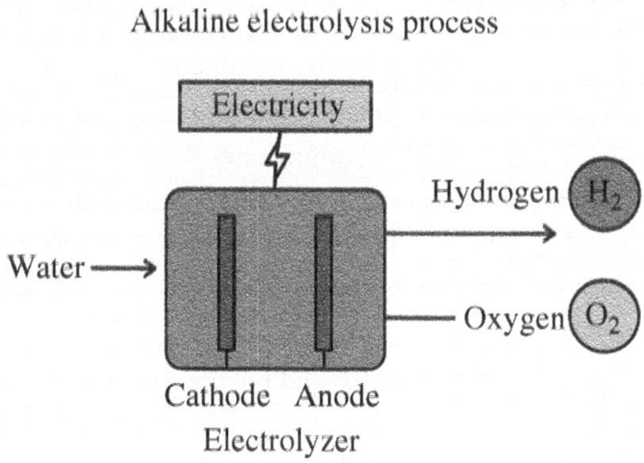


**Figure 2.11** Structure of Alkaline Electrolysis Process

molecules. The cathode is often made of a porous material such as nickel (Ni) or stainless steel, which allows for the efficient reduction of water molecules into hydrogen gas.

**Electrolysis Process:** When an electric current is passed through the electrolyte solution, water molecules near the anode undergo oxidation, releasing oxygen gas ($O_2$) and positively charged hydrogen ions (H+). This reaction occurs at the anode according to the following equation:

$$\text{Anode Reaction: } 2H_2O(l) \rightarrow O_2(g) + 4H + (aq) + 4e^-$$

Meanwhile, at the cathode, the positively charged hydrogen ions migrate toward the cathode where they undergo reduction, gaining electrons from the cathode and forming hydrogen gas ($H_2$). The overall cathode reaction can be represented as:

$$\text{Cathode Reaction: } 4H + (aq) + 4e^- \rightarrow 2H_2(g)$$

The net result of these reactions is the splitting of water into hydrogen and oxygen gases.

**Electrical Power Input:** The efficiency of the electrolysis process depends on the amount of electrical energy supplied to the cell. The electrical power input required for electrolysis is directly proportional to the amount of hydrogen gas produced. Higher current densities and voltages generally result in higher rates of hydrogen production, but they also lead to increased energy consumption and operating costs.

**Temperature and Pressure:** The temperature and pressure conditions within the electrolysis cell also influence its performance. Generally, higher temperatures can increase the rate of reaction kinetics and improve electrolysis efficiency. Additionally, applying pressure to the system can enhance the solubility of gases in the electrolyte solution, particularly for the generated hydrogen gas.

**Product Separation:** Once the electrolysis process is complete, the hydrogen and oxygen gases need to be separated from each other and from the electrolyte solution. This can be achieved through various methods such as gas diffusion, membrane separation, or physical separation techniques. The separated hydrogen gas can then be collected and utilized for various industrial applications, including fuel cell technology, chemical synthesis, and energy storage.

Alkaline electrolysis is a versatile and efficient method for producing hydrogen gas from water using electricity. It offers a clean and sustainable pathway for hydrogen production, particularly when powered by renewable energy sources such as solar or wind power. However, despite its advantages, alkaline electrolysis still faces challenges related to cost, efficiency, and scalability, which ongoing research and technological advancements aim to address.

> ***Key: Photobiological water splitting:*** *A process using biological organisms and sunlight to split water molecules into hydrogen and oxygen.*

### 2.3.1.2 Proton Exchange Membrane (PEM) Electrolysis:

PEM electrolysis is a method for producing hydrogen gas by splitting water molecules using electricity, similar to alkaline electrolysis. However, PEM electrolysis employs a different type of electrolyte membrane, known as a PEM, which separates the anode and cathode compartments while allowing the transfer of protons (hydrogen ions). Figure 2.12 shows the structure of PEM electrolysis.

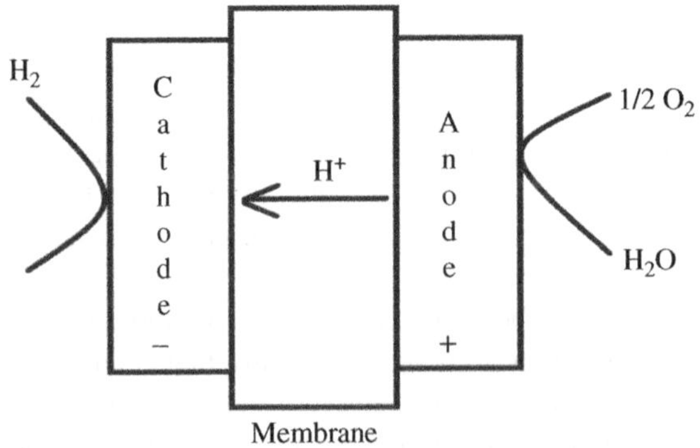


**Figure 2.12** Structure Proton Exchange Membrane of Alkaline Electrolysis Process

Following are the detailed explanation of the working principles and applications of PEM electrolysis:

**Electrolysis Cell Configuration:** A PEM electrolysis cell consists of two electrodes—an anode and a cathode separated by a PEM. The anode and cathode compartments are filled with water or a water-based electrolyte solution. The PEM acts as a barrier between the two compartments, allowing only the passage of protons ($H^+$) while blocking the flow of electrons.

**Electrode Materials:** Similar to alkaline electrolysis, the choice of electrode materials is crucial for PEM electrolysis. Typically, the anode is made of a corrosion-resistant material such as titanium coated with a catalyst like iridium oxide ($IrO_2$) or platinum (Pt) to facilitate the oxidation of water molecules. The cathode is often made of a porous material like carbon, which allows for the efficient reduction of water molecules into hydrogen gas.

**Electrolysis Process:** When an electric current is passed through the PEM electrolysis cell, water molecules near the anode undergo oxidation, releasing oxygen gas ($O_2$), protons ($H^+$), and electrons ($e^-$). The anode reaction can be represented as:

$$\text{Anode Reaction: } 2H_2O(l) \rightarrow O_2(g) + 4H + (aq) + 4e^-$$

Meanwhile, the protons produced at the anode migrate through the PEM to the cathode compartment. At the cathode, these protons combine with electrons from the external circuit and reduce to form hydrogen gas ($H_2$). The cathode reaction can be represented as:

$$\text{Cathode Reaction: } 4H + (aq) + 4e^- \rightarrow 2H_2(g)$$

The net result of these reactions is the splitting of water into hydrogen and oxygen gases.

**Proton Exchange Membrane (PEM):** The PEM plays a crucial role in facilitating the transport of protons from the anode to the cathode while preventing the mixing of gases and electrolyte solutions. It is typically made of a solid polymer electrolyte material, such as a perfluorosulfonic acid (PFSA) polymer, which provides high proton conductivity and chemical stability.

**Efficiency and Operating Conditions:** PEM electrolysis offers several advantages over alkaline electrolysis, including higher efficiency, faster response times, and the ability to operate at lower temperatures and pressures. The PEM technology allows for precise control of hydrogen production rates and enables dynamic operation to match varying electricity supply from renewable sources.

#### 2.3.1.3 Solid Oxide Electrolysis Cells (SOECs): Technology and Efficiency

SOECs are electrochemical devices that convert electricity and water into hydrogen and oxygen gas. They operate based on similar principles to solid oxide fuel cells (SOFCs), but in reverse, utilizing high-temperature electrolysis rather than power generation. Figure 2.13 shows the structure of SOECs of electrolysis process.

Following are the detailed explanation of SOEC technology and efficiency:

**Cell Configuration:** SOECs consist of a solid oxide electrolyte, typically made of materials like yttria-stabilized zirconia (YSZ) or ceria ($CeO_2$), sandwiched between porous electrodes. The electrolyte separates the cathode and anode compartments and conducts oxygen ions ($O_2{}^-$) from the cathode to the anode.

**Operating Temperature:** SOECs operate at high temperatures, typically between 600 and 900 °C. At these temperatures, the solid oxide electrolyte becomes sufficiently conductive to oxygen ions, facilitating the electrolysis process. High operating temperatures also enhance reaction kinetics, leading to improved efficiency and performance.

**Electrode Materials:** The electrodes of SOECs are typically made of mixed ionic–electronic conductors (MIECs) such as perovskite oxides (e.g., lanthanum strontium manganite, LSM) or nickel–cermet materials. These electrode materials facilitate the catalysis of electrochemical reactions occurring at the electrode–electrolyte interfaces.

**Electrolysis Process:** When an electric current is applied to the SOEC, oxygen ions ($O_2{}^-$) migrate through the solid oxide electrolyte from the cathode to the anode under the influence of the applied voltage. At the cathode, oxygen ions combine with electrons from the external circuit and water molecules to form oxygen gas:

$$\text{Cathode Reaction: } O_2{}^- + 2H_2O + 4e^- \rightarrow 2OH^-$$

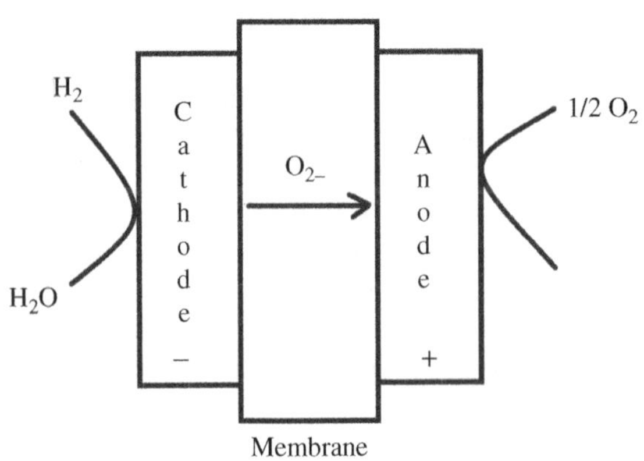


**Figure 2.13** Structure of Solid Oxide Electrolysis Cells of Electrolysis Process

At the anode, oxygen ions are oxidized to oxygen gas while releasing electrons into the external circuit:

Anode Reaction: $1/2O_2 + 2e^- \rightarrow O_2^-$

The overall electrolysis reaction is the splitting of water into hydrogen and oxygen gases:

Overall Reaction: $H_2O \rightarrow H_2 + 1/2O_2$

**Efficiency**

SOECs offer high efficiency for hydrogen production due to their ability to operate at high temperatures, which enhances reaction kinetics and reduces over potentials. The high operating temperatures also enable the use of waste heat from industrial processes or other sources to improve overall system efficiency. SOECs can achieve electrical-to-hydrogen conversion efficiencies of over 80%, making them competitive with other electrolysis technologies. Furthermore, SOECs can be coupled with renewable energy sources such as solar or wind power to produce hydrogen with zero GHG emissions, contributing to sustainable energy solutions.

> ***Key: Thermochemical water splitting:*** *A method of hydrogen production using high-temperature chemical reactions to split water.*

## 2.3.2 Electrolysis Equipment

### 2.3.2.1 Electrolyzer: Components and Design

**Anion Exchange Membrane Electrolyzer**

An anion exchange membrane electrolyzer (AEM electrolyzer) is a device used to produce hydrogen through the electrolysis of water. It operates on the principle of separating water molecules into hydrogen and oxygen gas by passing an electric current through water, using a special membrane that facilitates the transport of anions (negatively charged ions).

**Key Components of an AEM Electrolyzer**

**Anion Exchange Membrane (AEM):** The core component of the AEM electrolyzer is the AEM. This membrane selectively allows the passage of anions, such as hydroxide ions ($OH^-$), while blocking cations and other impurities. The AEM is usually made from a polymer material with functional groups that can bind anions and facilitate their transport.

**Electrodes**

**Cathode (Negative Electrode):** At the cathode, water molecules gain electrons (reduction) and are converted into hydrogen gas ($H_2$) and hydroxide ions ($OH^-$).

**Anode (Positive Electrode):** At the anode, hydroxide ions lose electrons (oxidation) and form oxygen gas ($O_2$) and water.

**Electrolyte:** The electrolyte in an AEM electrolyzer is typically an alkaline solution, such as potassium hydroxide (KOH), which enhances the conductivity and facilitates the movement of hydroxide ions.

**How an AEM Electrolyzer Works**

**Water Splitting:** When an electric current is applied to the electrolyzer, water molecules at the cathode are split into hydrogen gas ($H_2$) and hydroxide ions ($OH^-$):

$$2H_2O + 2e^- \rightarrow H_2 + 2OH^-$$

**AEM Function:** The hydroxide ions ($OH^-$) generated at the cathode move through the AEM toward the anode.

**Oxygen Formation:** At the anode, the hydroxide ions release electrons (are oxidized) to form oxygen gas ($O_2$) and water:

$$2OH^- \rightarrow O_2 + 2H_2O + 4e^-$$

**Gas Separation:** The hydrogen gas produced at the cathode is collected, while oxygen gas is collected at the anode, and the two gases are kept separate by the membrane.

**Advantages of AEM Electrolyzers**

- **Cost-Effectiveness:** AEM electrolyzers can use nonprecious metal catalysts (like nickel or iron) rather than precious metals (like platinum), reducing overall costs.
- **Flexibility:** They can operate under alkaline conditions, making them compatible with a wide range of electrolytes and reducing material degradation.
- **Environmental Benefits:** AEM electrolyzers produce green hydrogen if powered by renewable electricity, providing a sustainable alternative to fossil fuel-based hydrogen production.

**Challenges**

- **Membrane Durability:** The AEM material must be highly durable, resistant to chemical degradation, and capable of maintaining high ionic conductivity over time.
- **Efficiency Improvements:** Research is ongoing to improve the overall efficiency, as AEM electrolyzers are currently less efficient than PEM electrolyzers.

**Applications**

- **Hydrogen Production for Clean Energy:** AEM electrolyzers can produce hydrogen for fuel cells, which power vehicles, industrial processes, and electricity generation.

- **Decarbonization Efforts:** They play a crucial role in efforts to decarbonize industries by providing a clean hydrogen source for various applications.

AEM electrolyzers offer a promising technology for cost-effective and environmentally friendly hydrogen production. They operate under alkaline conditions, using a specialized membrane that selectively transports anions, and provide a pathway for sustainable energy production through green hydrogen. Table 2.9 shows the comparison of different types of electrolysis.

**Table 2.9** Comparison of Different Types of Electrolysis

| Aspect | Alkaline Electrolysis | Proton Exchange Membrane (PEM) Electrolysis | Solid Oxide Electrolysis | Anion Exchange Membrane (AEM) Electrolysis |
|---|---|---|---|---|
| Electrolyte | Potassium hydroxide (KOH) or sodium hydroxide (NaOH) | Solid polymer membrane (Nafion or similar) | Ceramic (yttria-stabilized zirconia) | AEM |
| Operating temperature | 60–90 °C | 50–80 °C | 600–1,000 °C | 40–90 °C |
| catalyst | Nickel-based (nonprecious metals) | Platinum or iridium (precious metals) | Nickel or cerium-based (non-precious metals) | Nickel, cobalt, or nonprecious metals |
| Efficiency | 60–70% | 70–80% | 80–90% | 60–80% |
| Energy requirement | Moderate | High (due to requirement for pure water) | Low (due to high operating temperature) | Moderate to high |
| Current density | Low (0.2–0.4 $A/cm^2$) | High (1–2 $A/cm^2$) | Very high (0.3–1.0 $A/cm^2$ at high temperatures) | Moderate (0.5–1 $A/cm^2$) |
| Durability | High (long lifespan) | Moderate to High (can degrade due to membrane wear) | Moderate (affected by thermal cycling) | Developing (current issues with membrane stability) |

(Continued)

**Table 2.9** (Continued)

| Aspect | Alkaline Electrolysis | Proton Exchange Membrane (PEM) Electrolysis | Solid Oxide Electrolysis | Anion Exchange Membrane (AEM) Electrolysis |
|---|---|---|---|---|
| Startup time | Slow | Fast | Slow | Moderate |
| Cost | Low (due to inexpensive materials) | High (due to expensive catalysts and membranes) | Moderate (complex materials and high-temperature requirements) | Moderate to low (depends on membrane development) |
| Water purity requirement | Low (can use tap water) | High (requires pure or deionized water) | Low (steam is typically used) | Low (can use lower purity water) |
| Scalability | High | Moderate to high | Low to moderate | Developing (Potential for high scalability) |
| Applications | Industrial hydrogen production, chlor-alkali industry | Small to medium-scale hydrogen production, Fuel cells | Industrial, potentially for high-temperature applications | Small- to medium-scale hydrogen production, Fuel cells |
| Maturity | Mature (long commercial history) | Mature (growing commercial use) | Developing (research stage) | Emerging (in early development stage) |

## 2.4 Materials Selection for Electrodes and Membranes

Selecting appropriate materials for electrodes and membranes in electrolyzer is crucial for ensuring efficient and durable operation. Following are some considerations for materials selection:

➢ **Electrodes**

**Anode:** Anodes in electrolyzer are typically subjected to oxidation reactions and harsh chemical environments. Key considerations for anode materials include corrosion resistance, catalytic activity, and electrical conductivity.

Common materials for anodes include:

**Titanium:** Titanium is often used as a substrate material for anodes due to its excellent corrosion resistance and mechanical properties. It is typically coated with catalyst materials such as iridium oxide ($IrO_2$) or platinum (Pt) to enhance catalytic activity.

**Ruthenium (Ru), Iridium (Ir), and Platinum (Pt):** These noble metals are commonly used as catalysts on the surface of anodes to facilitate the oxygen evolution reaction (OER) during water electrolysis. They exhibit high catalytic activity and stability under harsh operating conditions.

**Cathode:** Cathodes in electrolyzers are responsible for reduction reactions and hydrogen evolution. Cathode materials should possess good electrical conductivity, stability in alkaline or acidic environments, and catalytic activity for hydrogen evolution. Common materials for cathodes include:

**Nickel (Ni) and Stainless Steel:** These materials are often used as substrates for cathodes due to their good electrical conductivity and mechanical properties. They can be coated with catalyst materials like nickel or platinum to enhance hydrogen evolution activity.

**Platinum (Pt) and Nickel (Ni):** Platinum is a commonly used catalyst for hydrogen evolution due to its high catalytic activity and stability. Nickel is also used as a catalyst or alloyed with other metals to improve hydrogen evolution kinetics.

➢ **Membranes**

**Proton Exchange Membranes (PEMs):**

- PEM materials should possess high proton conductivity, chemical stability, and mechanical strength. They should also have low electronic conductivity to prevent shunting of electrons between the electrodes.
- Common materials for PEMs include PFSA polymer membranes like Nafion and hydrocarbon-based membranes like sulfonated polyetheretherketone (SPEEK).

**Solid Oxide Electrolyte Membranes:**

- Solid oxide electrolyte membranes should exhibit high oxygen ion conductivity at elevated temperatures and good chemical stability in reducing and oxidizing atmospheres.
- Common materials for solid oxide electrolyte membranes include YSZ, ceria ($CeO_2$), and doped lanthanum gallate ($LaGaO_3$).

**Considerations for Material Selection**

- **Compatibility:** Electrode and membrane materials should be compatible with the electrolyte and operating conditions of the electrolyzer to prevent degradation or failure.

- **Cost:** Material costs play a significant role in the overall cost-effectiveness of electrolyzer systems. Selecting materials that balance performance with affordability is essential.
- **Performance:** Materials should exhibit high catalytic activity, conductivity, and stability under operating conditions to ensure efficient and durable electrolyzer performance.
- **Scalability:** Materials should be readily available and suitable for large-scale manufacturing processes to enable widespread deployment of electrolyzer technology.

➢ **Electrolysis as a Chemical Process**

Electrolysis is a chemical process that uses electricity to drive a nonspontaneous chemical reaction. Electrolysis is a chemical process that involves the decomposition of a compound by passing an electric current through it. This process is commonly used to extract elements from compounds, purify metals, and produce various chemical compounds. Let's break down the key components and steps involved in electrolysis:

**Electrolyte:** The substance undergoing electrolysis, typically an ionic compound dissolved in a solvent (usually water). The electrolyte must contain ions that can move freely to conduct electricity. Common electrolytes include salts, acids, and bases.

**Electrodes:** These are conductive materials (usually metals like platinum, graphite, or inert metals) that are immersed in the electrolyte. There are two types of electrodes:

**Anode:** The electrode where oxidation (loss of electrons) occurs. It attracts negatively charged ions (anions).

**Cathode:** The electrode where reduction (gain of electrons) occurs. It attracts positively charged ions (cations).

**Electric Current:** A power source (such as a battery or power supply) is connected to the electrodes, creating a potential difference (voltage) between them. When the circuit is completed, electrons flow from the negative terminal (cathode) to the positive terminal (anode) through the external circuit. Within the electrolyte, positive ions move toward the cathode (negative electrode) and negative ions move toward the anode (positive electrode).

**Redox Reactions:** At the electrodes, chemical reactions take place. These are typically redox (reduction–oxidation) reactions, where one species loses electrons (oxidation) and another gains electrons (reduction). The specific reactions depend on the electrolyte and the substances present.

**Product Formation:** As a result of the redox reactions, new substances are formed at the electrodes. The substance produced at the cathode is typically the reduced form of the cation in the electrolyte, while the substance produced at the anode is typically the oxidized form of the anion in the electrolyte.

**Ion Migration:** As the electric current flows through the electrolyte, ions migrate toward the electrodes. This movement of ions maintains electrical neutrality within the solution.

**Gas Evolution:** In some cases, gases may be produced at the electrodes as by-products of electrolysis. For example, hydrogen gas is often produced at the cathode during the electrolysis of water.

> ***Key: Alkaline electrolysis:*** *Electrolysis carried out in an alkaline electrolyte solution, often using potassium hydroxide.*

## ➢ Renewable Energy Sources for Electrolysis

Renewable energy sources play a crucial role in powering electrolysis for green hydrogen production, ensuring that the process is environmentally sustainable and does not contribute to GHG emissions. Figure 2.14 illustrates the types of renewable energy sources for electrolysis.

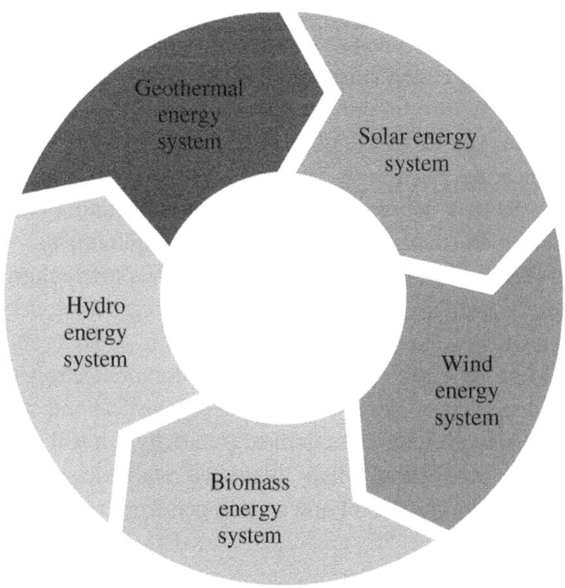


**Figure 2.14** Types of Renewable Energy Sources for Electrolysis

Some of the primary renewable energy sources utilized for electrolysis include:

**Solar Energy:** Solar power, harnessed through PV panels or concentrated solar power (CSP) systems, can provide electricity for electrolysis. PV panels convert sunlight directly into electricity, while CSP systems use mirrors or lenses to concentrate sunlight onto a small area, generating heat that can be converted into electricity. Solar energy is abundant, particularly in regions with high solar irradiance, making it an ideal renewable energy source for green hydrogen production.

**Wind Energy:** Wind power involves harnessing the kinetic energy of wind to generate electricity, typically through wind turbines. Wind farms located in areas with strong and consistent wind patterns can produce significant amounts of electricity to power electrolysis for green hydrogen production. Wind energy is particularly advantageous for electrolysis as it can complement solar energy by providing electricity during periods of low solar irradiance.

**Hydropower:** Hydropower utilizes the energy of flowing or falling water to generate electricity through hydroelectric dams or run-of-river systems. Hydropower is a reliable and established renewable energy source, providing consistent electricity generation regardless of weather conditions. Electrolysis plants located near hydropower facilities can benefit from access to clean and renewable electricity, making hydropower an attractive option for green hydrogen production.

**Geothermal Energy:** Geothermal energy harnesses heat from the Earth's subsurface to generate electricity or directly provide heat for various applications. Geothermal power plants can produce continuous and reliable electricity, making them suitable for powering electrolysis for green hydrogen production. Additionally, geothermal heat can be used directly in high-temperature electrolysis processes, improving efficiency and reducing energy requirements.

**Biomass Energy:** Biomass energy involves using organic materials such as agricultural residues, forestry waste, or dedicated energy crops to generate electricity or heat through processes like combustion, gasification, or anaerobic digestion. While biomass is a renewable energy source, its use for electrolysis may not be as common compared to other sources due to logistical and sustainability considerations. However, biomass-derived electricity can still contribute to green hydrogen production when integrated with electrolysis processes. Figure 2.15 shows the process of green hydrogen from renewable energy generation.

These renewable energy sources offer diverse options for powering electrolysis and producing green hydrogen sustainably. By utilizing electricity generated from

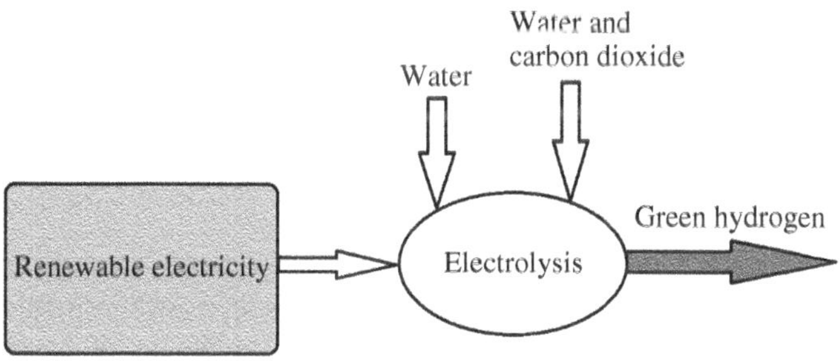


**Figure 2.15** Green Hydrogen from Renewable Energy Generation

renewable sources, electrolysis can produce hydrogen without emitting GHGs, thereby supporting efforts to mitigate climate change and transition to a cleaner and more sustainable energy future. Additionally, advancements in renewable energy technologies and improvements in electrolysis efficiency continue to drive down costs and increase the viability of green hydrogen production as a key component of the global energy transition.

➤ **Advancements in Electrolysis Technology**

Advancements in electrolysis technology have been instrumental in driving the efficiency, scalability, and cost-effectiveness of green hydrogen production. These advancements span various aspects of electrolysis systems, including the electrolyzer design, materials, efficiency, and integration with renewable energy sources. Some notable advancements include:

**Electrolyzer Efficiency:** Advances in electrolyzer design and materials have led to significant improvements in efficiency, reducing energy consumption and production costs. Innovations such as high-performance electrodes, advanced catalysts, and improved membrane materials have enhanced the electrolysis process, allowing for higher conversion efficiencies and faster reaction rates.

**PEM Electrolysis:** PEM electrolysis technology has seen substantial advancements, particularly in terms of efficiency, durability, and cost reduction. Thin-film membranes with enhanced proton conductivity, durable catalysts, and innovative cell designs have contributed to improved performance and reliability of PEM electrolyzer. These advancements have made PEM electrolysis competitive with other electrolysis technologies and well-suited for both small-scale and large-scale green hydrogen production.

**Alkaline Electrolysis:** Alkaline electrolysis, a mature and well-established technology, has also benefited from advancements in materials and system optimization. Improved electrode materials, such as nickel-based alloys, and advancements in cell design have increased the efficiency and durability of AELs, making them more cost-effective and reliable for industrial-scale hydrogen production.

**Solid Oxide Electrolysis Cells (SOECs):** SOEC technology, which operates at high temperatures and can achieve high efficiencies, has seen advancements in materials, cell design, and system integration. Development of novel ceramic electrolytes, stable electrode materials, and efficient heat management systems has improved the performance and longevity of SOECs, making them increasingly viable for large-scale green hydrogen production and co-electrolysis of water and carbon dioxide ($CO_2$).

**Integrated Systems and Power-to-Gas:** Advancements in electrolysis technology have facilitated the integration of electrolysis systems with renewable energy sources and grid infrastructure. Power-to-gas (P2G) systems, which utilize surplus renewable electricity to produce hydrogen through electrolysis, offer grid balancing services and energy storage solutions. Integrated P2G systems, coupled with electrolyzers and renewable energy generation assets, enable efficient utilization of renewable energy resources and contribute to grid stability and decarbonization efforts.

**Modular and Scalable Designs:** Modular electrolyzer designs and standardized components have made electrolysis systems more flexible and scalable, allowing for easy deployment and expansion of green hydrogen production capacity. Modular electrolyzer stacks, containerized systems, and standardized interfaces enable rapid deployment and integration with existing infrastructure, facilitating the adoption of green hydrogen technology in various applications and industries.

**Cost Reduction:** Advancements in materials, manufacturing processes, and system integration have contributed to significant cost reductions in electrolysis technology. Economies of scale, increased production volumes, and improved manufacturing techniques have lowered the capital and operational costs of electrolyzers, making green hydrogen more competitive with conventional hydrogen production methods and fossil fuels.

Advancements in electrolysis technology have been instrumental in driving the growth of green hydrogen production, enabling efficient and cost-effective utilization of renewable energy resources and contributing to the transition toward a sustainable and low-carbon energy future. Continued research, innovation, and collaboration across academia, industry, and government are essential to further accelerate the development and deployment of electrolysis technology and unlock the full potential of green hydrogen as a clean and versatile energy carrier.

## 2.5 Thermochemical Water Splitting

Thermochemical water splitting is a process that involves breaking down water molecules ($H_2O$) into their constituent elements, hydrogen ($H_2$) and oxygen ($O_2$), using heat as the primary energy input. Unlike traditional electrolysis, which uses electricity to split water molecules, thermochemical water splitting relies on high temperatures to drive chemical reactions that separate hydrogen and oxygen. The process typically involves several steps, each governed by specific chemical reactions. Commonly, metal oxides or other compounds act as catalysts or reactants in these steps. The overall reaction can be represented as:

$$2H_2O(\text{liquid}) + \text{Energy} \rightarrow 2H_2(g) + O_2(g)$$

The energy input required for thermochemical water splitting can come from various sources, including concentrated solar energy, nuclear energy, or waste heat from industrial processes. High temperatures are necessary to provide the activation energy for the chemical reactions involved in breaking the strong bonds in water molecules. Figure 2.16 shows the thermochemical water splitting cycle for hydrogen production.

One advantage of thermochemical water splitting is its potential for efficient hydrogen production without direct GHG emissions, particularly when coupled with renewable or low-carbon heat sources. However, challenges remain, including finding suitable catalysts and materials that can withstand the high temperatures and harsh chemical environments involved in the process, as well as optimizing the overall efficiency and cost-effectiveness.

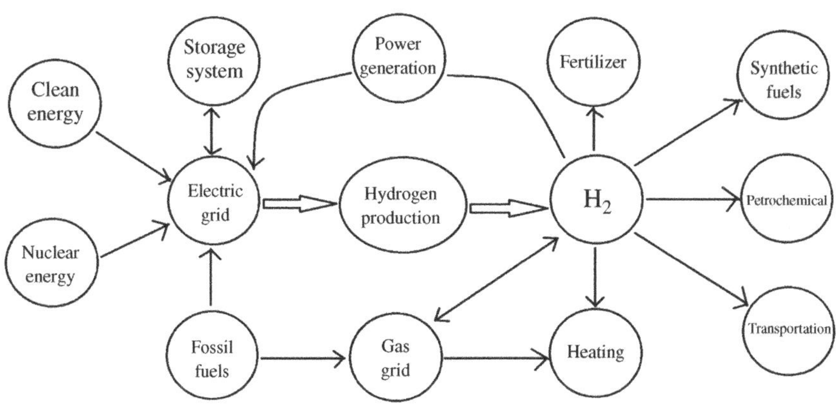


**Figure 2.16** Thermochemical Water Splitting Cycle for Hydrogen Production

R&D efforts continue to improve the efficiency and scalability of thermochemical water splitting technologies, aiming to make hydrogen production via this method more economically viable and environmentally sustainable for various applications, including energy storage, transportation, and industrial processes.

Thermochemical water splitting holds significant importance in the context of sustainable energy production for several reasons:

**Carbon Neutrality:** Thermochemical water splitting has the potential to produce hydrogen without directly emitting GHGs. By using renewable heat sources such as concentrated solar energy or nuclear power, thermochemical processes can avoid carbon emissions associated with traditional fossil fuel-based hydrogen production methods like SMR.

**Energy Storage:** Hydrogen produced through thermochemical water splitting can serve as an efficient means of energy storage, especially for renewable energy sources like solar and wind that are intermittent in nature. Excess energy generated during peak production periods can be used to produce hydrogen via thermochemical processes, which can then be stored and later converted back into electricity or used as a fuel for transportation and industrial applications when energy demand is high.

**Versatility:** Hydrogen produced via thermochemical water splitting can be utilized in a variety of applications, including fuel cells for EVs, industrial processes (such as ammonia production and refining), and power generation. Its versatility makes it a valuable component of a sustainable energy system, capable of decarbonizing multiple sectors of the economy.

**Resource Flexibility:** Thermochemical water splitting can utilize a wide range of heat sources, including solar, nuclear, and waste heat from industrial processes. This flexibility allows for the integration of thermochemical hydrogen production into existing energy infrastructure and enables the utilization of diverse energy resources, enhancing energy security and resilience.

**Research and Innovation:** Continued research and innovation in thermochemical water splitting technologies can lead to advancements in efficiency, cost reduction, and scalability. This, in turn, can drive down the overall cost of hydrogen production and accelerate its adoption as a clean energy carrier, contributing to the transition toward a sustainable and low-carbon energy future.

In summary, thermochemical water splitting plays a crucial role in sustainable energy production by providing a carbon-neutral pathway for hydrogen production, enabling energy storage and flexibility, and fostering innovation in clean energy technologies. Its integration into the broader energy system can contribute significantly to mitigating climate change and advancing global efforts toward a more sustainable and resilient energy future.

> ***Key: Proton exchange membrane electrolysis:*** *Electrolysis conducted with a PEM separating the electrodes and electrolytes.*

The thermodynamics of water splitting reactions in thermochemical processes are governed by the Gibbs free energy change ($\Delta G$) associated with the overall reaction. For water splitting, the overall reaction can be represented as:

$$2H_2O(\text{liquid}) \rightarrow 2H_2(g) + O_2(g)$$

The Gibbs free energy change for this reaction can be calculated using the standard Gibbs free energy of formation ($\Delta G°f$) of the reactants and products. The standard Gibbs free energy change for the reaction at a given temperature ($T$) and pressure ($P$) is given by the equation:

$$\Delta G = \leftarrow \Delta G^\circ + RT\ln\left(\frac{P(H_2)^2 \cdot P(O_2)}{P(H_2O)^2}\right)$$

where $\Delta G^\circ$ is the standard Gibbs free energy change of reaction.
$R$ is the gas constant (8.314 J/mol·K).
$T$ is the temperature in Kelvin.

$P(H_2)$, $P(O_2)$, and $P(H_2O)$ are the partial pressures of hydrogen, oxygen, and water vapor, respectively. In a thermochemical water splitting process, high temperatures are typically required to drive the endothermic reaction, breaking the bonds of water molecules. The reaction is often carried out in multiple steps, each with its own thermodynamics, to facilitate the overall water splitting process. These steps may involve the use of various catalysts or reactive materials to promote specific reactions and enhance overall efficiency.

The thermodynamics of thermochemical water splitting reactions are essential for understanding the energy requirements, feasibility, and efficiency of the process. By analyzing the Gibbs free energy change and considering factors such as temperature, pressure, and reaction kinetics, researchers can optimize reaction conditions and design effective thermochemical water splitting systems for hydrogen production.

Thermochemical water splitting involves a series of high-temperature chemical processes aimed at breaking down water molecules ($H_2O$) into hydrogen ($H_2$) and oxygen ($O_2$). These processes typically occur in multiple steps, each governed by specific chemical reactions. Following are some key high-temperature chemical processes involved in thermochemical water splitting:

**Oxidation Reaction (Endothermic)**

$$2H_2O(\text{liquid}) = 2H_2(g) + O_2(g)$$

This is the overall reaction of water splitting. It is endothermic, meaning it requires input energy, usually in the form of heat, to break the bonds in water molecules.

**Thermal Decomposition**

$$2H_2O(g) = 2H_2(g) + O_2(g)$$

Water vapor is decomposed into hydrogen and oxygen gases at high temperatures, typically above 1,000 °C. This process is energy-intensive and requires heat as the primary energy input.

**Redox Reactions**

Thermochemical water splitting often involves redox (reduction–oxidation) reactions between metal oxides and water to facilitate the splitting of water molecules. These reactions typically occur in multiple steps, involving the oxidation of the metal oxide by water to produce hydrogen and the reduction of the oxide to regenerate the active catalyst.

$$\text{Reduction: } Fe_2O_3 + 3H_2 = 2Fe + 3H_2O$$

$$\text{Hydrogen Generation: } Fe + H_2O = FeO + H_2$$

Once water molecules are decomposed, hydrogen gas is generated. Depending on the specific thermochemical process used, hydrogen may need to be separated from other gases produced during the water splitting reaction, such as oxygen or carbon dioxide.

**Oxygen Generation**

Alongside hydrogen, oxygen gas is also produced during the water splitting process. Separating oxygen from hydrogen is crucial to obtain high-purity hydrogen for various applications. Oxygen can often be released as a by-product or utilized in other industrial processes.

These high-temperature chemical processes require careful control of reaction conditions, including temperature, pressure, and the choice of catalysts or reactive materials, to optimize efficiency and minimize energy consumption. R&D efforts continue to focus on improving the kinetics, selectivity, and scalability of these processes to enable cost-effective and sustainable hydrogen production via thermochemical water splitting.

### 2.5.1 Thermochemical Water Splitting Cycles

Thermochemical water splitting cycles are multi-step processes that utilize high temperatures to decompose water molecules into hydrogen and oxygen. These cycles often involve a series of chemical reactions, some of which may be endothermic (requiring input energy) and others exothermic (releasing energy). Two prominent examples of thermochemical water splitting cycles are the sulfur–iodine (S—I) cycle and the copper–chlorine (Cu—Cl) cycle. Following an overview of each:

1) **Sulfur–Iodine (S—I) Cycle**

   **Overview:** The S—I cycle is a three-step thermochemical process for water splitting that utilizes sulfur and iodine compounds as reactive intermediates. It operates at high temperatures (typically 700–900 °C) and is often coupled with nuclear or solar heat sources. Figure 2.17 shows the structure of sulfur–iodine cycle.

   **Bunsen Reaction:** The Bunsen reaction is a chemical reaction that describes water, sulfur dioxide, and iodine reacting to form sulfuric acid and hydrogen iodide:

   $$2H_2O + SO_2 + I_2 = H_2SO_4 + 2HI$$

   **Sulfuric Acid Decomposition:** Sulfuric acid ($H_2SO_4$) is decomposed into sulfur dioxide ($SO_2$), oxygen ($O_2$), and water ($H_2O$) at high temperatures:

   $$2H_2SO_4 = 2SO_2 + 2H_2O + O_2$$

   **Sulfuric Acid Regeneration:** Sulfur dioxide reacts with iodine to regenerate sulfuric acid:

   $$2SO_2 + 2I_2 + 2H_2O = 2H_2SO_4 + 4HI$$

   **Hydrogen Production:** Hydroiodic acid (HI) is decomposed into hydrogen ($H_2$) and iodine ($I_2$):

   $$2HI = H_2 + I_2$$

   **Advantages:** The S—I cycle offers the potential for efficient hydrogen production without carbon emissions. It can utilize various heat sources and has been extensively studied for its suitability in large-scale hydrogen production.

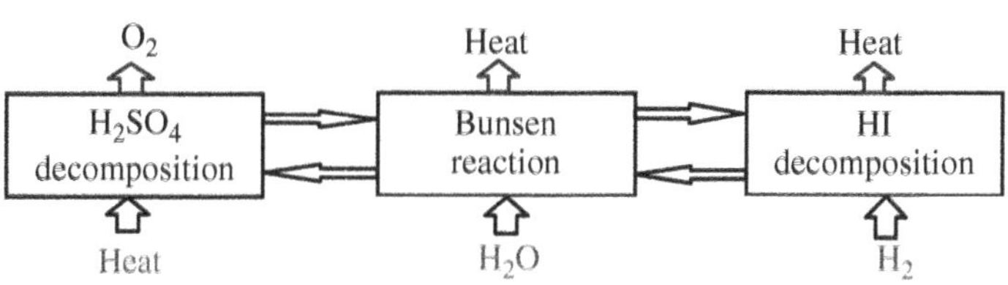


**Figure 2.17** Structure of Sulfur–Iodine Cycle

2) **Copper–Chlorine (Cu—Cl) Cycle**

**Overview:** The Cu—Cl cycle is another thermochemical water splitting cycle that involves multiple chemical reactions using copper and chlorine compounds. It operates at high temperatures (above 700 °C) and is suitable for integration with concentrated solar energy.

**Oxidation of Copper Chloride:** Copper chloride (CuCl) is oxidized by water vapor to form copper oxychloride (CuOCl) and hydrogen chloride (HCl):

$$2CuCl_2 + H_2O = Cu_2OCl_2 + 2HCl$$

**Hydrolysis of Copper Oxychloride:** Copper oxychloride undergoes hydrolysis in steam to produce copper oxide (CuO), hydrogen chloride, and oxygen:

$$Cu_2OCl_2 + H_2O = 2CuO + 2HCl + 1/2O_2$$

**Thermal Decomposition of Copper Oxide:** Copper oxide is thermally decomposed to regenerate copper chloride and oxygen:

$$2CuO = Cu_2O + 1/2O_2$$

**Advantages:** The Cu—Cl cycle offers the potential for high-temperature hydrogen production with efficient utilization of solar energy. It has been investigated for its scalability and suitability for large-scale hydrogen production facilities.

Both the S—I and Cu—Cl cycles are actively researched and developed for their potential to provide sustainable and efficient pathways for hydrogen production. However, challenges remain in terms of materials compatibility, reaction kinetics, and system integration, requiring further R&D to realize their full potential in commercial applications.

Table 2.10 illustrates a comparison of different thermochemical water splitting cycles, including the sulfur–iodine (S—I) cycle, the copper–chlorine (Cu—Cl) cycle, and the hybrid sulfur (HyS) cycle, highlighting their key features and advantages. This comparison highlights the different approaches and characteristics of thermochemical water splitting cycles, each with its own advantages and challenges. While the sulfur–iodine cycle is well-studied and offers high efficiency, the copper–chlorine cycle utilizes abundant copper and chlorine compounds and efficient solar energy utilization. The hybrid sulfur cycle combines feature of sulfuric acid decomposition and metal oxide reduction, offering potential advantages in efficiency and heat source flexibility. Ongoing R&D efforts aim to address challenges and optimize these cycles for large-scale, sustainable hydrogen production.

***Key: Solid oxide electrolysis cells:*** *Electrolysis cells operating at high temperatures with a solid oxide electrolyte, enabling efficient hydrogen production.*

**Table 2.10** Comparison of S—I and Cu—Cl Cycle

| Cycle | Sulfur–Iodine (S—I) | Copper–Chlorine (Cu—Cl) |
|---|---|---|
| Principle | Utilizes sulfuric acid decomposition, sulfur dioxide, and iodine reactions | Involves oxidation and hydrolysis of copper compounds |
| Operating temperature | 700–900 °C | Above 700 °C |
| Heat source | Nuclear, concentrated solar | Concentrated solar |
| Reaction steps | Three-step process involving sulfuric acid decomposition, regeneration, and hydrogen production | Multiple steps involving oxidation, hydrolysis, and thermal decomposition of copper compounds |
| Hydrogen production efficiency | High | High |
| Advantages | • Well-studied and researched<br>• High efficiency<br>• Potential for large-scale hydrogen production | • Utilizes abundant copper and chlorine compounds<br>• Efficient utilization of solar energy<br>• Suitable for large-scale facilities |
| Challenges | • Corrosion issues due to sulfuric acid<br>• Complex chemical processes<br>• Potential for sulfur compound emissions | • Corrosion and material compatibility issues<br>• Thermal stability of copper compounds<br>• Complexity of reaction steps |
| Current status | Well-researched and developed, with ongoing optimization efforts | Actively researched, with progress toward scalability and commercialization |

Water splitting thermochemical cycles are processes that use a sequence of chemical reactions to split water into hydrogen and oxygen using heat as the energy input. In these cycles, various materials can be used as redox pairs to facilitate the water splitting reaction. Some common materials include:

**Metal Oxides and Metals:** Metal oxides such as cerium oxide ($CeO_2$), iron oxide ($Fe_2O_3$), and manganese oxide ($MnO_2$) can serve as the oxidizing agent, while metals such as zinc (Zn), iron (Fe), or cerium (Ce) can serve as the reducing agent. For example, in the cerium-based cycle, $CeO_2$ is reduced to $CeO_{2-x}$ by high-temperature steam and then the reduced cerium oxide reacts with water to produce hydrogen and regenerate $CeO_2$.

**Sulfur-Based Compounds:** Compounds containing sulfur, such as sulfur dioxide ($SO_2$) and hydrogen sulfide ($H_2S$), can participate in redox reactions. For instance, in the sulfur–iodine cycle, sulfur dioxide is oxidized to sulfuric acid, while iodine is reduced to form hydrogen iodide.

**Iodine-Based Compounds:** Iodine and iodine-based compounds like iodine pentoxide ($I_2O_5$) and hydrogen iodide (HI) are used in cycles like the sulfur–iodine cycle. In these cycles, iodine is reduced to hydrogen iodide, which can then react with water to produce hydrogen and regenerate iodine.

**Perovskite-Type Materials:** Perovskite-type materials, such as barium cerate ($BaCeO_3$) or strontium titanate ($SrTiO_3$), have been investigated for their potential use in thermochemical water splitting cycles. These materials can undergo redox reactions under high temperatures to facilitate water splitting.

**Metal Halides:** Metal halides, particularly those of alkali metals and alkaline earth metals, can serve as redox pairs. For example, calcium chloride ($CaCl_2$) and calcium bromide ($CaBr_2$) have been studied for their use in water splitting cycles.

**Metal Hydrides:** Some metal hydrides, such as sodium borohydride ($NaBH_4$), have been explored for their potential in water splitting cycles. These materials can release hydrogen upon hydrolysis or react with water to produce hydrogen.

These are just a few examples, and there are many other materials and combinations being researched for their suitability in thermochemical water splitting cycles. The choice of redox pairs depends on factors such as thermodynamic feasibility, kinetics of the reactions, stability of the materials, and economic viability.

Selecting suitable materials for thermochemical reactors in water splitting thermochemical cycles involves considering several important criteria to ensure the reactor's efficiency, durability, and safety. Following are some key criteria:

**Chemical Compatibility:** Materials must be compatible with the reactive species involved in the thermochemical cycle. They should resist corrosion, degradation, and unwanted chemical reactions when exposed to high-temperature gases, steam, and reactive intermediates such as metal oxides, halides, or sulfur compounds.

**Thermal Stability:** The materials should maintain their structural integrity and mechanical strength at high temperatures encountered during the thermochemical cycle. They should have high melting points and thermal conductivity to withstand thermal stresses and prevent deformation or failure. Figure 2.18 illustrates the criteria for selecting suitable materials for the thermochemical process.

**Gas Permeability:** Reactor materials should allow for efficient gas transport, including the diffusion of reactants and products through the reactor's structure. Porous materials or membranes with controlled pore sizes can facilitate gas permeation while maintaining structural stability.

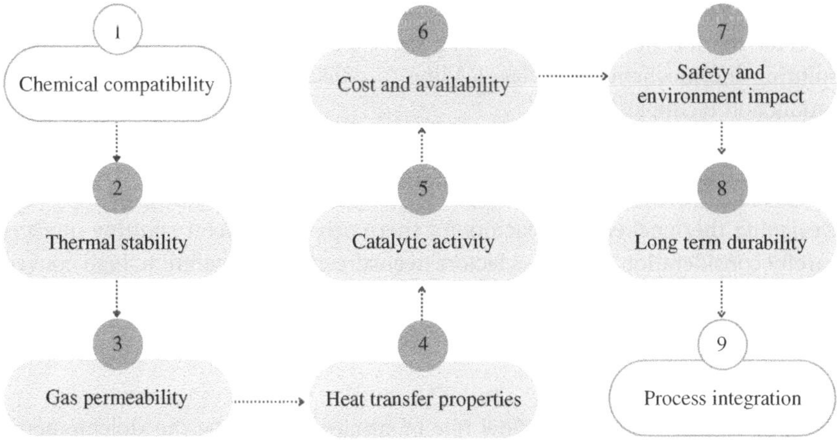


**Figure 2.18** Criteria for Selecting Suitable Materials

**Heat Transfer Properties:** Efficient heat transfer is essential for maintaining uniform temperature distributions within the reactor and promoting the desired chemical reactions. Materials with high thermal conductivity can enhance heat transfer rates and minimize temperature gradients, leading to improved reactor performance.

**Catalytic Activity:** In some cases, reactor materials may also act as catalysts to enhance reaction rates and selectivity. Catalysts should exhibit high catalytic activity, stability, and compatibility with the reactor environment to facilitate the thermochemical reactions effectively.

**Cost and Availability:** The cost and availability of materials are important considerations for large-scale implementation of thermochemical water splitting technologies. Economically viable materials should be readily available, cost-effective, and scalable for industrial applications.

**Safety and Environmental Impact:** Materials used in thermochemical reactors should meet safety standards and regulations to ensure the protection of operators, the environment, and surrounding communities. They should not release toxic gases or contaminants during operation and should be environmentally sustainable throughout their lifecycle.

**Long-Term Durability:** The durability and lifespan of reactor materials are crucial for minimizing maintenance and replacement costs over time. Materials should exhibit resistance to degradation mechanisms such as thermal cycling, mechanical stress, chemical attack, and erosion.

**Process Integration:** The choice of reactor materials should be compatible with the overall process design and integration of the thermochemical water splitting cycle within the energy system. Materials should facilitate ease of operation, maintenance, and control while ensuring optimal system performance.

By considering these criteria, engineers and researchers can select suitable materials for thermochemical reactors that meet the specific requirements of water splitting thermochemical cycles, leading to efficient and sustainable hydrogen production technologies.

### 2.5.2 Design Considerations for Thermochemical Reactors

Designing thermochemical reactors for thermochemical water splitting involves careful consideration of various factors to ensure efficient operation, high conversion rates, and safety. Following are the mathematical interpretations of the thermochemical reactors for thermochemical water splitting:

➢ **Mass Balance Equations**

Let Fi represent the molar flow rate of species $i$, and $v_i$ be the stoichiometric coefficient.

For a general thermochemical reaction:

$$\sum_i v_i A_i \rightleftharpoons \sum_j v_j B_j$$

The mass balance in a steady-state reactor is given by:

$$\frac{dC_i}{dt} = r_i$$

where $C_i$ is the concentration of species $i$ and $r_i$ is the reaction rate of species $i$.

For a flow reactor,

$$\frac{dF_i}{dV} = r_i \tag{2.1}$$

where $V$ is the reactor volume.

➢ **Reaction Kinetics**

The reaction rate follows the Arrhenius equation:

$$r = k(T)C_A^m C_B^m \tag{2.2}$$

where

$k(T) = Ae^{-E_a/RT}$ is the temperature-dependent rate constant.

$A$ is the preexponential factor.
$E_a$ is the activation energy.
$R$ is the universal gas constant.
$T$ is the absolute temperature.
$m$ and $n$ are reaction orders.

➢ **Energy Balance Equation**

The enthalpy balance for the reactor is:

$$\sum_{in} \dot{H}_i - \sum_{out} \dot{H}_j + Q \ = 0 \tag{2.3}$$

where $H_i = F_i C_p \Delta T$ is the enthalpy flow rate, and $Q$ is the heat supplied.

For an adiabatic reactor,

$$\sum_i v_i \Delta H_i = 0$$

where $\Delta H_i$ is the enthalpy of reaction.

- **Thermodynamic Efficiency**

The efficiency of a thermodynamic cycle is:

$$\eta = \frac{\Delta G_{H_{2O}} \longrightarrow H_2 + 1/2O_2}{Q_{input}} \tag{2.4}$$

where $\Delta G = \Delta H - T\Delta S$ is the Gibbs free energy change, and $Q_{input}$ is the thermal energy supplied.

For ideal conditions: $\eta_{ideal} = \dfrac{\Delta G}{\Delta H} \approx 0.75 - 0.85$ for efficient cycles.

➢ **Heat Transfer Consideration**

The reactor must maintain a temperature $T$, modeled as:

$$Q = UA\Delta T \tag{2.5}$$

where $U$ is the overall heat transfer coefficient, $A$ is the surface area, and $\Delta T$ is the temperature difference between the heat source and the reactor.

For high-temperature solar-driven reactors:

$$Q_{solar} = \eta_{optical} IA \tag{2.6}$$

where $\eta_{optical}$ is the optimal efficiency and $I$ is the solar irradiance.

➢ **Reactor Design Parameters**

To optimize hydrogen production, the residence time ($\tau$) in a continuous flow reactor is

$$\tau = \frac{V}{F_{in}} \tag{2.7}$$

For maximum hydrogen yield, the equilibrium constant $K_{eq}$ must be maximized:

$$K_{eq} = e^{-\Delta G/RT} \tag{2.8}$$

Following are some key design considerations:

**Reaction Kinetics:** Understanding the kinetics of the thermochemical reactions involved in water splitting is crucial for reactor design. Reactor geometry, residence time, and temperature profiles should be optimized to maximize reaction rates while minimizing energy losses.

**Temperature Control:** Thermochemical reactions typically require high temperatures, which must be carefully controlled to ensure optimal reaction kinetics and avoid thermal gradients within the reactor. Reactor design should incorporate efficient heating and temperature monitoring systems, such as resistive heating elements, concentrated solar energy, or combustion heat sources.

**Reactant Mixing:** Effective mixing of reactants is essential for achieving uniform reaction conditions and maximizing conversion rates. Reactor designs should incorporate mechanisms for thorough mixing, such as stirring systems, baffles, or fluidized beds, to enhance mass transfer and reaction kinetics.

**Heat Transfer:** Efficient heat transfer is critical for maintaining uniform temperature distributions within the reactor and promoting the desired thermochemical reactions. Reactor designs should maximize heat transfer rates through mechanisms such as heat exchangers, thermal insulation, or fluid circulation systems.

**Material Selection:** Selecting appropriate materials for reactor construction is essential for ensuring compatibility with the reactive species and operating conditions. Materials should exhibit high thermal stability, corrosion resistance, and mechanical strength to withstand high temperatures and reactive environments.

**Pressure Control:** Some thermochemical reactions may require elevated pressures to achieve optimal conversion rates. Reactor designs should incorporate pressure vessels capable of safely handling the desired operating pressures while minimizing leakage and maintaining system integrity.

**Product Separation:** Efficient separation of reaction products, such as hydrogen and oxygen, is essential for downstream processing and utilization. Reactor designs should include provisions for product separation, such as membranes, condensers, or separation columns, to achieve high purity and yield.

**Scale-Up Considerations:** Reactor designs should be scalable to accommodate varying production capacities and enable cost-effective implementation at different scales, from laboratory prototypes to industrial-scale facilities. Considerations such as reactor volume, throughput, and equipment size should be optimized for each scale.

**Safety Systems:** Thermochemical reactors must incorporate safety systems to mitigate the risks associated with high temperatures, reactive species, and potential equipment failures. Safety features such as pressure relief valves, gas detection systems, emergency shutdown procedures, and containment measures should be implemented to ensure operator safety and prevent environmental hazards.

**Energy Integration:** Integrating thermochemical reactors with energy sources, such as CSP, waste heat recovery, or renewable electricity, can enhance overall system efficiency and sustainability. Reactor designs should consider opportunities for energy integration and optimization to minimize energy consumption and environmental impact. By addressing these design considerations, engineers and researchers can develop thermochemical reactors tailored to specific thermochemical water splitting cycles, enabling efficient and sustainable hydrogen production for various applications.

Thermochemical water splitting involves high-temperature reactions that necessitate efficient heat exchange mechanisms and carefully designed reactor configurations to ensure optimal performance. Following are the common heat exchange mechanisms and reactor configurations used in thermochemical water splitting processes:

➢ **Heat Exchange Mechanisms**

**Direct Heating:** In this mechanism, heat is directly supplied to the reactor using external heat sources such as concentrated solar energy, combustion of fuels, or electrical resistance heating. Direct heating is often used in fixed-bed reactors or fluidized-bed reactors where the reactants pass through a heated zone to facilitate the thermochemical reactions.

**Indirect Heating:** Indirect heating involves transferring heat from an external source to the reactor using heat exchange fluids such as molten salts, thermal oils, or steam. Heat exchangers are employed to transfer heat from the heat exchange fluid to the reactor. This mechanism is commonly used in tubular reactors or heat exchanger reactors, allowing for precise control of temperature and thermal management.

**Radiative Heating:** Radiative heating utilizes electromagnetic radiation, typically from concentrated solar energy systems, to directly heat the reactor. Reactor walls are often coated with absorptive materials to efficiently absorb solar radiation and transfer heat to the reactants. Radiative heating can be combined with other heat exchange mechanisms for enhanced efficiency.

**Thermal Cycling:** Some thermochemical water splitting cycles involve alternating high- and low-temperature steps. Heat exchange mechanisms are used to efficiently transfer heat between these steps, allowing for effective thermal cycling and energy recovery within the reactor system.

➢ **Reactor Configurations**

**Fixed-Bed Reactors:** Fixed-bed reactors consist of a stationary bed of solid catalyst or reactant particles through which the reactant gases flow. Heat is supplied to the bed to facilitate the thermochemical reactions. Fixed-bed reactors offer simplicity in design and operation but may suffer from issues such as heat and mass transfer limitations and bed channeling.

**Fluidized-Bed Reactors:** Fluidized-bed reactors employ a bed of solid particles that are fluidized by a gas stream. The reactants are introduced into the fluidized bed, where they react with the solid catalyst particles. Fluidized-bed reactors offer excellent mixing and heat transfer characteristics, making them suitable for high-temperature reactions.

**Tubular Reactors:** Tubular reactors consist of a cylindrical tube through which the reactants flow. Heat is supplied to the reactor either directly or indirectly, and the reactants undergo thermochemical reactions as they pass through the tube. Tubular reactors offer good heat transfer and are well-suited for continuous operation and scalability.

**Membrane Reactors:** Membrane reactors integrate catalytic reaction and separation processes within a single unit. They utilize selective membranes to separate reaction products or intermediates, allowing for enhanced reaction rates and product purity. Membrane reactors can be configured in various geometries, such as tubular or flat-sheet configurations.

**Heat Exchanger Reactors:** Heat exchanger reactors combine the functions of heat exchange and reaction within a single unit. They utilize heat exchanger surfaces to transfer heat between the reactor and the heat source or heat sink. Heat exchanger reactors offer efficient thermal management and temperature control, particularly in thermochemical cycles involving thermal cycling.

By selecting appropriate heat exchange mechanisms and reactor configurations, engineers can design thermochemical water splitting systems that achieve efficient heat transfer, precise temperature control, and optimal reactor performance, leading to high yields of hydrogen production.

The energy requirements for driving thermochemical water splitting processes depend on several factors, including the thermodynamics of the specific reaction pathways, the operating conditions of the process, and the efficiency of heat and energy transfer within the system. Following are the breakdown of the main energy requirements involved:

**Endothermic Reaction Energy**: Thermochemical water splitting involves endothermic reactions where energy is required to break the bonds of water molecules and generate hydrogen and oxygen. The energy input required for these reactions depends on the enthalpy change ($\Delta H$) associated with the specific reaction pathway. This energy input typically comes in the form of heat supplied to the reactor.

**Heat Input:** The primary energy input for thermochemical water splitting is heat, which provides the necessary activation energy for the endothermic reactions to occur. The amount of heat required depends on factors such as the reaction enthalpies, reaction kinetics, temperature levels, and heat losses within the reactor system. Efficient heat transfer mechanisms and heat recovery systems can help minimize energy losses and improve overall energy efficiency.

**Temperature and Pressure:** Higher temperatures and pressures generally facilitate thermochemical reactions, reducing the energy input required to achieve the desired conversion rates. However, operating at elevated temperatures and pressures may also increase energy consumption for heat generation and system maintenance.

**Energy for Heat Generation:** The energy used to generate heat for thermochemical water splitting processes can come from various sources, including fossil fuels, renewable energy sources such as CSP, nuclear energy, or waste heat recovery systems. The choice of heat source impacts the overall energy efficiency, environmental footprint, and sustainability of the process.

**Energy for Auxiliary Systems:** Energy is also required to power auxiliary systems such as pumps, compressors, fans, and controls necessary for maintaining reactor conditions, handling reactants and products, and ensuring safe operation. These energy requirements should be accounted for in the overall energy balance of the system.

**Energy for Separation and Purification:** Depending on the specific thermochemical cycle and reactor configuration, additional energy may be required for separation, purification, and storage of hydrogen and oxygen products. Energy-intensive separation processes such as membrane separation, cryogenic distillation, or PSA contribute to the overall energy requirements of the hydrogen production system.

Optimizing the energy efficiency of thermochemical water splitting processes involves careful design of reactor systems, selection of appropriate reaction pathways, efficient heat management strategies, integration of renewable energy sources, and implementation of advanced separation technologies to minimize energy consumption and maximize hydrogen production yield.

Improving the overall process efficiency of thermochemical water splitting is crucial for enhancing the economic viability, sustainability, and scalability of hydrogen production. Following are the several strategies for improving efficiency across various aspects of thermochemical water splitting processes:

**Selection of Optimal Reaction Pathways:** Identify and optimize thermochemical reaction pathways with favorable thermodynamics and kinetics. Focus on reactions with high conversion rates, low activation energies, and minimal byproduct formation to maximize hydrogen yield and minimize energy losses.

**Integration of Renewable Energy Sources:** Utilize renewable energy sources such as CSP, wind, biomass, or geothermal energy to provide heat for thermochemical reactions. Integration with renewable energy sources reduces GHG emissions, enhances sustainability, and mitigates reliance on fossil fuels.

**Efficient Heat Transfer Mechanisms:** Implement efficient heat transfer mechanisms to supply heat to thermochemical reactors. Utilize heat exchangers, thermal insulation, and advanced heat transfer fluids to minimize heat losses, improve temperature control, and enhance overall energy efficiency.

**Heat Recovery and Recycle:** Capture and recycle waste heat generated during thermochemical reactions and heat exchange processes. Utilize heat recovery systems, such as recuperates or regenerators, to recover and reuse thermal energy, reducing energy consumption and improving process efficiency.

**Process Integration and Optimization:** Integrate thermochemical water splitting processes with other energy systems and industrial processes to maximize resource utilization and energy efficiency. Implement process optimization techniques, such as pinch analysis or process simulation software, to identify opportunities for energy savings and improve overall process efficiency.

**Advanced Reactor Design:** Develop and deploy advanced reactor designs tailored to specific thermochemical water splitting cycles. Utilize fluidized-bed reactors, membrane reactors, or multi-stage reaction systems to improve mixing, heat transfer, and reaction kinetics, leading to higher conversion rates and energy efficiency.

**Catalyst Development and Optimization:** Invest in R&D of efficient catalysts tailored to thermochemical water splitting reactions. Develop catalysts with high activity, selectivity, and stability to enhance reaction rates and minimize energy requirements.

**Hybrid Process Integration:** Explore hybrid approaches that combine thermochemical water splitting with other hydrogen production methods, such as electrolysis or photo electrochemical water splitting. Integration of complementary technologies allows for synergistic benefits, improved energy efficiency, and increased flexibility in hydrogen production.

**Resource Efficiency and Waste Minimization:** Optimize resource utilization and minimize waste generation throughout the thermochemical water splitting process. Implement closed-loop systems, recycle reactants and by-products, and minimize energy-intensive separation processes to improve overall process efficiency and sustainability.

**Scale-Up and Commercialization:** Scale up thermochemical water splitting processes from laboratory prototypes to pilot- and commercial-scale facilities. Address scalability challenges, optimize process economics, and collaborate with industry partners to accelerate deployment and commercialization.

By implementing these strategies, researchers, engineers, and policymakers can enhance the overall process efficiency of thermochemical water splitting, leading to cost-effective, sustainable, and scalable hydrogen production for a wide range of applications.

## 2.6 Cost Analysis and Energy Efficiency Comparisons for Green Hydrogen

Green hydrogen produced via the electrolysis of water using renewable electricity is widely recognized as a cornerstone of the clean energy transition. Unlike gray hydrogen, which relies on fossil fuels such as natural gas, and blue hydrogen, which incorporates CCS, green hydrogen eliminates direct carbon emissions during production. However, the economic feasibility of green hydrogen is a central challenge, given its currently higher production costs compared to conventional hydrogen sources. To understand the path toward cost parity and scalability, it is essential to analyze the capital and operational expenditures, conversion efficiencies, impacts of renewable electricity prices, economies of scale, and benchmark costs through the levelized cost of hydrogen (LCOH) metric. The production of green hydrogen primarily relies on electrolyzers, which convert water into hydrogen and oxygen using electricity. The three major electrolyzer technologies, AELs, PEM electrolyzers, and solid oxide electrolyzers, differ in maturity, cost structure, and performance.

➢ **Capital Expenditures**

Capital expenditure (CAPEX) in green hydrogen projects is dominated by the cost of electrolyzers, followed by balance-of-plant (BOP) components such as compressors, purification systems, power electronics, storage tanks, and auxiliary infrastructure. Following are the details of the CAPEX:

**Alkaline Electrolyzers (AELs):** Mature technology with relatively low capital cost, typically in the range of $500–$1,000 per kW. They have longer lifespans (60,000–90,000 hours) but lower dynamic load response, making them less suited for intermittent renewable energy.

**PEM Electrolyzers:** More compact, responsive, and capable of higher current densities. CAPEX ranges from $900 to $1,500 per kW due to the use of costly materials like platinum group metals (PGMs) for catalysts and advanced membranes.

**SOEC Electrolyzers:** Operating at high temperatures (600–900 °C) allows them to achieve high efficiencies, but they remain in the early stages of commercialization. CAPEX is expected to be high initially (>$2,000 per kW) but with strong potential for reduction as the technology matures.

**Infrastructure Costs** also vary depending on the hydrogen application. For industrial clusters, shared pipelines and centralized compression can reduce CAPEX per unit of hydrogen compared to distributed refueling stations.

➢ **Operating Expenditures**

Operating expenditure (OPEX) consists mainly of electricity costs (typically 60–75% of total production cost), maintenance, and water purification. Following are the details of the OPEX:

**Electricity Cost Share:** Given that producing 1 kg of hydrogen via electrolysis requires around 50–55 kWh of electricity, the price of electricity directly dictates hydrogen cost competitiveness.

**Maintenance:** PEM systems require more frequent stack replacements, increasing OPEX. In contrast, alkaline systems have lower maintenance costs but may incur performance penalties in renewable-driven applications.

**Water Supply:** While water consumption is modest (about 9 L/kg of $H_2$), ultra-pure water is required, adding treatment costs.

CAPEX reduction through mass production and supply chain maturity, combined with OPEX optimization via cheap renewable electricity, is critical to scaling green hydrogen.

### 2.6.1 Impact of Renewable Electricity Price on Electrolysis Cost

Electricity price is the single largest determinant of green hydrogen cost. Given the electricity consumption of ~50 kWh/kg $H_2$, the impact is straightforward:

- At $0.02/kWh: Electricity cost per kg $H_2 \approx$ $1.00
- At $0.05/kWh: Electricity cost per kg $H_2 \approx$ $2.50
- At $0.10/kWh: Electricity cost per kg $H_2 \approx$ $5.00

This linear relationship means:

- Low-cost renewable energy regions (e.g., solar-rich deserts, high-wind coasts) have a massive advantage.
- Co-location of electrolyzers with renewable plants can reduce grid charges, curtailment losses, and transmission costs.
- Off-peak renewable energy pricing (dynamic tariffs) can be leveraged for intermittent hydrogen production.

Regions with stable renewable generation can achieve higher capacity factors (>50%), reducing fixed costs per kg $H_2$. In contrast, high intermittency reduces electrolyzer utilization and increases per-unit costs unless storage or hybrid renewable sources are integrated. Like solar PV and wind energy, green hydrogen production benefits from economies of scale and learning-by-doing.

**Scaling Effects:** Large-scale plants (>100 MW) can achieve:

- Lower per-unit equipment cost due to bulk purchasing.
- Shared infrastructure for water treatment, compression, and storage.
- Reduced financing costs through stable, large offtake agreements.

**Learning Curve:** Historical evidence from renewable energy shows cost reductions of 15–25% for each doubling of installed capacity. The IRENA projects that electrolyzer CAPEX could fall by **60–70% by 2030** as manufacturing scales and supply chains mature.

**Technology Standardization:** Standardized, modular electrolyzer units can reduce installation complexity, permitting time, and engineering costs. Countries with early large-scale deployment (e.g., Australia, Saudi Arabia, Chile) will drive down costs faster, potentially becoming hydrogen export hubs.

Green hydrogen's cost trajectory will follow a familiar renewable energy pattern: initially high costs driven by CAPEX and expensive renewable electricity, followed by rapid reductions through scale, innovation, and supply chain maturity. The convergence of low renewable electricity costs, high electrolyzer efficiency, and economies of scale will be the tipping point for green hydrogen's mass adoption.

**Short-Term Focus (2025–2030):** Co-location with renewables, cost-sharing through industrial hubs, and early government subsidies.

**Mid-Term Focus (2030–2040):** Large-scale deployment, global hydrogen trade routes, advanced storage and transport solutions.

**Long-Term Focus (2040–2050):** Mature, cost-competitive hydrogen economy with integration into all major hard-to-abate sectors.

If these cost and efficiency challenges are addressed, green hydrogen could transform from a niche decarbonization tool into the backbone of a global zero-carbon energy system. Figures 2.19–2.22 illustrate capital and operating cost by technology, energy conversion efficiency, impact of renewable electricity price, and economies of scale, respectively.

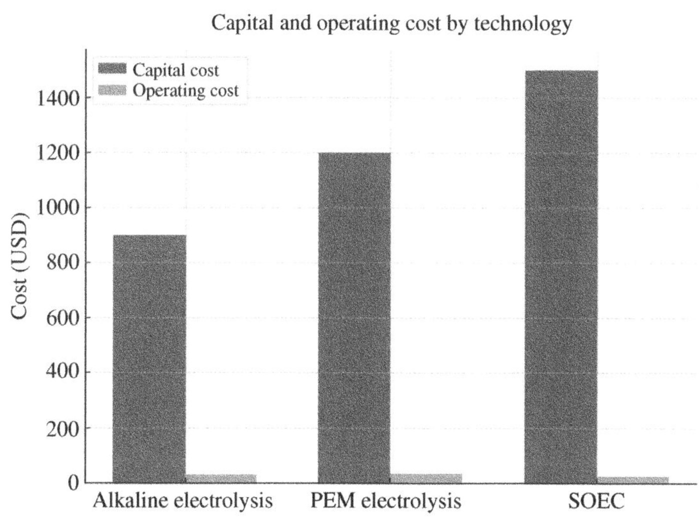


**Figure 2.19** Capital and Operating Cost by Technology

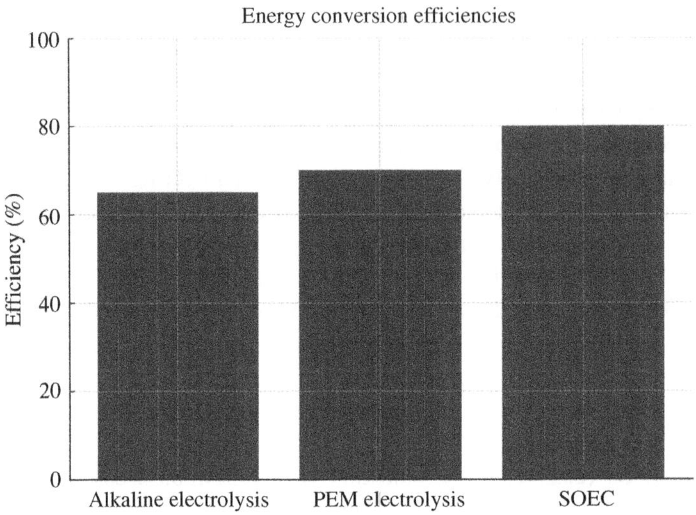


**Figure 2.20** Energy Conversion Efficiency

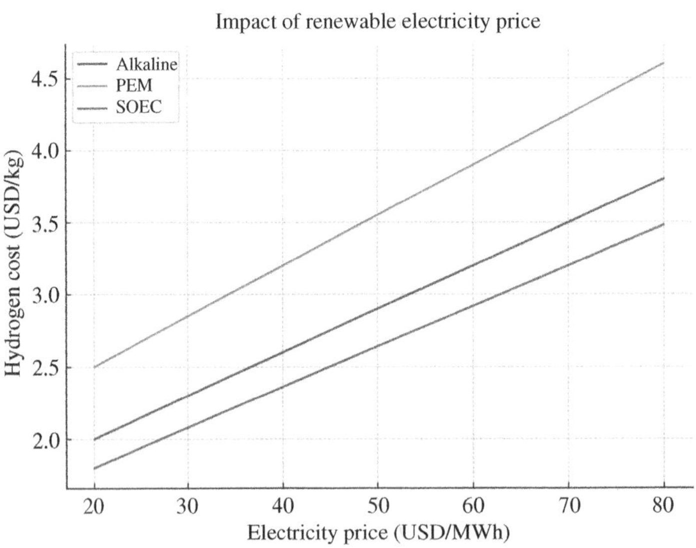


**Figure 2.21** Impact of Renewable Electricity Price

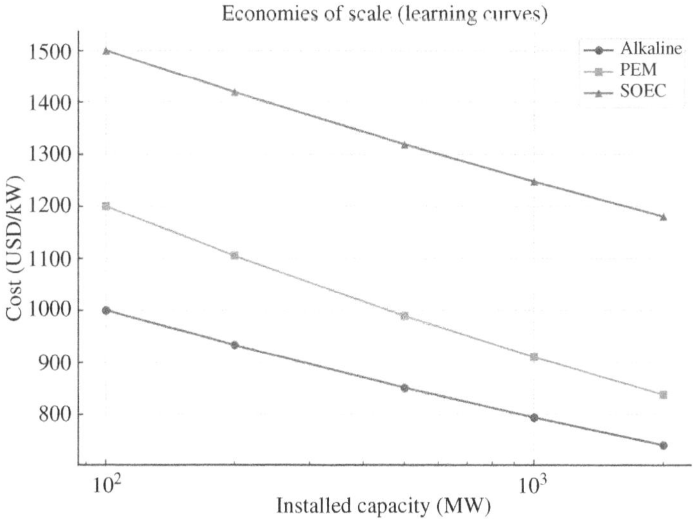


**Figure 2.22** Economies of Scale

## 2.7 Lifecycle Assessment of Hydrogen Production Pathways

Hydrogen has emerged as a pivotal energy carrier in the global transition toward decarbonization and clean energy. Unlike fossil fuels, hydrogen itself is a clean fuel, as its combustion produces only water as a by-product. However, the environmental benefits of hydrogen are highly dependent on the method of its production. Currently, hydrogen can be produced through various pathways, including SMR, biomass gasification, and electrolysis powered by renewable or conventional electricity. To evaluate the true environmental footprint of these pathways, researchers and policymakers employ a structured approach called *lifecycle assessment (LCA)*. LCA provides a comprehensive framework for analyzing the environmental impacts associated with all stages of a product's life, from raw material extraction to end-of-life disposal. When applied to hydrogen, LCA offers valuable insights into GHG emissions, resource consumption, energy efficiency, and ecological impacts across different production routes. This section explores in detail the methodologies for conducting LCA; the distinction between cradle-to-grave and cradle-to-gate assessments; comparative LCA results of SMR, biomass, and electrolysis pathways; the sensitivity of outcomes to feedstock, grid mix, and technological assumptions; and the standards and tools commonly employed in LCA studies, such as GREET and ISO 14040.

### 2.7.1 Methodologies for Conducting LCA

LCA is governed by standardized methodologies that ensure consistency, transparency, and comparability across studies. At its core, an LCA consists of four main phases as defined in the ISO 14040/14044 framework: goal and scope definition, lifecycle inventory (LCI), lifecycle impact assessment (LCIA), and interpretation. In the context of hydrogen production, the goal and scope definition phase typically establishes the system boundaries, functional units, and key environmental indicators to be analyzed. For example, a study may define the functional unit as "1 kg of hydrogen produced" and set boundaries that include feedstock extraction, processing, hydrogen conversion, storage, transportation, and end-use combustion.

The LCI phase involves collecting quantitative data on energy inputs, raw material requirements, emissions, and wastes across each stage of the production chain. For hydrogen produced via SMR, this would include natural gas extraction, methane reforming, $CO_2$ emissions from the reforming process, and potential emissions from transportation and storage. For biomass pathways, inventory data must cover biomass cultivation, harvesting, transportation, gasification, and subsequent hydrogen purification. For electrolysis, inventory involves electricity generation, water input, electrolyzer efficiency, and emissions associated with the electricity grid.

The LCIA translates inventory data into environmental impact categories. These categories often include GWP, acidification potential, eutrophication potential, ozone depletion, and resource depletion. For hydrogen pathways, GWP tends to be the most scrutinized metric, given the importance of hydrogen in mitigating climate change. The interpretation stage synthesizes findings, identifies uncertainties, and highlights areas where technological improvements or policy interventions could yield significant benefits.

Several methodologies exist for conducting LCAs with varying levels of detail and complexity. Attributional LCA focuses on describing the average environmental impacts associated with a given system, while consequential LCA examines the broader consequences of changes in demand or supply, such as how large-scale hydrogen adoption might alter electricity grid dynamics or biomass markets. Hybrid approaches, which combine process-based LCA with input–output analysis, also provide a more holistic view by incorporating indirect economic activities linked to hydrogen production.

#### ➢ Cradle-to-Grave Versus Cradle-to-Gate Assessments

A critical aspect of hydrogen LCA studies is the definition of system boundaries, which significantly influence results and interpretations. Two widely used approaches are *cradle-to-grave* and *cradle-to-gate* assessments. Cradle-to-grave LCAs evaluate the environmental impacts across the entire lifecycle of hydrogen, starting from raw material extraction and ending with its ultimate use or disposal. For hydrogen, this includes feedstock extraction, production, compression/liquefaction, storage, distribution, end-use combustion in fuel cells or turbines, and any residual impacts. This approach provides the most comprehensive view, enabling a true comparison of hydrogen against other energy carriers such as gasoline, diesel, or natural gas. Figure 2.23 illustrates system boundaries in LCA.

In contrast, cradle-to-gate LCAs restrict the analysis to upstream processes, ending at the production plant gate where hydrogen is produced and ready for distribution. This approach is particularly useful when the objective is to compare production technologies independent of the variability introduced by downstream applications. For example, whether hydrogen is used in a steelmaking plant, an FCEV, or a power generation turbine, cradle-to-gate studies allow policymakers and investors to focus on the production stage, which often contributes the most significant share of emissions. The choice between cradle-to-grave and cradle-to-gate assessments depends on the research question at hand. Policymakers seeking to understand hydrogen's role in decarbonizing transport might prefer cradle-to-grave studies, whereas technology developers focused on optimizing electrolyzers or biomass gasifiers may rely on cradle-to-gate analyses. Importantly, both approaches must carefully define functional units and allocation methods, particularly when dealing with co-products such as oxygen in electrolysis or syngas by-products in biomass pathways.

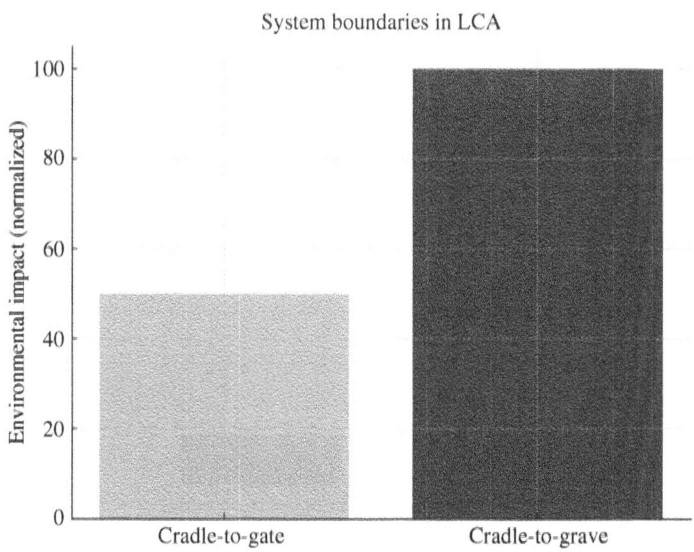


**Figure 2.23** System Boundaries in LCA

### ➢ LCA of SMR, Biomass, and Electrolysis Compared

Hydrogen production pathways differ widely in their environmental impacts, and LCA studies provide a structured way to compare them. SMR is the dominant hydrogen production pathway today, accounting for over 70% of global hydrogen supply. However, its environmental footprint is substantial. A typical cradle-to-gate LCA for SMR shows that producing 1 kg of hydrogen emits around 9–12 kg of $CO_2$, depending on methane leakage rates during natural gas extraction and transport. When coupled with CCS, emissions can be reduced to 2–4 kg $CO_2$/kg $H_2$, though this introduces additional energy penalties and costs. The key challenge with SMR lies in fugitive methane emissions, which have a high GWP. Even small leakage rates (e.g., 1–3%) can significantly worsen its climate impact. Figure 2.24 illustrates GHG emission by hydrogen production pathway.

Biomass gasification represents an intermediate pathway with more favorable environmental characteristics. In theory, biomass is carbon-neutral since the $CO_2$ emitted during combustion is offset by $CO_2$ absorbed during plant growth. However, cradle-to-gate LCAs reveal that emissions depend heavily on cultivation practices, fertilizer use, land use change, and supply chain logistics. If biomass is sourced from agricultural residues or sustainably managed forests, net emissions can be close to zero or even negative when combined with carbon capture (BECCS). On the other hand, dedicated energy crops grown on cleared forestland

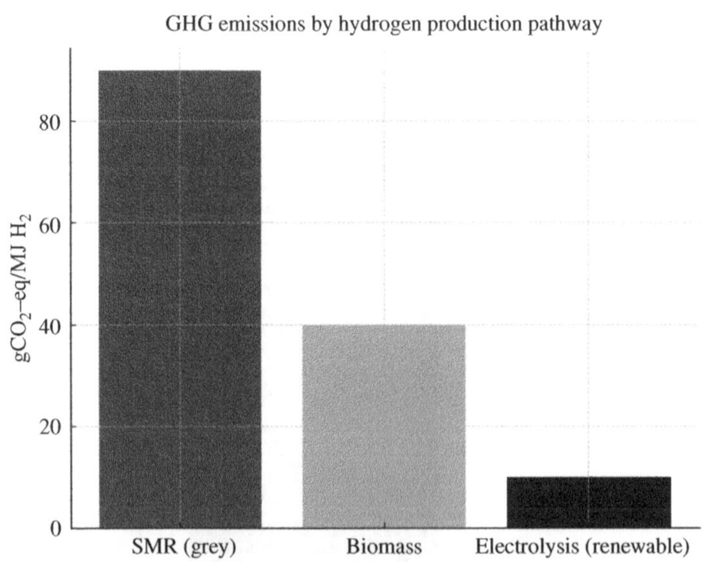


**Figure 2.24** GHG Emission by Hydrogen Production Pathway

can lead to substantial carbon debts that undermine climate benefits. Biomass hydrogen also requires careful consideration of noncarbon impacts such as eutrophication from fertilizer runoff and biodiversity loss. Electrolysis, particularly when powered by renewable energy, is considered the cornerstone of green hydrogen production. The environmental performance of electrolysis is primarily determined by the carbon intensity of the electricity input. If powered by coal-heavy grids, electrolysis can be more carbon-intensive than SMR. However, when electricity comes from solar, wind, or hydropower, cradle-to-gate LCAs typically show emissions below 1–2 kg $CO_2$/kg $H_2$. Beyond climate impacts, electrolysis also reduces air pollutant emissions and avoids methane leakage associated with natural gas supply chains. The efficiency of electrolyzers, currently ranging from 60% to 75%, plays a critical role in determining energy requirements and overall environmental impacts. Emerging technologies such as solid oxide electrolyzers promise higher efficiencies, though they face durability challenges.

**➤ Sensitivity to Feedstock, Grid Mix, and Technology Assumptions**

One of the defining features of hydrogen LCAs is their sensitivity to underlying assumptions. Small variations in feedstock quality, grid mix, and technological parameters can significantly alter results. Feedstock variability is particularly critical for biomass-based hydrogen. Residue-based biomass generally performs better than dedicated crops because it avoids land use change emissions and minimizes fertilizer requirements. However, supply chain logistics such as transportation distances and drying methods can offset these benefits. Similarly, for SMR, the carbon intensity depends on whether natural gas is sourced from conventional fields, shale gas, or liquefied natural gas (LNG). Shale gas extraction, for instance, is often associated with higher fugitive methane emissions. Figure 2.25 illustrates electrolysis sensitivity to grid mix.

The electricity grid mix is the most important factor for electrolysis. Electrolysis powered by grids with high shares of renewables such as Iceland or Norway can achieve near-zero emissions. Conversely, in coal-dependent grids such as parts of China or India, electrolysis can emit over 20 kg $CO_2$/kg $H_2$, making it less favorable than SMR. Temporal dynamics of renewable electricity also matter: if electrolysis is run during times of excess renewable generation (curtailment), the effective emissions can be near-zero. Technological assumptions around efficiency, capacity factor, and lifetime also introduce variability. Electrolyzer efficiency improvements directly reduce energy input and environmental impact per kilogram of hydrogen. Similarly, assumptions about carbon capture rates in SMR–CCS systems or durability of solid oxide electrolyzers significantly influence comparative results. Sensitivity analyses are, therefore, a vital part of hydrogen LCAs, ensuring transparency and guiding both technology development and policy interventions.

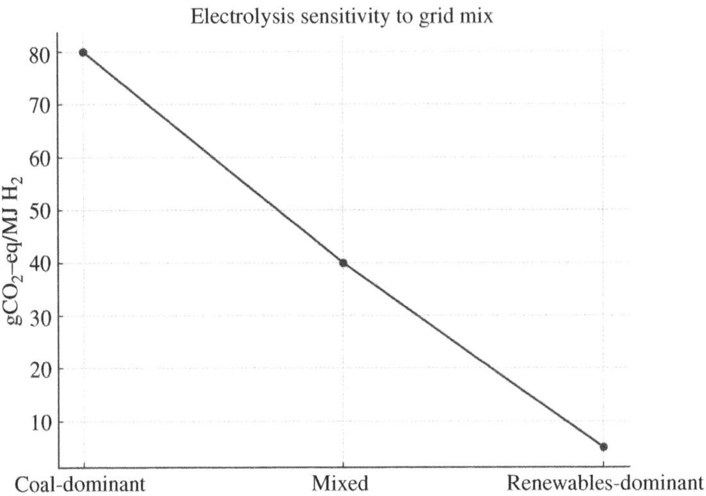


**Figure 2.25** Electrolysis Sensitivity to Grid Mix

### ➢ Tools and Standards

To ensure rigor and comparability, LCA studies of hydrogen production often rely on established tools and international standards. The ISO 14040/14044 standards provide the overarching framework for conducting LCAs, including definitions of goal and scope, requirements for inventory and impact assessment, and guidelines for interpretation. These standards ensure consistency across studies and enhance credibility for policymakers and industry stakeholders.

The GREET model (Greenhouse gases, Regulated Emissions, and Energy use in Transportation), developed by Argonne National Laboratory, is one of the most widely used tools for hydrogen LCA. GREET provides detailed datasets for different feedstocks, technologies, and regional grid mixes, enabling researchers to model cradle-to-gate and cradle-to-grave scenarios with relative ease. GREET is particularly valuable for comparing hydrogen with alternative fuels in transportation applications. Other tools include SimaPro, GaBi, and OpenLCA, which allow detailed modeling of LCIs and impact categories across a wide range of products and processes. These platforms support both attributional and consequential LCAs, making them adaptable to different research questions. The choice of tool often depends on data availability, the required level of detail, and the intended audience. International collaboration has also led to the development of hydrogen-specific guidelines under initiatives such as the International Partnership for Hydrogen and Fuel Cells in the Economy (IPHE). These guidelines aim to harmonize LCA methodologies and establish common benchmarks for green hydrogen certification. Figure 2.26 illustrates adoption of LCA tools and standards.

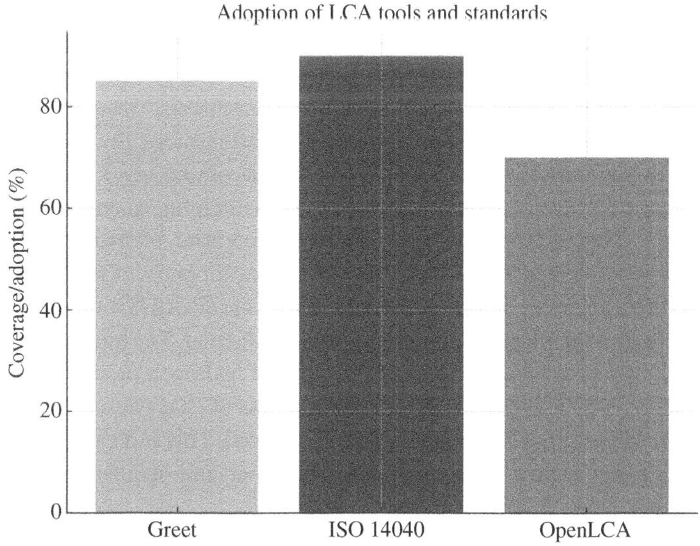


**Figure 2.26** Adoption of LCA Tools and Standards

LCA of hydrogen production pathways is an indispensable tool for understanding the true environmental implications of the hydrogen economy. By systematically analyzing inputs, outputs, and impacts across production chains, LCAs reveal that not all hydrogen is created equal. SMR remains carbon-intensive unless paired with effective carbon capture, biomass hydrogen can offer significant benefits if sustainably sourced, and electrolysis achieves the lowest emissions only when powered by renewables. Furthermore, results are highly sensitive to assumptions about feedstock, grid mix, and technology performance. Tools such as GREET, ISO 14040 standards, and advanced LCA software provide the necessary frameworks for conducting robust analyses. Ultimately, LCAs guide policymakers, investors, and researchers toward the most sustainable hydrogen pathways, ensuring that hydrogen fulfills its promise as a cornerstone of the global energy transition.

## 2.8 Environmental Footprint

Hydrogen is often celebrated as the "fuel of the future," particularly because it offers a pathway toward decarbonization in sectors such as power generation, steel production, transportation, and chemicals. However, the environmental footprint of hydrogen depends entirely on how it is produced. Traditional hydrogen production methods such as SMR generate large amounts of GHG emissions. In contrast, "green hydrogen," produced via water electrolysis powered by renewable energy, presents the lowest-carbon option. Nevertheless, the sustainability of hydrogen requires careful assessment not only of direct carbon emissions but also of other environmental dimensions such as water use, land requirements, waste generation, and the role of emerging technologies like CCUS. Furthermore, the certification frameworks and potential to earn carbon credits are becoming key to ensuring that green hydrogen becomes commercially viable while helping to combat climate change. This essay explores the environmental footprint of hydrogen in detail, structured into five key areas: GHG emissions per kilogram of hydrogen, comparison of carbon intensity across technologies, the role of CCUS, other environmental impacts, and certification plus carbon credit potential. The most important metric for evaluating the environmental footprint of hydrogen production is the carbon intensity, typically expressed as the kilograms of $CO_2$ equivalent (kg $CO_2$e) emitted per kilogram of hydrogen (kg $H_2$) produced. This indicator helps in comparing hydrogen technologies against one another and against conventional fuels.

**Gray Hydrogen (SMR Without CCUS):** SMR is the dominant method for hydrogen production today, accounting for ~70–75% of global hydrogen supply. However, it is also highly carbon-intensive. Producing 1 kg of hydrogen via SMR releases approximately 9–12 kg of $CO_2$ depending on efficiency, methane leakage rates, and natural gas quality. Methane leakage during extraction and transport can worsen this footprint significantly because methane has a GWP 28–34 times greater than $CO_2$ over 100 years.

**Blue Hydrogen (SMR + CCUS):** When combined with CCUS, the emissions from SMR can be reduced by 50–90%. Depending on the capture rate and efficiency, blue hydrogen emits between 2 and 5 kg of $CO_2$ per kg $H_2$. However, residual emissions are unavoidable, and upstream methane leakage often diminishes the actual gains.

**Green Hydrogen (Electrolysis with Renewable Energy):** Green hydrogen has the lowest emissions profile. If electrolysis is powered entirely by renewable electricity (solar, wind, hydro), emissions can be close to 0.1–1 kg $CO_2$e per kg $H_2$. The small footprint results mainly from the embodied carbon of building solar panels, wind turbines, electrolyzers, and storage infrastructure rather than from the hydrogen generation process itself.

**Biomass Gasification with CCS (Bio-Hydrogen):** Another emerging pathway is hydrogen production from biomass or biogas with integrated carbon capture. Since biomass absorbs $CO_2$ during its growth phase, this process can yield carbon-neutral or even carbon-negative hydrogen. Carbon intensity in this pathway ranges from −2 to 2 kg $CO_2$e/kg $H_2$, depending on capture efficiency, feedstock type, and supply chain logistics.

Thus, the GHG footprint varies dramatically across production routes, with gray hydrogen being the worst and green hydrogen being the cleanest option.

In the era of sustainability, CCUS has been positioned as a bridging technology to reduce emissions from fossil fuel-based hydrogen production until green hydrogen becomes mainstream.

**Carbon Capture Efficiency:** Current CCUS technologies can capture between 60% and 95% of $CO_2$ emissions from SMR or coal gasification plants. However, capturing 100% is practically impossible. Therefore, even "blue hydrogen" is not fully clean.

**Methane Leakage Issues:** Studies have shown that upstream methane leakage from natural gas systems, if greater than **3%**, can make blue hydrogen nearly as carbon-intensive as gray hydrogen. Thus, CCUS alone is insufficient without addressing methane leakage across the supply chain.

**Utilization Options:** Captured $CO_2$ can be used in EOR, carbonated beverages, or chemical production. However, most of these uses re-release $CO_2$ later. Permanent sequestration (deep saline aquifers, depleted oil reservoirs) offers greater climate benefits but raises concerns about leakage and monitoring.

**Economic Viability:** CCUS adds significant cost to hydrogen production—estimated at \$0.5–1/kg $H_2$ and requires large-scale infrastructure. Without strong carbon pricing or subsidies, CCUS adoption remains limited.

While CCUS can mitigate emissions in the short to medium term, it should be seen as a transition technology rather than a permanent solution, as only green hydrogen can achieve long-term net-zero goals.

As the hydrogen market grows, certification frameworks and carbon credit systems are essential to distinguish low-carbon hydrogen from conventional alternatives and to incentivize green hydrogen production.

**Certification Schemes:** Several certification systems are emerging globally:

- **CertifHy (EU):** Defines green hydrogen as hydrogen with less than 36.4 g $CO_2e$/MJ (~4.4 kg $CO_2$/kg $H_2$).
- **Guarantees of Origin (GoO):** Track renewable hydrogen production in Europe.
- **U.S. DOE Clean Hydrogen Standard:** Defines clean hydrogen as having ≤4 kg $CO_2e$/kg $H_2$ at the point of production.

These frameworks provide transparency and enable trade in green hydrogen.

**Carbon Credit Potential:** Green hydrogen can earn carbon credits by displacing gray hydrogen or fossil fuels in industries like steel, chemicals, and transport. The value of these credits depends on carbon pricing markets. For example, in the EU ETS (Emissions Trading System), carbon credits are currently priced around €80–100 per ton $CO_2$, making green hydrogen financially more competitive.

**Border Carbon Adjustments:** The EU's Carbon Border Adjustment Mechanism (CBAM) will soon require imports of carbon-intensive products to pay carbon tariffs. This creates a market advantage for certified green hydrogen.

The environmental footprint of hydrogen varies drastically depending on production technologies. Gray hydrogen is among the most carbon-intensive industrial processes, releasing up to 12 kg $CO_2e$ per kg $H_2$, whereas green hydrogen can reduce this footprint to near-zero if powered by renewables. Blue hydrogen, with CCUS, offers partial emissions reduction but suffers from residual $CO_2$, methane leakage, and cost challenges. Beyond carbon intensity, hydrogen production also affects water resources, land use, waste management, and local ecosystems. Certification frameworks and carbon credit mechanisms will play an essential role in promoting green hydrogen adoption and ensuring that environmental benefits are realized. Ultimately, the long-term hydrogen economy must be anchored in green hydrogen supported by renewable electricity expansion, efficient electrolysis technologies, sustainable resource management, and robust certification schemes. Only then can hydrogen fulfill its promise as a cornerstone of the global transition to net-zero.

## 2.9 Applications of Green Hydrogen

### 2.9.1 Transportation

Green hydrogen plays a pivotal role in the transformation of the transportation sector toward sustainability and decarburization. Following are some key roles that green hydrogen plays in transportation:

**Zero-Emission Fuel for FCVs:** Green hydrogen serves as a clean and sustainable fuel for FCVs. In FCVs, hydrogen is fed into a fuel cell where it reacts with oxygen to produce electricity, powering the vehicle's electric motor. FCVs emit only water vapor as a by-product, contributing to significant reductions in GHG emissions compared to conventional internal combustion engine vehicles.

**Long-Range Driving Capability:** Hydrogen fuel cell technology offers long-range driving capabilities, making it suitable for applications requiring extended travel distances, such as long-haul freight transport and intercity travel. FCVs equipped with hydrogen fuel cells can achieve driving ranges comparable to conventional vehicles, addressing range anxiety concerns associated with BEVs.

**Fast Refueling Times:** Hydrogen fueling stations offer fast refueling times, typically similar to conventional gasoline or diesel refueling times. This quick refueling capability enhances the convenience and usability of hydrogen-powered vehicles, making them suitable for a wide range of transportation applications, including commercial fleets and public transportation.

**Decarbonization of Heavy-Duty Transportation:** Green hydrogen plays a critical role in decarbonizing heavy-duty transportation sectors such as trucks, buses, and trains. These vehicles often require higher energy densities and longer ranges than passenger cars, making hydrogen fuel cells a compelling option for achieving zero-emission transport in these sectors.

**Maritime and Aviation Applications:** Green hydrogen can also be utilized to decarbonize maritime and aviation sectors, where electrification using battery technology may not be feasible due to weight and energy density constraints. Hydrogen fuel cells can power ships, ferries, and airplanes, offering zero-emission propulsion for long-distance travel and cargo transportation.

**Hydrogen Refueling Infrastructure:** Building a network of hydrogen refueling stations is essential for supporting the widespread adoption of hydrogen-powered vehicles. Green hydrogen production facilities can be co-located with renewable energy sources such as solar or wind farms, enabling on-site electrolysis and ensuring a sustainable and environmentally friendly hydrogen supply chain.

**Diversification of Energy Sources:** Hydrogen serves as a versatile energy carrier that can be produced from a variety of renewable energy sources, including solar, wind, hydroelectric, and biomass. This diversification of energy sources enhances energy security and resilience in the transportation sector, reducing dependence on fossil fuels and imported oil.

**Integration with Renewable Energy:** Green hydrogen production can be integrated with renewable energy sources to leverage surplus renewable electricity for electrolysis. By utilizing excess renewable energy that would otherwise be curtailed, hydrogen production can help balance supply and demand on the grid, contributing to grid stability and supporting the integration of variable renewable energy sources.

**Emissions Reduction Targets:** Green hydrogen is aligned with global emissions reduction targets and climate change mitigation efforts. By displacing fossil fuels in transportation, hydrogen fuel cells contribute to significant reductions in GHG emissions, air pollutants, and particulate matter, leading to cleaner air and improved public health outcomes.

In summary, green hydrogen plays a multifaceted role in the transportation sector, offering zero-emission fuel options for a wide range of vehicles and applications while contributing to the transition toward a sustainable and low-carbon transportation system.

- **Benefits of Green Hydrogen for Emissions Reduction**

  **Zero Emissions:** Green hydrogen production and utilization result in zero GHG emissions, helping to mitigate climate change and reduce air pollution.

  **Energy Security:** Hydrogen can be produced domestically from renewable sources, reducing dependence on imported fossil fuels and enhancing energy security.

  **Technological Flexibility:** Green hydrogen can be produced and utilized in a variety of transportation applications, offering flexibility and versatility in emissions reduction strategies.

  **Scalability:** The scalability of green hydrogen production technologies enables widespread deployment across transportation sectors, driving emissions reductions at scale.

- **Challenges and Considerations:**

  **Infrastructure Development:** Deployment of hydrogen infrastructure, including production facilities and refueling stations, requires significant investment and coordination among stakeholders.

  **Cost Competitiveness:** Despite cost reductions in electrolysis technology, green hydrogen production costs remain higher than conventional fuels, posing a challenge to widespread adoption.

  **Technological Maturity:** Further R&D is needed to improve the efficiency and performance of hydrogen FCVs, as well as electrolysis and hydrogen storage technologies.

  **Policy Support:** Policy frameworks and incentives are essential for fostering market demand, incentivizing investment, and overcoming barriers to adoption of green hydrogen in transportation.

Green hydrogen holds immense potential for reducing emissions in the transportation sector, offering a clean, sustainable, and versatile fuel option across various transportation modes. By leveraging renewable energy sources and advancing hydrogen production and utilization technologies, green hydrogen can play a critical role in decarbonizing transportation and achieving emissions reductions at scale. Adoption of green hydrogen in FCVs, heavy-duty transportation, maritime and aviation sectors, and last-mile delivery vehicles can contribute to significant emissions reductions, improve air quality, and enhance energy security. Addressing challenges such as infrastructure development, cost competitiveness, technological maturity, and policy support will be essential to unlock the full potential of green hydrogen in transforming the transportation sector toward sustainability and decarbonization. By fostering collaboration among industry stakeholders, policymakers, and researchers, we can accelerate the transition to a cleaner, greener future powered by green hydrogen in transportation.

### 2.9.2 Energy Storage

Green hydrogen, produced through the electrolysis of water using renewable energy sources, has emerged as a key player in the quest for sustainable energy storage solutions. As the world transitions toward a low-carbon future, the importance of energy storage technologies in balancing intermittent renewable energy generation and meeting energy demand is becoming increasingly evident.

**Seasonal Energy Storage:** Green hydrogen can serve as a long-term energy storage solution for storing excess renewable energy generated during periods of high supply and using it during periods of low supply. By converting surplus renewable electricity into hydrogen through electrolysis and storing it in tanks or underground caverns, green hydrogen enables seasonal energy balancing, enhancing grid stability and reliability.

**Grid Balancing and Peak Shaving:** Green hydrogen can be used for grid balancing and peak shaving, where excess electricity generated from renewable sources is used to produce hydrogen through electrolysis. Hydrogen can then be stored and utilized to generate electricity during peak demand periods or when renewable energy supply is insufficient, reducing reliance on fossil fuel-based power plants and improving grid stability.

**Transportation Fuel:** Green hydrogen can be used as a clean and sustainable fuel for FCVs, offering zero-emission transportation options for passenger cars, buses, trucks, and other heavy-duty vehicles. Hydrogen fuel cells convert hydrogen into electricity through chemical reactions with oxygen, providing an efficient and environmentally friendly alternative to conventional internal combustion engines.

**Industrial Applications:** Green hydrogen has various industrial applications, including feedstock for chemical processes, refining, metallurgy, and ammonia production. By replacing fossil fuels with green hydrogen in industrial processes, emissions can be significantly reduced, contributing to climate change mitigation and environmental sustainability.

➢ **Advantages of Green Hydrogen as an Energy Storage Medium**

**Versatility:** Green hydrogen is a versatile energy carrier that can be stored, transported, and utilized in various applications, offering flexibility and adaptability to different energy storage needs.

**Scalability:** The scalability of green hydrogen production and storage technologies enables deployment at various scales, from small-scale distributed systems to large-scale centralized facilities, supporting diverse energy storage requirements.

**Long-Term Storage:** Hydrogen can be stored for extended periods without significant degradation, making it suitable for long-term energy storage applications, such as seasonal energy storage and grid balancing.

**Zero Emissions:** Green hydrogen production and utilization result in zero GHG emissions, helping to mitigate climate change and reduce air pollution, particularly when powered by renewable energy sources.

➢ **Challenges and Considerations**

**Cost Competitiveness:** Despite cost reductions in electrolysis technology, green hydrogen production costs remain higher than conventional fossil fuels, posing a challenge to widespread adoption. Continued R&D is needed to improve cost competitiveness and drive down production costs.

**Infrastructure Development:** Deployment of hydrogen infrastructure, including production facilities, storage tanks, distribution networks, and refueling stations, requires significant investment and coordination among stakeholders. Addressing infrastructure barriers is essential to enable the widespread adoption of green hydrogen.

**Technological Innovation:** Further R&D is needed to advance electrolysis and hydrogen storage technologies, improve efficiency, reduce energy consumption, and enhance system integration. Innovation in materials, catalysts, and system design will be crucial to unlocking the full potential of green hydrogen as an energy storage medium.

**Policy Support:** Policy frameworks and incentives are essential for fostering market demand, incentivizing investment, and overcoming barriers to adoption of green hydrogen as an energy storage solution. Supportive policies can include renewable energy targets, carbon pricing mechanisms, financial incentives, and regulatory frameworks to promote the development and deployment of green hydrogen technologies.

Green hydrogen offers a promising solution for addressing the challenges of intermittent renewable energy generation and meeting energy storage needs in a sustainable and environmentally friendly manner. As the world seeks to transition toward a low-carbon energy future, the role of green hydrogen in energy storage will continue to grow, driving innovation, investment, and collaboration across sectors. By harnessing the potential of green hydrogen, societies can achieve a cleaner, greener, and more resilient energy system for future generations.

### 2.9.3 Industry

The industrial application of green hydrogen encompasses a wide range of sectors and processes where hydrogen serves as a clean, sustainable, and versatile energy carrier. As the world seeks to reduce GHG emissions and transition toward a low-carbon economy, the industrial sector plays a crucial role in adopting green hydrogen technologies. Following are some key industrial applications of green hydrogen:

**Ammonia Production:** One of the largest industrial applications of hydrogen is in the production of ammonia ($NH_3$) for fertilizer production. Green hydrogen can replace fossil fuel-derived hydrogen in the Haber–Bosch process, which combines nitrogen from the air with hydrogen to produce ammonia. By using green hydrogen instead of fossil fuels, ammonia production can become more sustainable and environmentally friendly, reducing carbon emissions and dependence on fossil fuels.

**Petrochemical Industry:** Hydrogen is widely used in the petrochemical industry for various processes, including hydrogenation, hydrocracking, and desulfurization. Green hydrogen can be utilized in these processes instead of hydrogen derived from fossil fuels, reducing carbon emissions and environmental impact. By integrating green hydrogen into petrochemical operations, companies can improve their sustainability profile and meet emissions reduction targets.

**Steel Production:** Green hydrogen has the potential to revolutionize the steel industry by replacing coal-derived coke in the blast furnace with hydrogen as a reducing agent. Direct reduction of iron ore using hydrogen can lead to significant reductions in carbon dioxide emissions associated with traditional steelmaking processes. Several pilot projects and research initiatives are underway to explore the feasibility of hydrogen-based steel production, known as "green steel."

**Refining Industry:** Hydrogen is essential in the refining industry for processes such as hydrocracking, hydro treating, and catalytic reforming. Green hydrogen can be used in refinery operations to reduce the carbon intensity of petroleum products and meet increasingly stringent environmental regulations. Refineries can integrate green hydrogen production facilities on-site, utilizing renewable energy sources to power electrolysis units.

**Chemical Manufacturing:** Hydrogen is a critical feedstock for various chemical manufacturing processes, including methanol production, hydrogenation reactions, and synthesis of organic compounds. Green hydrogen offers a sustainable alternative to fossil fuel-derived hydrogen, enabling chemical manufacturers to reduce their carbon footprint and enhance sustainability. Adoption of green hydrogen in chemical manufacturing can support the transition toward a circular economy by enabling the production of renewable and biodegradable materials.

**Power Generation:** Green hydrogen can be used as a fuel in gas turbines, fuel cells, and combustion engines for power generation. Hydrogen-based power generation offers flexibility and grid stability, particularly in conjunction with renewable energy sources such as solar and wind. Power plants equipped with hydrogen combustion technology can provide dispatchable, low-carbon electricity generation, helping to integrate intermittent renewable energy into the grid.

**Hydrogen Fueling Infrastructure:** The development of green hydrogen infrastructure, including production facilities, storage tanks, distribution networks, and refueling stations, is an industrial application in itself. Investment in hydrogen refueling infrastructure is essential to support the widespread adoption of hydrogen FCVs and other hydrogen-based transportation modes. Industrial players, energy companies, and government agencies are collaborating to deploy hydrogen infrastructure and accelerate the transition to a hydrogen economy.

**Industrial Heating and Heat Treatment:** Hydrogen is used in industrial processes for heating, annealing, brazing, and heat treatment of metals and materials. Green hydrogen can replace natural gas or other fossil fuels in industrial heating applications, reducing carbon emissions and environmental impact. Industrial facilities can invest in hydrogen-powered heating systems to improve energy efficiency, reduce emissions, and enhance sustainability.

The industrial application of green hydrogen offers significant opportunities for reducing carbon emissions, enhancing sustainability, and driving the transition toward a low-carbon economy. By replacing fossil fuels with green hydrogen in various industrial processes and applications, companies can achieve environmental goals, meet regulatory requirements, and contribute to global efforts to combat climate change. Collaboration among industry stakeholders, policymakers, and research institutions is essential to accelerate the adoption and deployment of green hydrogen technologies in the industrial sector.

## 2.10 Conclusion

Green hydrogen stands at the forefront of the global transition toward clean and sustainable energy systems. This chapter has explored the core production methods and enabling technologies that define the green hydrogen ecosystem, with a primary focus on water electrolysis powered by renewable energy sources. As nations strive to decarbonize critical sectors, green hydrogen offers a viable and scalable solution to reduce reliance on fossil fuels, particularly in hard-to-abate industries such as transportation, steel, and chemicals.

The development and deployment of advanced electrolyzers, innovations in storage and distribution, and supportive policy frameworks are crucial to making green hydrogen economically competitive and widely accessible. While challenges remain—including cost, infrastructure, and energy efficiency ongoing research, public–private partnerships, and global collaboration are accelerating progress. Ultimately, green hydrogen is not just a technological advancement; it is a cornerstone of a sustainable energy future. Understanding its production methods and technological landscape equips us to harness its full potential in building a low-carbon economy.

## 2.11 Case Study

### 2.11.1 Green Hydrogen in the EU

**Introduction**

Green hydrogen is a promising clean energy source produced through the electrolysis of water using renewable electricity, such as wind, solar, or hydroelectric power. Unlike "gray" hydrogen, derived from natural gas, or "blue" hydrogen, produced using natural gas with carbon capture, green hydrogen emits no carbon dioxide during its production process. As the world increasingly shifts toward sustainable energy solutions, green hydrogen presents an opportunity to reduce GHG emissions significantly, enabling countries to meet their climate goals and foster a sustainable economy.

**Background and Importance**

The Paris Agreement, adopted in 2015, set a global framework to limit global warming to below 2 °C, preferably to 1.5 °C. This ambitious target requires significant reductions in GHG emissions across all sectors. The energy sector, responsible for more than 70% of these emissions, is pivotal in achieving these goals. Green hydrogen is considered a vital component of the future energy mix due to its versatility, energy density, and ability to decarbonize "hard-to-abate" sectors such as heavy industry, long-haul transportation, and aviation.

**The Challenge**

The EU, a leader in climate action, aims to become the world's first climate-neutral continent by 2050. However, reaching this target requires deep decarbonization of all sectors, particularly those heavily reliant on fossil fuels, such as steel production, chemicals, transportation, and heating. Despite the extensive use of renewable energy, these sectors continue to face challenges due to technological and economic barriers. This gap highlighted the need for an alternative clean energy source capable of supporting the EU's ambitious targets.

**The Solution: Scaling Up Green Hydrogen**

1) **Investment and Policy Support:** In July 2020, the European Commission unveiled the "EU Hydrogen Strategy," which outlines a roadmap for creating a European Hydrogen Ecosystem. The strategy targets an installed capacity of at least 40 GW of renewable hydrogen electrolysis by 2030. Significant funding has been allocated from the EU Recovery Fund and other instruments, emphasizing support for research, innovation, and infrastructure.
2) **Public–Private Partnerships:** The EU established the "European Clean Hydrogen Alliance," bringing together more than 1,000 stakeholders from industry, academia, and government to collaborate on scaling up green hydrogen production, distribution, and end-use. The partnership aims to

foster a robust hydrogen market by facilitating large-scale projects, sharing best practices, and overcoming regulatory and technological barriers.

3) **Pilot Projects and Infrastructure Development:** Several pilot projects have been launched across Europe to demonstrate green hydrogen's technical and economic viability. For example, the "HyDeal Ambition" project aims to deliver 100% green hydrogen at a competitive price across Europe by 2030. This project involves the construction of electrolyzers, solar power plants, and a dedicated hydrogen pipeline network connecting production sites to industrial demand centers.
4) **Sector Coupling and Integration:** The EU strategy focuses on integrating green hydrogen into the broader energy system, promoting its use in sectors such as heavy industry (e.g., steel and chemicals), transport (e.g., hydrogen FCVs), and residential heating. The strategy promotes "sector coupling," where green hydrogen serves as a storage medium and enables the use of renewable energy across various applications.

**Results and Impact**

1) **Reduction in GHG Emissions:** The deployment of green hydrogen is expected to significantly reduce emissions across multiple sectors. For instance, the steel industry could cut up to 95% of its carbon emissions by using hydrogen in place of coal in blast furnaces.
2) **Economic Opportunities and Job Creation:** The green hydrogen sector is projected to create hundreds of thousands of jobs in Europe, particularly in engineering, manufacturing, and infrastructure development. The EU estimates that by 2050, hydrogen could supply up to 24% of its energy demand, creating a market worth over €820 billion.
3) **Energy Security and Independence:** By promoting the domestic production of green hydrogen, Europe aims to reduce its dependence on fossil fuel imports, particularly natural gas. This is particularly significant in the context of geopolitical tensions and supply disruptions.
4) **Innovation and Technological Leadership:** Europe's investments in green hydrogen research and innovation position it as a global leader in clean energy technology. Advances in electrolyzer efficiency, fuel cells, and hydrogen storage could set standards for the worldwide hydrogen economy.

**Challenges and Future Outlook**

1) **Cost Competitiveness:** Currently, green hydrogen is more expensive than gray or blue hydrogen, primarily due to the high costs of electrolyzers and renewable electricity. Scaling up production and technological advancements are critical to reducing costs.

2) **Infrastructure Development:** The lack of infrastructure for hydrogen production, storage, and distribution remains a significant barrier. Significant investments are needed to develop pipelines, refueling stations, and storage facilities.
3) **Regulatory and Safety Concerns:** Standardizing safety protocols, regulatory frameworks, and certification systems across countries is essential to ensure safe and efficient deployment of hydrogen technologies.
4) **Market Creation:** To foster a robust green hydrogen market, policies must be aligned to incentivize production and use across various sectors. This includes subsidies, carbon pricing mechanisms, and mandates for hydrogen blending in natural gas networks.

## 2.12 Exercise

**1** What are the distinguishing properties of hydrogen gas, and how do these properties make it suitable for various industrial applications?

**2** How does green hydrogen differ from other types like gray and blue hydrogen, and what are the implications of these differences for environmental sustainability?

**3** Can you explain the process of electrolysis and how it is used to produce green hydrogen? What role do renewable energy sources like solar and wind play in this process?

**4** In what ways does the production and utilization of green hydrogen contribute to reducing greenhouse gas emissions? Can you provide examples of sectors that could benefit from this reduction?

**5** What are some of the advancements in electrolysis technology that have contributed to the increasing viability and efficiency of green hydrogen production?

**6** How is green hydrogen being utilized in transportation, particularly in the development of fuel cells and hydrogen-powered vehicles? What are the potential benefits of this technology for reducing emissions in the transportation sector?

**7** Could you elaborate on the role of green hydrogen in industrial processes such as refining and ammonia production? How does its use in these sectors contribute to decarbonization efforts?

8 How does green hydrogen serve as a means of energy storage and grid stabilization? What advantages does it offer compared to other forms of energy storage in this regard?

9 How does steam methane reforming (SMR) work as a method for hydrogen production, and what are the key inputs and outputs of this process?

10 Can you explain the principles behind electrolysis as a method for hydrogen production, including the role of electricity and water in the process?

11 What is biomass gasification, and how does it differ from other methods of hydrogen production? What types of biomass can be used, and what are the environmental implications?

12 How does photo-biological water splitting utilize biological organisms to produce hydrogen from water and sunlight? What are the advantages and limitations of this approach?

13 What is thermochemical water splitting, and how does it utilize heat to produce hydrogen from water? What are the key reactions involved in this process, and what are its potential applications?

14 How do the efficiency and environmental impact of each hydrogen production method vary? Can you compare and contrast their sustainability, scalability, and economic feasibility?

15 What are the main challenges associated with scaling up each of these hydrogen production methods to meet the growing demand for clean energy? How are researchers and industry stakeholders addressing these challenges?

16 Considering factors such as energy efficiency, cost-effectiveness, and carbon footprint, which hydrogen production method(s) do you think hold the most promise for widespread adoption in the near future, and why?

17 What are the fundamental principles of electrolysis as a chemical process, and how does it facilitate the separation of molecules into their constituent elements?

18 Can you compare and contrast the operation and key features of alkaline electrolysis, proton exchange membrane (PEM) electrolysis, and solid oxide electrolysis cells (SOECs) as different types of electrolysis technologies?

19 How are electrolyzers designed, and what are the essential components involved in their operation? What role do materials selection and design play in optimizing electrolyzer performance?

20 What factors influence the energy input requirements and overall efficiency of electrolysis processes? How can efficiency be improved, and what strategies are employed to enhance energy utilization?

21 Why is the selection of water sources crucial for electrolysis, and what are the implications of water purity on the efficiency and longevity of electrolysis equipment? How can water be treated to ensure suitable purity levels?

22 How do renewable energy sources such as solar and wind power integrate with electrolysis, and what role do they play in driving the sustainability of hydrogen production? How are electrolysis systems synchronized with variable renewable energy generation?

23 What challenges and considerations arise when scaling up electrolysis technology for large-scale hydrogen production? How do engineers address these challenges, and what are the economic implications of scaling up electrolysis plants?

## 2.13 Objective Questions

1 What is hydrogen primarily known as?
**a)** A metal
**b)** A nonmetal
**c)** A noble gas
**d)** An alkali metal

2 Which of the following is NOT a property of hydrogen gas?
**a)** Colorless and odorless
**b)** Highly flammable
**c)** Heavier than air
**d)** Low density

3 What is one significant application of hydrogen in various industries?
**a)** Food preservation
**b)** Fertilizer production
**c)** Textile manufacturing
**d)** Pharmaceutical synthesis

**4** What is green hydrogen?
**a)** Hydrogen extracted from fossil fuels
**b)** Hydrogen produced using renewable energy sources
**c)** Hydrogen with a greenish hue
**d)** Hydrogen with environmentally friendly properties

**5** How does green hydrogen differ from gray and blue hydrogen?
**a)** Green hydrogen is produced using fossil fuels.
**b)** Gray hydrogen is produced using renewable energy.
**c)** Blue hydrogen emits carbon dioxide during production.
**d)** Blue hydrogen is not suitable for industrial applications.

**6** Which method is commonly used for the production of green hydrogen?
**a)** Steam methane reforming
**b)** Biomass gasification
**c)** Electrolysis
**d)** Natural gas reforming

**7** What are the environmental benefits of green hydrogen?
**a)** It reduces ozone depletion.
**b)** It contributes to air pollution.
**c)** It reduces dependence on fossil fuels.
**d)** It increases greenhouse gas emissions.

**8** What role does green hydrogen play in reducing greenhouse gas emissions?
**a)** It increases emissions due to production inefficiencies.
**b)** It traps greenhouse gases in the atmosphere.
**c)** It serves as a clean energy alternative, emitting only water vapor.
**d)** It accelerates the depletion of the ozone layer.

**9** In which sector does green hydrogen have the potential to decarbonize various processes?
**a)** Agriculture
**b)** Construction
**c)** Transportation
**d)** Entertainment

**10** Which renewable energy sources are commonly used for electrolysis in green hydrogen production?
**a)** Geothermal and tidal energy
**b)** Solar and wind energy

c) Biomass and hydroelectric energy
d) Nuclear and fossil fuel energy

11 What advancements have been made in electrolysis technology for green hydrogen production?
a) Increased energy consumption
b) Decreased efficiency
c) Enhanced scalability and cost-effectiveness
d) Reduced compatibility with renewable energy sources

12 What are some applications of green hydrogen in transportation?
a) Gasoline-powered cars
b) Hydrogen-powered vehicles and fuel cells
c) Electric bicycles
d) Diesel trucks

13 In which industrial processes can green hydrogen be used?
a) Paper production
b) Cement manufacturing
c) Refining and ammonia production
d) Textile dyeing

14 How can green hydrogen contribute to energy storage and grid stabilization?
a) By increasing grid instability
b) By providing a reliable backup power source
c) By accelerating the depletion of fossil fuel reserves
d) By emitting harmful pollutants into the atmosphere.

15 Which of the following is a method commonly used for hydrogen production that involves the reaction of methane with steam?
a) Electrolysis
b) Biomass gasification
c) Steam methane reforming
d) Photo-biological water splitting

16 What is the primary source of hydrogen in steam methane reforming?
a) Natural gas
b) Solar energy
c) Wind energy
d) Biomass

**17** Electrolysis is a process that involves the decomposition of water into hydrogen and oxygen using:
**a)** Chemical reactions
**b)** Heat
**c)** Electricity
**d)** Pressure

**18** Biomass gasification is a method of hydrogen production that involves:
**a)** Reacting biomass with oxygen at high temperatures
**b)** Heating biomass to produce syngas, which is then processed to obtain hydrogen
**c)** Exposing biomass to sunlight to split water molecules
**d)** Treating biomass with chemicals to extract hydrogen gas

**19** Which method of hydrogen production involves the use of biological organisms and sunlight to split water molecules and produce hydrogen?
**a)** Steam methane reforming
**b)** Electrolysis
**c)** Biomass gasification
**d)** Photo-biological water splitting

**20** Thermochemical water splitting is a method of hydrogen production that utilizes:
**a)** Solar energy
**b)** Geothermal energy
**c)** Chemical reactions at high temperatures
**d)** Electrical currents

## 2.14 Glossary

1) **Green Hydrogen:** Hydrogen produced using renewable energy sources (such as solar, wind, hydroelectric, or biomass) through water electrolysis, without generating greenhouse gas emissions.
2) **Electrolysis:** A chemical process that uses electricity to split water ($H_2O$) into hydrogen ($H_2$) and oxygen ($O_2$). It is the primary method for producing green hydrogen.
3) **Alkaline Electrolysis:** An electrolysis method that uses an alkaline solution (such as potassium hydroxide) as the electrolyte. It is a mature and widely used technology for large-scale hydrogen production.

4) **Proton Exchange Membrane (PEM) Electrolysis:** An electrolysis technique that uses a solid polymer electrolyte (PEM) to conduct protons while separating gases. Known for high efficiency and compact design, it is well-suited for renewable energy integration.
5) **Solid Oxide Electrolysis Cells (SOECs):** High-temperature electrolysis technology that operates using ceramic solid oxide cells to convert water and electricity into hydrogen and oxygen. Offers high efficiency but is still under development.
6) **Renewable Energy Sources:** Natural sources of energy that are replenished constantly, such as solar, wind, hydroelectric, and biomass, used to power electrolysis for green hydrogen production.
7) **Grid Compatibility:** The ability of green hydrogen production systems to integrate efficiently with existing electrical grids, ensuring stable energy supply and optimal utilization of renewable energy.
8) **Sector Coupling:** The integration of various energy-consuming sectors (such as transport, industry, and heating) with renewable energy and hydrogen systems to increase efficiency and reduce carbon emissions.
9) **Electrolyzer:** A device that performs the electrolysis process, converting electricity and water into hydrogen and oxygen.
10) **Greenhouse Gas Emissions (GHG):** Gases such as carbon dioxide ($CO_2$) and methane ($CH_4$) that trap heat in the atmosphere, contributing to global warming. Green hydrogen aims to reduce such emissions.
11) **Thermochemical Water Splitting:** A hydrogen production method involving high-temperature chemical reactions to split water into hydrogen and oxygen, often considered alongside electrolysis in clean energy research.
12) **System Reliability:** The ability of hydrogen production technologies to operate consistently and dependably under varying conditions, ensuring continuous hydrogen supply.
13) **Capital and Operational Costs:** Financial investments required for establishing (capital cost) and running (operational cost) hydrogen production facilities. Technological advancements seek to minimize these.
14) **Policy Support:** Government initiatives, subsidies, and incentives designed to encourage investment, development, and deployment of green hydrogen technologies.
15) **Regulatory Frameworks:** Rules and guidelines established by authorities to standardize, monitor, and promote safe and sustainable adoption of hydrogen technologies.
16) **Market Mechanisms:** Economic systems such as carbon credits, renewable energy certificates, and subsidies that support the commercialization and competitiveness of green hydrogen.

17) **Stakeholders:** Entities involved in the green hydrogen ecosystem, including governments, industries, research institutions, investors, and consumers, who contribute to its growth and adoption.

## References


1 Hwangbo, S., Lee, I. B., & Han, J. (2016). Multi-period stochastic mathematical model for the optimal design of integrated utility and hydrogen supply network under uncertainty in raw material prices. *Energy*, 114, 418–430.

2 Hwangbo, S., Lee, I. B., & Han, J. (2017). Mathematical model to optimize design of integrated utility supply network and future global hydrogen supply network under demand uncertainty. *Applied Energy*, 195, 257–267.

3 Ingason, H. T., Pall Ingolfsson, H., & Jensson, P. (2008). Optimizing site selection for hydrogen production in Iceland. *International Journal of Hydrogen Energy*, 33, 3632–3643.

4 Jagannath, A., & Almansoori, A. (2014). Modeling of hydrogen networks in a refinery using a stochastic programming approach. *Industrial and Engineering Chemistry Research*, 53, 19715–19735.

5 Jiang, H., Qi, B., du, E., Zhang, N., Yang, X., Yang, F., & Wu, Z. (2022. https://doi.org/10.1109/TIA.2021.3117748). Modeling hydrogen supply chain in renewable electric energy system planning. *IEEE Transactions on Industry Applications*, 58, 2780–2791.

6 Joffe, D., Hart, D., & Bauen, A. (2004). Modelling of hydrogen infrastructure for vehicle refuelling in London. *Journal of Power Sources*, 131, 13–22.

7 Johnson, N., & Ogden, J. (2012). A spatially-explicit optimization model for long-term hydrogen pipeline planning. *International Journal of Hydrogen Energy*, 37, 5421–5433.

8 Kamarudin, S. K., Daud, W. R. W., Yaakub, Z., Misron, Z., Anuar, W., & Yusuf, N. N. A. N. (2009). Synthesis and optimization of future hydrogen energy infrastructure planning in peninsular Malaysia. *International Journal of Hydrogen Energy*, 34, 2077–2088.

9 Khalilnejad A, Abbaspour A, Sarwat AI. Multi-level optimization approach for directly coupled photovoltaic electrolyser system. *International Journal of Hydrogen Energy* 2016; 41: 11884–11894; Kazi MK, Eljack F, El-Halwagi MM, Haouari M. Green hydrogen for industrial sector decarbonization: costs and impacts on hydrogen economy in Qatar. *Computers and Chemical Engineering* 2021; 145, 107144.

10 Kim, M., & Kim, J. (2016). Optimization model for the design and analysis of an integrated renewable hydrogen supply (IRHS) system: application to Korea's hydrogen economy. *International Journal of Hydrogen Energy*, 41, 16613–16626.

**11** Khare, V. (2023). Prediction of tidal energy resources at Gulf of Kutch and Gulf of Khambhat by data mining. *Cleaner Energy Systems*, 6, 100090.

**12** Khare, V. (2023). Design, optimization, and data analysis of solar-tidal hybrid renewable energy system for Hurawalhi, Maldives. *Cleaner Energy Systems*, 6, 100088.

**13** Khare, V. (2023). Assessment of hydro energy potential from rain fall data set in India through data analysis. *Advances in Electrical Engineering, Electronics and Energy*, 6, 100290.

**14** Khare, V., & Chaturvedi, P. (2023). Design, control, reliability, economic and energy management of microgrid: A review. *Advances in Electrical Engineering, Electronics and Energy*, 5, 100239.

**15** Khare, V., & Bhuiyan, M. A. (2023). Perspective of renewable energy in the BRICS country. *Advances in Electrical Engineering, Electronics and Energy*, Elsevier, 5, 100250.

**16** Khare, V., & Chaturvedi, P. (2023). Solar energy system concept change from trending technology: A comprehensive review. *Advances in Electrical Engineering, Electronics and Energy*, 4, 100183.

**17** Khare, V., & Chaturvedi, P. (2023). Design and operational management of sustainable multi- networks green energy system for study area at India. *Materials Today: Proceedings* https://doi.org/10.1016/j.matpr.2023.05.575.

**18** Kim, M., & Kim, J. (2017). An integrated decision support model for design and operation of a wind-based hydrogen supply system. *International Journal of Hydrogen Energy*, 42, 3899–3915.

**19** Kim, J., & Moon, I. (2008). Strategic design of hydrogen infrastructure considering cost and safety using multiobjective optimization. *International Journal of Hydrogen Energy*, 33, 5887–5896.

**20** Kim, J., Lee, Y., & Moon, I. (2008). Optimization of a hydrogen supply chain under demand uncertainty. *International Journal of Hydrogen Energy*, 33, 4715–4729.

**21** Li, Z., Gao, D., Chang, L., Liu, P., & Pistikopoulos, E. N. (2008). Hydrogen infrastructure design and optimization: a case study of China. *International Journal of Hydrogen Energy*, 33, 5275–5286.

# 3

# The Role of Hydrogen in Energy Transition

## 3.1 Introduction to Energy Transition

The global energy landscape is undergoing a profound transformation, driven by the urgent need to reduce carbon emissions, enhance energy security, and transition toward more sustainable energy systems. This shift, known as the energy transition, involves the gradual replacement of fossil fuels with cleaner and renewable energy sources such as solar, wind, and bioenergy. However, to achieve deep Decarbonization (DE) carbonization across various sectors, including power generation, transportation, and industry an integrated approach, is required.

Among the emerging solutions, hydrogen has gained significant attention as a versatile energy carrier capable of addressing some of the key challenges in the transition to a low-carbon economy. Hydrogen offers a promising pathway to store and transport energy efficiently while enabling sectoral integration and decarbonization of hard-to-abate industries. As nations strive to meet ambitious climate targets, the role of hydrogen in the energy transition has become increasingly vital. The subsequent discussion explores the role of hydrogen in energy transition, highlighting its production methods, applications, challenges, and potential in shaping a cleaner and more sustainable future [1].

Energy transition refers to the global shift from traditional fossil fuel-based energy systems such as coal, oil, and natural gas toward sustainable, renewable, and low-carbon energy sources, such as wind, solar, hydropower, and bioenergy. This transformation encompasses changes in energy production, distribution, and consumption patterns aimed at achieving energy efficiency, environmental sustainability, and economic growth. Energy transition is a multifaceted and dynamic process driven by technological advancements, environmental concerns, and socioeconomic considerations. Figure 3.1 shows the different elements of the energy transition.

*Hydrogen Technologies in Decarbonizing Industries and Energy Systems*, First Edition.
Vikas Khare, Sanjeet K. Dwivedi, Monica Bhatia, and K. K. Pant.

Published 2026 by John Wiley & Sons, Inc.

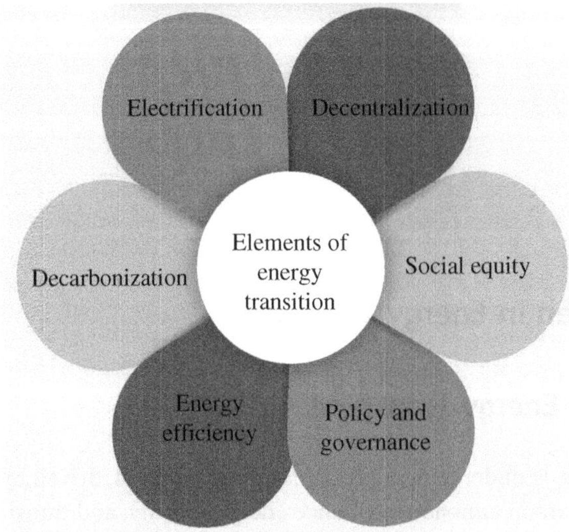


**Figure 3.1** Elements of Energy Transition

It involves:

1) **Decarbonization:** Reducing carbon emissions to combat climate change by replacing fossil fuels with renewable energy sources.
2) **Decentralization:** Transitioning from centralized energy systems to distributed generation models, empowering communities and individuals to produce and consume energy locally.
3) **Electrification:** Increasing the use of electricity in sectors such as transportation and heating by leveraging renewable energy technologies.
4) **Energy Efficiency:** Improving energy systems and technologies to reduce waste and optimize energy use.
5) **Policy and Governance:** Enacting laws, regulations, and incentives to encourage investment in clean energy and innovation.
6) **Social Equity:** Ensuring equitable access to energy and addressing socioeconomic disparities during the transition [2].

➢ **Key Drivers of Energy Transition**

- **Environmental Concerns:** Mitigating the effects of climate change and reducing GHG emissions.
- **Technological Innovation:** Advances in renewable energy technologies, storage solutions, and smart grids.
- **Economic Opportunities:** Creating jobs in the clean energy sector and reducing reliance on imported fuels.

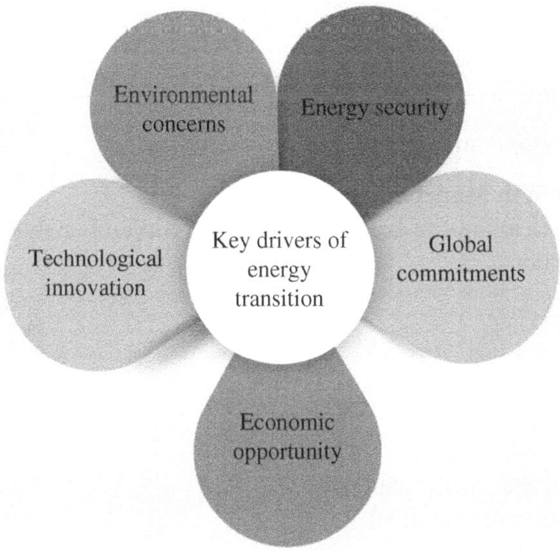


**Figure 3.2** Key Drivers of Energy Transition

- **Energy Security:** Ensuring a reliable and resilient energy supply.
- **Global Commitments:** Achieving targets set by international agreements, such as the Paris Agreement. Figure 3.2 shows the key drivers of energy transition.

➢ **Mathematical Interpretation of Energy Transition**

1) **Energy Supply and Demand Dynamics**

The fundamental equation governing energy balance in a system is:

$$S(t) + I(t) = D(t) + L(t) \tag{3.1}$$

where

$S(t)$ = Energy supply at time "$t$" (sum of all energy sources).
$I(t)$ = Imported energy.
$D(t)$ = Energy demand.
$L(t)$ = Energy losses.

For an energy transition scenario, we decompose supply into fossil and renewable components:

$$S(t) = S_f(t) + S_r(t) \tag{3.2}$$

where

$S_f(t)$ = Fossil fuel-based energy generation.
$S_r(t)$ = Renewable energy generation.

The goal is to minimize $S_f(t)$ and $S_r(t)$ increases to meet demand.

2) **Renewable Energy Growth Model**
Renewable energy growth follows an exponential adoption curve, often modeled using a logistic function:

$$S_r(t) = \frac{S_{r,max}}{1 + e^{-k(t-t_0)}} \tag{3.3}$$

where

$S_{r,max}$ = Maximum renewable capacity.
$k$ = Growth rate of renewable energy adoption.
$t_0$ = Inflection point (mid-transition time).

This function ensures that renewable energy follows a slow rate, rapid growth, and saturation phase.

3) **Decarbonization Rate**
To measure how fast fossil fuel energy is replaced, we define the decarbonization rate:

$$\lambda(t) = -\frac{dS_f}{dt} \tag{3.4}$$

where a higher $\lambda(t)$ means faster reduction in fossil fuel use. This is often constrained by economic and policy decisions.

4) **Economic Cost of Transition**
The total cost of transitioning to renewables includes capital investment $C_r$, operational cost $C_0$, and decommissioning fossil fuel infrastructure $C_f$:

$$C_{total} = \int_{t_0}^{T} \left(C_r(S_r) + C_0(S_r) + C_f\left(S_f\right)\right)dt \tag{3.5}$$

Optimization techniques such as integer programming or dynamic programming can be used to minimize cost while maximizing energy security.

5) **Carbon Emission Reduction Model**
Carbon emissions are proportional to fossil fuel use:

$$E(t) = \alpha S_f(t) \tag{3.6}$$

where

$E(t)$ = Carbon emission.
$\alpha$ = Emission factor (kg $CO_2$ per unit energy).

$$\text{To meet net-zero goals: } \int_0^T E(t)dt \leq E_{max} \tag{3.7}$$

where $E_{max}$ is the carbon budget.

A carbon tax or cap and trade system can be incorporated by adding a penalty function:

$$\text{Penalty} = \beta E(t)$$

where $\beta$ is the carbon tax rate.

6) **Energy Storage and Grid Stability**

Renewable energy is intermittent, so we introduce storage $B(t)$ with charging and discharging dynamics:

$$\frac{dB}{dt} = S_r(t) - D(t) - L(t) \tag{3.8}$$

Constraints:

$$0 \leq B(t) \leq B_{max}$$

where $B_{max}$ is the maximum storage capacity.

An optimization problem arises:

$$\min_{S_r, B} \int_0^T (C_r + C_0 + C_f)dt \tag{3.9}$$

subject to supply demand balance and emission limits.

➢ **Significance of Energy Transition**

Energy transition is vital for building a sustainable future by:

- Reducing the environmental impact of energy production and consumption.
- Promoting economic resilience through diversification of energy sources.
- Supporting global climate goals to limit global warming to 1.5 or 2 °C.
- Enhancing energy access and security for all [3].

This transition is reshaping global energy landscapes and fostering a shift toward cleaner, greener, and more equitable energy systems.

Hydrogen plays a pivotal role in the global energy transition, acting as a versatile and sustainable energy carrier that can help decarbonize hard-to-abate sectors and integrate renewable energy into the energy system. Its unique properties and potential applications make it an essential component of a cleaner, more sustainable energy future. Table 3.1 shows the key contributions of hydrogen to the energy transition. Table 3.2 shows the types of hydrogen in the energy transition. Table 3.3 shows the challenges for hydrogen in energy transition.

Hydrogen is a cornerstone of the energy transition, offering solutions for decarbonizing energy-intensive sectors and enabling a sustainable, low-carbon future. With continuous innovation, investment, and collaboration, hydrogen can unlock its full potential as a clean energy carrier and drive the global shift toward sustainability [4].

**Table 3.1** Key Contributions of Hydrogen to the Energy Transition

| Aspect | Contribution |
|---|---|
| Decarbonizing industry | Hydrogen can replace fossil fuels in industrial processes such as steel production, cement manufacturing, and refining. |
| Clean transportation | Hydrogen fuel cell vehicles (FCVs), especially for heavy-duty transport like trucks, buses, and ships, reducing emissions. |
| Energy storage | Hydrogen acts as a long-term energy storage solution, balancing the intermittent supply of renewable energy sources. |
| Power generation | Hydrogen can be used in fuel cells or burned in turbines to generate electricity without carbon emissions. |
| Replacing fossil fuels | Green hydrogen (produced using renewable energy) provides a clean alternative to natural gas for heating and power. |
| Grid flexibility | Hydrogen supports energy grids by storing excess renewable energy and providing backup power during peak demand. |
| Fuel for aviation | Hydrogen can power aircraft through direct combustion or as a component of synthetic fuels, decarbonizing air travel. |

**Table 3.2** Types of Hydrogen in the Energy Transition

| Type | Description | Relevance to Energy Transition |
|---|---|---|
| Green hydrogen | Produced using renewable energy sources (solar, wind) via electrolysis | The most sustainable option, crucial for achieving net-zero goals |
| Blue hydrogen | Produced from natural gas with carbon capture and storage (CCS) | Transition fuel for reducing emissions in the near term |
| Gray hydrogen | Produced from fossil fuels without CCS, leading to carbon emissions | Not sustainable; being replaced by green and blue hydrogen |

**Table 3.3** Challenges for Hydrogen in Energy Transition

| Challenge | Explanation | Possible Solutions |
|---|---|---|
| High production costs | Green hydrogen is expensive due to the cost of renewable energy and electrolysis technology. | Invest in scaling production, innovation, and government incentives. |
| Infrastructure gaps | Limited infrastructure for hydrogen storage, transport, and refueling | Develop pipelines, storage facilities, and fueling stations. |
| Energy efficiency | Conversion losses during production, storage, and use reduce overall efficiency. | Focus on technological advancements to optimize efficiency. |
| Safety concerns | Hydrogen is highly flammable and requires specialized handling and storage solutions. | Implement stringent safety standards and training programs. |
| Policy and regulation | Lack of clear global policies and regulatory frameworks | Develop harmonized regulations and international collaboration. |

***Key: Energy transition:*** *The global shift from fossil fuel-based energy systems to renewable and low-carbon energy sources such as wind, solar, and hydrogen, aimed at achieving sustainability and mitigating climate change.*

### 3.1.1 Historical Context

The energy transition, as understood today, is not a singular event but a series of gradual shifts in energy systems driven by evolving technological, economic, environmental, and social needs. Its history reflects humanity's adaptation to new challenges and opportunities, moving from reliance on traditional energy sources to more efficient and sustainable alternatives. Figure 3.3 shows the historical context of energy transition.

1) **Early Energy Sources and Pre-Industrial Era (Before 18th Century)**
   - **Biomass and Manual Labor:** The earliest energy systems were based on human labor, animal power, and biomass (wood, peat, and dung). These energy sources were sufficient for agrarian societies but limited in scale and efficiency.
   - **Wind and Water Power:** Windmills and watermills were introduced, harnessing natural forces for grinding grain, pumping water, and other mechanical tasks.

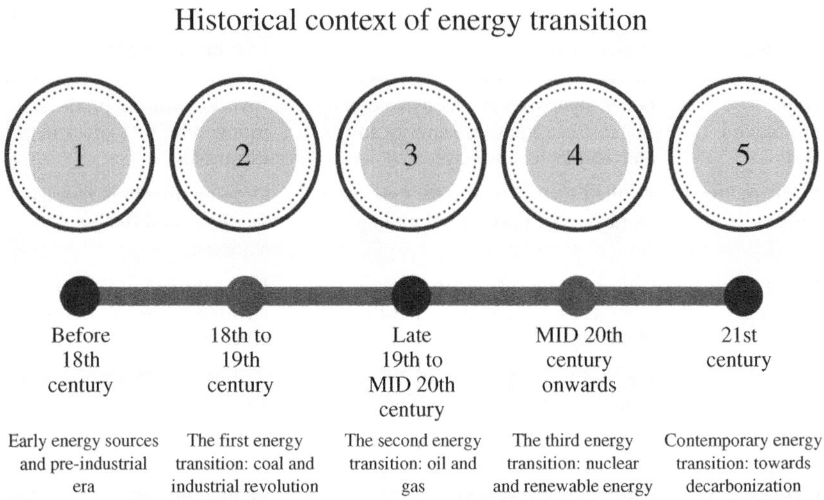


**Figure 3.3** Historical Context of Energy Transition [5]

- **Constraints:** Energy use was localized, and the lack of advanced technologies hindered large-scale energy production and distribution.

2) **The First Energy Transition: Coal and the Industrial Revolution (18th–19th Centuries)**
   - **Introduction of Coal:** Coal became the dominant energy source during the Industrial Revolution, replacing biomass. Its high energy density and availability fueled innovations in steam engines, manufacturing, and transportation.
   - **Mechanization:** Factories powered by coal revolutionized production systems, enabling mass production and urbanization.
   - **Economic and Social Changes:** The shift to coal-driven economies created industrial cities, expanded trade, and reshaped societal structures.
   - **Environmental Impact:** This transition marked the beginning of large-scale carbon emissions, laying the groundwork for modern environmental challenges.
3) **The Second Energy Transition: Oil and Gas (Late 19th to Mid-20th Centuries)**
   - **Discovery of Oil:** The drilling of the first commercial oil well in 1859 (Pennsylvania, USA) introduced a new era of energy. Oil became essential for transportation and industrial applications.
   - **Rise of Gas:** Natural gas emerged as a cleaner alternative to coal for heating and electricity generation [6].

- **Electricity Revolution:** The development of electricity systems, powered initially by coal and later by gas and oil, transformed energy accessibility and usage.
- **Global Impact:** This period witnessed rapid technological advancements, urbanization, and the rise of consumerism.

4) **The Third Energy Transition: Nuclear and Renewable Energy (Mid-20th Century Onward)**
   - **Nuclear Power:** Post-World War II, nuclear energy was introduced as a powerful and low-carbon energy source. It contributed significantly to electricity generation but faced public skepticism due to safety concerns (e.g., Chernobyl and Fukushima disasters).
   - **Emergence of Renewables:** Technologies harnessing solar, wind, hydropower, and geothermal energy gained traction as awareness of environmental issues grew.
   - **Oil Crises of the 1970s:** The oil embargoes underscored the vulnerabilities of fossil fuel dependence and spurred investments in alternative energy sources.
   - **Environmental Movements:** The rise of global environmental awareness and policies such as the Kyoto Protocol and Paris Agreement emphasized the need for sustainable energy systems.
5) **Contemporary Energy Transition: Toward Decarbonization (21st Century)**
   - **Climate Change Mitigation:** The urgent need to address global warming has accelerated the shift toward renewable energy and energy efficiency.
   - **Technological Innovations:** Advances in energy storage, smart grids, and hydrogen technologies are enabling deeper integration of renewables.
   - **Global Commitments:** Countries and organizations are adopting net-zero targets, promoting green hydrogen, and phasing out coal-fired plants.
   - **Just Transition:** Emphasis is placed on equitable energy systems that address social and economic disparities during the transition [7].

The historical context of the energy transition illustrates humanity's ingenuity and resilience in addressing energy needs. The current phase is a continuation of this evolution, driven by the pressing need to mitigate climate change and achieve sustainability. Learning from past transitions will be crucial in navigating the complexities of the modern energy landscape.

> ***Key: Driver of energy transition:*** *Key forces that accelerate the transition to clean energy, including climate change concerns, technological advancements, policy support, economic viability, and public awareness.*

### 3.1.2 Drivers of Energy Transition

The energy transition, defined as the shift from fossil fuel-based energy systems to renewable and sustainable energy sources, is propelled by a combination of technological advancements, environmental imperatives, economic considerations, and societal demands.

Following are the key drivers of the energy transition:

1) **Environmental and Climate Concerns**
   - **Climate Change Mitigation:** Rising global temperatures and extreme weather events underscore the need to reduce GHG emissions.
   - **Carbon Reduction Targets:** Commitments under international agreements like the Paris Agreement drive countries to adopt cleaner energy sources.
   - **Pollution Control:** Concerns about air and water pollution from fossil fuels are pushing governments and industries toward renewable energy solutions. Figure 3.4 illustrates drivers of energy transition.
2) **Technological Advancements**
   - **Renewable Energy Technologies:** Innovations in solar, wind, and hydropower technologies have increased efficiency and reduced costs, making them viable alternatives to fossil fuels.
   - **Energy Storage Solutions:** Advances in battery technologies, such as lithium-ion and solid-state batteries, facilitate reliable storage and integration of renewable energy [8].

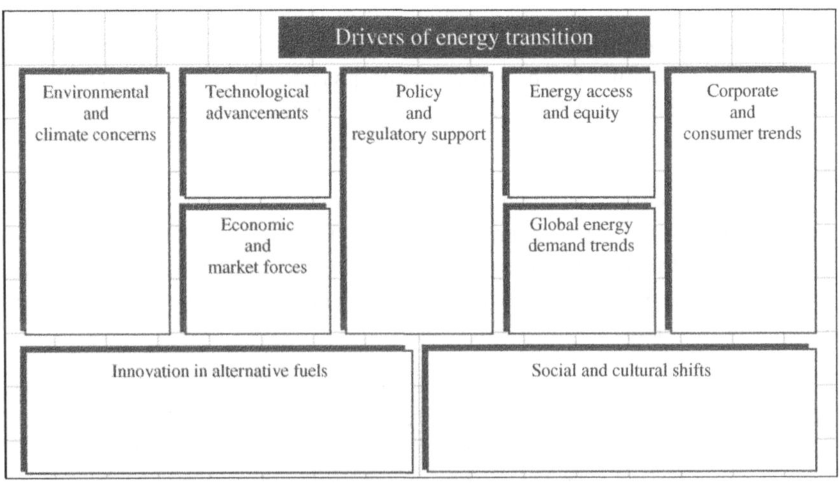


**Figure 3.4** Drivers of Energy Transition

- **Smart Grids and Digitalization:** Smart grids, AI, and IoT technologies enable efficient energy management and distribution, accommodating variable renewable energy sources.

3) **Economic and Market Forces**
   - **Cost Competitiveness:** The declining costs of renewable energy systems, including solar and wind, make them more economically attractive than fossil fuels in many regions.
   - **Energy Security:** Dependence on imported fossil fuels exposes countries to geopolitical risks. Transitioning to domestically produced renewable energy enhances energy security.
   - **Job Creation:** The renewable energy sector has the potential to create significant employment opportunities, driving economic growth and innovation.
4) **Policy and Regulatory Support**
   - **Incentives and Subsidies:** Governments worldwide provide tax credits, subsidies, and grants to promote renewable energy adoption.
   - **Carbon Pricing:** Mechanisms like carbon taxes and cap-and-trade systems encourage industries to transition to cleaner energy sources.
   - **Renewable Energy Targets:** National and regional policies mandate specific renewable energy shares in the energy mix, accelerating the transition.
5) **Energy Access and Equity**
   - **Universal Energy Access:** Renewable energy technologies, particularly decentralized solutions like solar home systems, enable energy access in remote and underserved regions.
   - **Energy Justice:** The transition aims to address energy inequities by ensuring affordable and reliable energy for all, including marginalized communities [9].
6) **Corporate and Consumer Trends**
   - **Corporate Sustainability Goals:** Many corporations are committing to net-zero targets and investing in renewable energy to align with ESG (Environmental, Social, and Governance) criteria.
   - **Consumer Awareness:** Growing consumer demand for environmentally friendly products and services pressures companies to adopt cleaner energy practices.
   - **Green Branding:** Companies use renewable energy adoption as part of their branding strategy to attract eco-conscious consumers.
7) **Global Energy Demand Trends**
   - **Rising Energy Consumption:** Increasing global energy demand, driven by urbanization and industrialization, necessitates sustainable energy solutions to prevent resource depletion.

- **Energy Efficiency:** Improving energy efficiency in buildings, transportation, and industries complements the transition by reducing overall energy demand.

8) **Innovation in Alternative Fuels**
   - **Hydrogen Economy:** Green hydrogen is emerging as a versatile and clean energy carrier for hard-to-decarbonize sectors like heavy industry and transportation.
   - **Bioenergy:** Advances in biofuels and biogas production provide sustainable energy alternatives for transportation and heating [10].
9) **Social and Cultural Shifts**
   - **Public Support for Sustainability:** Increased awareness of climate issues drives public demand for renewable energy solutions and green policies.
   - **Youth Activism:** Movements like Fridays for Future amplify the urgency of transitioning to sustainable energy systems.
10) **Geopolitical Shifts**
   - **Global Competitiveness:** Countries leading in renewable energy adoption and innovation gain economic and strategic advantages on the global stage.
   - **Reduced Fossil Fuel Influence:** The transition diminishes the geopolitical power of fossil fuel-rich nations, redistributing influence toward renewable energy leaders.

The energy transition is driven by a convergence of environmental, technological, economic, and societal factors. Addressing climate change, enhancing energy security, and meeting rising energy demand sustainably are critical imperatives. These drivers not only shape the trajectory of the transition but also underscore its inevitability in achieving a resilient and equitable energy future.

## 3.2 Fundamentals of Hydrogen Storage

Hydrogen storage is a crucial aspect of hydrogen energy systems, enabling the practical use of hydrogen as a clean, sustainable, and efficient energy carrier. Hydrogen, as a versatile fuel, has applications in a wide range of fields, including transportation, energy generation, and industrial processes. However, due to its low volumetric energy density in gaseous form, effective storage solutions are essential to facilitate its widespread adoption.

➢ **Why Hydrogen Storage Is Important**

The global energy system is undergoing a dramatic transformation driven by the twin imperatives of decarbonization and sustainability. In this transition, hydrogen has emerged as a clean, versatile, and high-energy-density carrier that can bridge the gap between renewable energy generation and its practical, large-scale use. While hydrogen itself holds immense promise, the ability to store it safely, efficiently, and at scale is the true key to unlocking its potential in the energy system. Hydrogen storage is not merely a technical detail but rather a foundational requirement that underpins its application across electricity generation, transportation, industry, and heating. Without robust storage solutions, hydrogen cannot fulfill its envisioned role in shaping a sustainable energy future [11].

One of the most pressing challenges in modern energy systems is the mismatch between energy supply and demand. Renewable energy sources such as solar and wind are inherently intermittent and unpredictable, often producing electricity at times when demand is low or failing to generate power during periods of peak consumption. Hydrogen storage provides a crucial solution to this challenge. Excess renewable electricity can be used to produce hydrogen through electrolysis, which splits water into hydrogen and oxygen. The hydrogen produced can then be stored for hours, days, or even months, effectively decoupling energy generation from energy use. In this way, hydrogen acts as a long-duration energy storage medium, ensuring a reliable and continuous supply of energy even when renewable output fluctuates.

Energy grids are complex networks that require constant balancing of input and output to maintain stability. As renewable penetration increases, grid operators face new difficulties in maintaining frequency, voltage, and load management. Hydrogen storage offers an invaluable tool for stabilizing the grid. By converting surplus electricity into hydrogen during times of low demand and then converting it back into electricity through fuel cells or turbines when demand rises, hydrogen provides both short-term flexibility and long-term reliability. Unlike batteries, which are effective for short-term balancing but limited in duration and scalability, hydrogen can store energy for extended periods and in much larger quantities, making it especially important for seasonal balancing in renewable-heavy grids.

Hydrogen is particularly crucial in addressing the energy needs of sectors where direct electrification is difficult or inefficient. Heavy industries such as steelmaking, cement production, and chemical manufacturing rely heavily on high-temperature heat and fossil-based feedstocks, making them challenging to decarbonize. Hydrogen, stored and transported efficiently, can serve as both a fuel and a raw material for these industries. For example, stored hydrogen can replace coal in steelmaking through direct reduction processes or serve as a feedstock in ammonia and methanol production. In these contexts, hydrogen storage ensures a continuous and dependable supply of the gas, enabling industries to transition toward low-carbon processes without disruption.

The transportation sector accounts for a significant share of global carbon emissions, and hydrogen offers a pathway to reduce its environmental impact, particularly in segments where batteries face limitations. Heavy-duty vehicles, long-haul trucks, ships, and aircraft require energy-dense fuels that can provide long ranges and quick refueling times. Hydrogen, stored as compressed gas or liquefied hydrogen, meets these requirements effectively. The development of hydrogen refueling infrastructure depends on reliable and scalable storage solutions, ensuring that hydrogen is available where and when it is needed. Moreover, large-scale storage allows for centralized hydrogen production, which can then be distributed to refueling stations, further reducing emissions and enhancing transportation efficiency.

Unlike conventional batteries, which are best suited for short-term storage, hydrogen excels in long-term, seasonal storage applications. For example, in regions with strong solar energy resources, electricity generation peaks during summer, while demand for heating rises in winter. Hydrogen storage bridges this seasonal gap by capturing surplus summer electricity in the form of hydrogen and releasing it as heat or power during the colder months. This characteristic makes hydrogen storage a linchpin in achieving 100% renewable energy systems, where mismatches between seasonal generation and consumption must be effectively managed. Underground storage in salt caverns and depleted gas fields is especially promising for large-scale hydrogen storage, offering vast capacities at relatively low costs [12].

Hydrogen storage also contributes to energy security and system resilience. Unlike fossil fuels, which are often subject to geopolitical risks, price volatility, and supply chain disruptions, hydrogen can be produced domestically using local renewable resources. Large-scale hydrogen storage ensures that countries maintain a strategic reserve of clean energy that can be mobilized during supply shortages or emergencies. This reduces reliance on imported fuels and enhances national energy independence. In the face of natural disasters, cyberattacks, or extreme weather events, stored hydrogen can also act as a backup energy source, keeping critical infrastructure running and improving resilience against shocks.

Hydrogen storage also supports the integration of CCUS technologies. When combined with captured carbon dioxide, stored hydrogen can be used to produce synthetic fuels such as e-methane, e-diesel, or e-kerosene, which are compatible with existing infrastructure and can decarbonize sectors like aviation and shipping. The feasibility of these processes depends heavily on a steady and reliable supply of hydrogen, which, in turn, necessitates effective storage systems. In this way, hydrogen storage enables not only clean energy use but also the creation of carbon-neutral fuels, aligning with circular economy principles.

There are several technological pathways for storing hydrogen, each with distinct advantages and challenges. Compressed hydrogen gas storage is the most mature and widely used method, offering flexibility for small- to medium-scale applications such as FCVs. Liquefied hydrogen storage, although energy-intensive due to the need for cryogenic temperatures, is attractive for applications requiring high energy density, such as aerospace or shipping. Solid-state storage using metal hydrides, chemical carriers like ammonia, or liquid organic hydrogen carriers (LOHCs) provides additional options for safer and denser storage, though many of these technologies are still in the development stage. The diversity of storage technologies allows hydrogen to be tailored to specific applications, but all require further innovation and cost reduction to become widely viable.

The role of hydrogen storage in the energy system cannot be separated from economic and infrastructural considerations. While hydrogen itself is abundant and clean, storing it remains expensive and technologically challenging. Infrastructure for storage, transport, and distribution is still limited, creating a "chicken-and-egg" problem where demand and supply must evolve in tandem. Large-scale investments in hydrogen storage facilities are necessary to create economies of scale and reduce costs. Moreover, policy support, regulatory frameworks, and international cooperation are crucial to accelerate infrastructure deployment. As these hurdles are overcome, hydrogen storage will become more cost-effective and accessible, enhancing its role in the global energy transition.

The environmental case for hydrogen storage is compelling. By enabling the large-scale integration of renewables, hydrogen storage reduces dependence on fossil fuels and cuts GHG emissions. Furthermore, unlike traditional energy reserves such as coal or natural gas, hydrogen does not produce harmful pollutants when used. Stored hydrogen, when converted back to electricity in fuel cells, emits only water vapor as a by-product. This clean cycle makes hydrogen storage an essential tool for meeting global climate goals, improving air quality, and supporting sustainable development.

Hydrogen storage is not an optional complement but a non-negotiable pillar of the future energy system. Its ability to balance intermittent renewable generation, stabilize grids, decarbonize hard-to-abate sectors, support clean transportation, provide seasonal storage, and enhance energy security underscores its central importance. While challenges related to cost, efficiency, and infrastructure remain, ongoing technological advances and policy initiatives are steadily addressing these barriers. The strategic deployment of hydrogen storage will determine the pace and scale of the global energy transition. Without it, hydrogen's potential as a clean energy carrier cannot be fully realized; with it, the vision of a carbon-neutral, resilient, and sustainable energy system becomes not just possible, but inevitable.

***Key: Hydrogen storage:*** The process and technologies used to store hydrogen in various forms (gaseous, liquid, or solid-state) to ensure safe, efficient, and scalable energy use and distribution.

### ➢ Temporary Versus Long-Term Storage Applications

Hydrogen storage is a critical enabler of the hydrogen economy, but not all storage needs are the same. Depending on the application, hydrogen may need to be stored for a few hours or days (temporary storage) or for weeks to months (long-term storage). The difference lies not only in the duration but also in the scale, technology, and purpose of storage. Understanding the distinction between temporary and long-term hydrogen storage is essential for building efficient energy systems that can balance short-term operational needs with long-term strategic requirements.

#### Temporary Hydrogen Storage

Temporary storage refers to the short-duration holding of hydrogen, typically for hours to a few days. This type of storage is essential for balancing daily fluctuations in hydrogen supply and demand. For example, in renewable energy systems, hydrogen can be produced during periods of excess solar or wind generation and stored temporarily until it is needed for electricity generation later the same day or within a few days.

At smaller scales, temporary storage is widely used in FCVs, where compressed hydrogen tanks act as mobile storage units. Similarly, hydrogen refueling stations (HRS) rely on temporary storage tanks to ensure a steady supply of fuel to vehicles throughout the day. In industrial settings, hydrogen produced on-site may be stored temporarily before being consumed in processes such as refining, glass production, or electronics manufacturing.

Technologies for temporary storage include compressed hydrogen gas cylinders at 350–700 bar and small cryogenic tanks for liquefied hydrogen. These systems are relatively well-developed, compact, and designed for frequent charge–discharge cycles. However, they are generally expensive per unit of hydrogen stored and unsuitable for very large-scale or seasonal applications.

#### Applications of Temporary Storage

1) Transportation: Hydrogen-powered cars, buses, trucks, and trains require temporary storage tanks for onboard fuel supply.
2) Refueling Infrastructure: Filling stations hold hydrogen temporarily in high-pressure tanks before dispensing it to vehicles.
3) Daily Grid Balancing: Renewable energy stored as hydrogen during the day can be used in fuel cells or turbines to generate electricity at night.
4) Industrial Processes: On-site storage ensures an uninterrupted supply for industries requiring continuous hydrogen use.

#### Long-Term Hydrogen Storage

In contrast, long-term storage involves holding hydrogen for weeks, months, or even an entire season. The goal here is not just to manage daily fluctuations but also to bridge seasonal mismatches between energy supply and demand.

Renewable-rich regions often face this challenge: solar power peaks in the summer, while energy demand for heating peaks in winter. Long-term hydrogen storage allows surplus energy produced in one season to be stored and used in another.

Large-scale, long-term storage relies on geological formations, such as salt caverns, depleted natural gas reservoirs, and aquifers, which can hold massive amounts of hydrogen at relatively low cost. Another option is liquid hydrogen ($LH_2$) storage in large cryogenic tanks, though energy requirements for liquefaction remain high. Emerging technologies like ammonia or LOHCs also show promise for long-term storage and global hydrogen transport.

Unlike temporary storage systems, long-term storage must prioritize capacity, durability, and cost-effectiveness over compactness or high discharge rates. These storage methods are designed to handle seasonal variability and provide backup reserves for energy security.

### Applications of Long-Term Storage

1) **Seasonal Energy Storage**: Capturing surplus summer solar or spring wind energy and releasing it during winter or during prolonged low-renewable periods.
2) **Grid Resilience**: Acting as a strategic reserve of hydrogen to ensure supply during disruptions such as natural disasters, geopolitical crises, or energy price shocks.
3) **Industrial Supply Chains**: Large-scale, centralized hydrogen storage supporting steel, ammonia, and methanol production across extended timeframes.
4) **Export and Import**: Long-term hydrogen storage technologies like ammonia and LOHCs enable global trade, allowing countries to store hydrogen for shipping overseas.

### Comparison Between Temporary and Long-Term Storage

- **Duration**: Temporary storage covers hours to days; long-term storage spans weeks to months or seasons.
- **Scale**: Temporary storage is small to medium scale (cylinders, tanks), while long-term storage is large to massive scale (caverns, reservoirs, carriers).
- **Technology**: Temporary relies on compressed or liquefied hydrogen tanks, whereas long-term relies on geological storage, cryogenic tanks, or chemical carriers.
- **Applications**: Temporary is suited for vehicles, refueling stations, and daily balancing; long-term is for seasonal energy management, industrial reserves, and strategic security.
- **Cost**: Temporary storage has higher costs per unit but is flexible and mobile, while long-term storage is more cost-efficient at scale but requires infrastructure investment.

Both temporary and long-term hydrogen storage are indispensable in the energy system, but they serve fundamentally different purposes. Temporary storage ensures the smooth operation of vehicles, industries, and short-term grid balancing, while long-term storage addresses seasonal mismatches, energy security, and large-scale industrial needs. A robust hydrogen economy will depend on a hybrid approach, leveraging both types of storage to meet diverse requirements. As technology advances and costs decline, integrating short-term and long-term hydrogen storage will play a pivotal role in building a sustainable, resilient, and decarbonized energy future.

**➢ Thermodynamic and Kinetic Considerations of Hydrogen Storage**

Hydrogen storage is governed by a delicate balance between thermodynamics and kinetics. While thermodynamics determines whether hydrogen can be stored and released under specific conditions of temperature and pressure, kinetics defines how quickly these processes occur in practice. To make these concepts more precise, researchers use mathematical models and equations that describe equilibrium conditions, energy requirements, and reaction rates. These mathematical interpretations provide deeper insight into the feasibility and efficiency of hydrogen storage systems, and they guide the design of new materials and technologies.

**Thermodynamic Interpretations**

Thermodynamics of hydrogen storage can be understood through the concepts of enthalpy ($\Delta H$), entropy ($\Delta S$), and Gibbs free energy ($\Delta G$). The fundamental condition for a reaction to be spontaneous is that the Gibbs free energy change is negative, expressed mathematically as:

$$\Delta G = \Delta H - T\Delta S$$

For hydride formation, if $\Delta G$ is negative at a given temperature, hydrogen will be absorbed by the material. Conversely, for hydrogen release, $\Delta G$ must become positive, which usually requires raising the temperature. This explains why materials with high enthalpy of formation need heating to desorb hydrogen, while those with low enthalpy release hydrogen more readily but may not store it stably. The relationship between equilibrium pressure, temperature, and enthalpy can be captured by the van't Hoff equation:

$$\ln P = -\frac{\Delta H}{R} \cdot \frac{1}{T} + \frac{\Delta S}{R} \tag{3.10}$$

Here, $P$ is the equilibrium hydrogen pressure, $T$ is the temperature in Kelvin, and $R$ is the universal gas constant. This equation shows that at higher temperatures, equilibrium pressure increases, making hydrogen release more favorable. It also highlights why certain hydrides, like magnesium hydride ($MgH_2$), require high temperatures (>300 °C) to reach usable pressures for hydrogen release. In compressed gas storage, the energy required for compression can be described using the work of compression equation:

$$W = \frac{nRT}{\gamma - 1}\left[\left(\frac{P_2}{P_1}\right)^{\frac{\gamma-1}{\gamma}} - 1\right] \tag{3.11}$$

where $n$ is the number of moles of hydrogen, $P_1$ and $P_2$ are the initial and final pressures, respectively, $T$ is the temperature, and $\gamma$ is the heat capacity ratio. This equation quantifies the significant energy investment required to compress hydrogen to high pressures for temporary storage.

**Kinetic Interpretations**

Kinetics in hydrogen storage describes how fast hydrogen molecules are absorbed or released by a material. These processes are often modeled by **Arrhenius-type equations**, which capture the role of activation energy. The reaction rate constant $k$ is given by:

$$k = Ae^{-E_a/RT} \tag{3.12}$$

where $A$ is the preexponential factor (frequency of successful collisions), $E_a$ is the activation energy, $R$ is the gas constant, and $T$ is the temperature. A high activation energy means slow kinetics, even if the reaction is thermodynamically favorable. For example, $MgH_2$ has a favorable enthalpy but very high activation energy for hydrogen diffusion, making it slow to release hydrogen without catalysts.

Diffusion processes within solids can also be described mathematically. Hydrogen diffusion into a solid hydride often follows **Fick's law of diffusion**:

$$J = -D\frac{dc}{dx} \tag{3.13}$$

Here, $J$ is the diffusion flux, $D$ is the diffusion coefficient, and $dc/dx$ is the concentration gradient. The diffusion coefficient $D$ itself is temperature-dependent and follows an Arrhenius relationship:

$$D = D_0 e^{-E_d/RT} \tag{3.14}$$

where $D_0$ is the preexponential factor and $E_d$ is the activation energy for diffusion. These equations demonstrate why nanostructuring (reducing diffusion distances) and catalyst addition (lowering activation energy) improve hydrogen absorption/desorption rates.

**Thermodynamic–Kinetic Trade-Off**

Mathematical models help explain the well-known trade-off between thermodynamic stability and kinetic performance. A hydride with a very negative enthalpy of formation ($\Delta H$) will be stable but requires high temperatures to desorb hydrogen, as predicted by the van't Hoff relation. On the other hand, a material with a small $\Delta H$ may release hydrogen at low temperatures but cannot retain it for long-term storage. Similarly, Arrhenius-based models reveal that even thermodynamically favorable reactions can be kinetically slow if activation energies are high. This dual mathematical framework helps in identifying the "sweet spot" where $\Delta H$, $\Delta S$, and $E_a$ are optimized for practical hydrogen storage.

In real-world applications, these thermodynamic and kinetic equations are used to design and test storage systems. For example, in FCVs, the van't Hoff equation predicts whether hydrogen will be released at the moderate temperatures of onboard systems, while Arrhenius-based kinetic models determine how quickly refueling can be achieved. For seasonal energy storage in geological caverns, thermodynamic stability is prioritized, as hydrogen must remain stored for months without significant leakage, while kinetics is less critical since release does not need to be instantaneous. In portable devices, however, fast kinetics at low temperatures is essential, which requires materials with low activation energy and optimized diffusion pathways.

The thermodynamic and kinetic behavior of hydrogen storage systems can be rigorously described through mathematical models, which not only explain observed performance but also guide the development of improved materials. Thermodynamic equations, such as Gibbs free energy and van't Hoff relations, clarify the conditions under which hydrogen can be stored or released, while kinetic equations, such as Arrhenius and Fick's laws, reveal the speed and mechanisms of these processes. Together, they highlight the constant trade-off between stability and reactivity that defines hydrogen storage technology. By applying these mathematical frameworks, scientists and engineers can design storage systems that meet the diverse needs of transportation, grid storage, and industrial applications, pushing hydrogen closer to becoming the cornerstone of a sustainable energy future.

### ➢ Hydrogen Physical Properties: Volumetric Versus Gravimetric Density

Hydrogen, as the lightest element in the periodic table, exhibits unique physical properties that significantly impact its storage and utilization in energy systems. Unlike conventional fuels, where density values align more closely with energy requirements, hydrogen poses both opportunities and challenges due to its low molecular mass and high energy content per unit mass. To understand these implications, researchers often distinguish between gravimetric density (energy per unit mass) and volumetric density (energy per unit volume). These two perspectives are critical in evaluating storage feasibility and determining the most appropriate applications of hydrogen in transportation, stationary energy systems, or portable devices.

Because hydrogen is diatomic ($H_2$) with a molar mass $M = 2.016$ g/mol, its gravimetric energy density is outstanding about 120 MJ (LHV), or 142 MJ (HHV). The challenge is volumetric energy density, which follows directly from the gas density $\rho$ at a given pressure $P$ and temperature $T$. For gases, a practical expression is

$$\rho = \frac{PM}{ZRT}$$

where $R$ is the universal gas constant and $Z$ is the compressibility factor (deviation from ideal gas). Volumetric energy density is then $E_v = \rho \times E_g$, where $E_g$ is the gravimetric energy density (e.g., 120 MJ/kg LHV). At ambient conditions (1 bar, 25 °C) and $Z \approx 1$, hydrogen's density is only ~0.08 kg/m$^3$, giving $E_v \approx 0.01$ MJ/L, which is why uncompressed hydrogen is impractical for compact storage even though it is superb by mass. The same framework explains why $LH_2$ and high-pressure gas (350–700 bar) are used: both routes raise $\rho$, and therefore $E_v$, into application-worthy ranges.

Hydrogen's molecular simplicity also sets the thermodynamic cost to "pack" it tightly. For compression, the (idealized) isentropic work per mole is

$$W_m = \frac{\gamma}{\gamma - 1} RT \left[ \left( \frac{P_2}{P_1} \right)^{(\gamma-1)/\gamma} - 1 \right] \tag{3.15}$$

with $\gamma \approx 1.4$ for diatomic gases. Single-stage, isentropic compression from $1 \rightarrow 700$ bar at 25 °C yields an upper-bound work near a few 104 J/mol, i.e., tens of MJ/kg; multi-stage compression with intercooling typically reduces this to roughly 10–15% of the stored LHV. For liquefaction, energy consumption is even larger (roughly 30–40% of LHV) because hydrogen must be cooled to 20 K, and cryogenic boil-off then eats into delivered energy unless tanks are exquisitely insulated. Mathematically, you can see the volumetric win immediately: $LH_2$ has $\rho \approx 70$ kg/m$^3$ $\Rightarrow E_v \approx 8.4$ MJ/L (LHV), but you "pay" with the liquefaction work and cryogenic losses. These numbers come into sharper relief versus alternatives. Gasoline's gravimetric energy density is ~44 MJ/kg and its density ~0.74 kg/L, so $E_v \approx 32$ MJ/L. Modern Li-ion batteries are far lower by mass 140–250 Wh/kg = 0.5–0.9 MJ/kg but decent by volume, 250–700 Wh/L = 0.9–2.5 MJ/L. Hydrogen thus dominates on mass, but gasoline beats it on volume, and batteries land in between volumetrically while trailing gravimetrically. This is why the application dictates the storage mode: aircraft and heavy trucks value mass; cars and buildings often hit volume limits first. Even within hydrogen options, the math clarifies the weight–volume trade-off. Consider 700-bar compressed $H_2$ at 25 °C. Real-gas effects matter, so include $Z$. Using with $\rho = \frac{PM}{ZRT}$

$P = 700$ bar $= 7.0 \times 10^7$ Pa, $T = 298$ K, $M = 0.002016$ kg/mol, $R = 8.314$ J/(mol K), and a representative $Z = 1.25$ at high pressure, we get

$$\rho = \frac{7.0 \times 10^7 \times 0.002016}{1.25 \times 8.314 \times 298} = \frac{1.4112 \times 10^5}{3.097 \times 10^3} = 45.6\ \text{kg/m}^3$$

Multiplying by LHV gives the volumetric energy density

$$E_v \approx 45.6\ \text{kg/m}^3 \times 120\ \text{MJ/kg} = 5,470\ \text{MJ/m}^3 = 5.47\ \text{MJ/L}$$

For $LH_2$, take $\rho \approx 70.8\ \text{kg/m}^3$at 20 K, $E_v \approx 70.8 \times 120 = 8{,}496\ \text{MJ/m}^3 = 8.50$ MJ/L. Both trail gasoline's 32 MJ/L, but they crush it by mass: 120 MJ/kg versus 44 MJ/kg.

A short design-scale example makes the engineering implications concrete. Suppose a fuel cell car needs 5 kg of onboard $H_2$ (a common target). With 700-bar storage at $\rho \approx 45.6\ \text{kg/m}^3$, the required tank volume is $V = \frac{m}{\rho} = \frac{5}{45.6} = 0.1096\ \text{m}^3 = 110\ \text{L}$.

That 5 kg carries $5 \times 120 = 600$ MJ of chemical energy (LHV). To carry the same energy as gasoline, using $E_v \approx 32$ MJ/L, you'd need

$$V_{gasoline} = \frac{600}{32} = 18.8\ \text{L}$$

which is why gasoline tanks are much smaller for a given range. By contrast, to match the **mass**, gasoline at 44 MJ/kg would need $600/44 \approx 13.6$ kg 600/44, so hydrogen wins handily on weight but loses on volume exactly the gravimetric/volumetric trade-off.

Finally, look at the efficiency penalties from packing hydrogen more tightly. Using the isentropic compression expression above with $\gamma = 1.4$ and $T = 298$ K, a one-shot $1 \rightarrow 700$ bar calculation gives a theoretical work of roughly.

$$W_m = \frac{1.4}{0.4} RT\left[(700)^{0.2857} - 1\right] = 3.5 \times 8.314 \times 298 \times (\sim 6.5 - 1)$$
$$\approx 8.66 \times 10^3\ \text{J/mol} \times 5.5 \approx 4.8 \times 10^4 \text{J/mol}$$

Per kilogram (about 496 mol), that's $\sim$ 24 MJ/kg, close to 20% of LHV an upper bound since real systems use multi-stage compression with intercooling to bring this down toward 10–15%. For liquefaction, typical specific energy is around 10–12 kWh/kg (≈36–43 MJ/kg), i.e., 30–36% of LHV, which is why $LH_2$ is reserved for use cases that prize volumetric density enough to justify the energy cost (e.g., aerospace or very long-range transport). These calculations capture the heart of hydrogen's physics: magnificent energy per kilogram, modest to low energy per liter unless you invest work to raise density and that investment shows up as efficiency loss and system complexity.

Hydrogen exists primarily in the diatomic molecular form ($H_2$). With a molecular weight of only 2 g/mol, hydrogen boasts the highest gravimetric energy density of any fuel about 120 MJ/kg, nearly three times higher than gasoline (~44 MJ/kg). This exceptional gravimetric density makes hydrogen an attractive energy carrier for reducing overall fuel weight in applications like aviation or space exploration. However, the same property contributes to its extremely low volumetric density, since a large volume is required to contain significant amounts of energy at ambient temperature and pressure. For instance, at 1 bar and 25 °C, hydrogen gas has a volumetric energy density of only about 0.01 MJ/L, compared to gasoline's ~ 34 MJ/L. This disparity means that while hydrogen is excellent from a weight perspective, it presents formidable challenges when considering compact, portable storage solutions.

When compared to natural gas and batteries, hydrogen exhibits both advantages and drawbacks. Natural gas, primarily composed of methane ($CH_4$), has a much higher volumetric density at ambient conditions, which is why it has become the fuel of choice in many industrial and residential applications. Methane's molecular structure allows it to be stored more compactly, even without extreme compression or liquefaction. Hydrogen, on the other hand, requires either compression (350–700 bar) or liquefaction (at −253 °C) to approach comparable volumetric energy densities.

On the other side of the comparison, hydrogen's gravimetric energy density far surpasses that of modern batteries, such as lithium-ion, which typically offer only 0.5–0.9 MJ/kg. This is why hydrogen fuel cells are often viewed as superior for long-range transportation, where minimizing weight is critical. However, batteries have much higher volumetric density (about 0.9–2.6 MJ/L), making them more practical for compact energy storage, such as in electric cars or portable electronics. This contrast underscores hydrogen's central storage challenge: balancing its excellent mass-based efficiency with its poor volume-based efficiency.

The choice of hydrogen storage technology often depends on the relative importance of weight versus volume. In aerospace applications, where every kilogram of payload is critical, the gravimetric advantage of hydrogen is indispensable. This explains why $LH_2$ has long been the fuel of choice for rockets, despite the immense engineering complexity required for cryogenic storage. In contrast, for automobiles or residential applications, volumetric density becomes the limiting factor, as vehicles and storage tanks must remain compact while still providing sufficient energy for long ranges or daily use.

For example, compressed hydrogen at 700 bar achieves a volumetric density of about 5.6 MJ/L, still significantly lower than gasoline but sufficient for FCVs that emphasize efficiency. Meanwhile, metal hydrides or other solid-state storage methods are being explored as a compromise, providing higher volumetric density at the expense of added weight. This constant trade-off illustrates the practical dilemma: achieving high storage capacity in both gravimetric and volumetric terms is extremely challenging, and optimizing one usually worsens the other.

To overcome the volumetric limitations, hydrogen must be stored either in compressed gaseous form or as a cryogenic liquid. However, both approaches face significant thermodynamic and engineering challenges. Compressing hydrogen to 700 bar consumes approximately 10–15% of the energy content of the stored hydrogen, representing a substantial efficiency penalty. Moreover, high-pressure tanks require advanced materials such as carbon–fiber composites to withstand repeated cycling, increasing cost, and complexity.

Liquefaction, while improving volumetric density to about 8.5 MJ/L, introduces even greater inefficiencies. The process of cooling hydrogen to −253 °C consumes around 30–40% of its energy content, and boil-off losses can occur during storage and transportation due to heat ingress. Furthermore, hydrogen's small molecular size increases the risk of leakage and permeation, even in well-sealed systems, complicating both safety and efficiency. These challenges emphasize that while hydrogen has immense potential as a clean energy vector, its physical properties impose constraints that demand sophisticated engineering solutions to balance practicality, cost, and efficiency.

Hydrogen's gravimetric versus volumetric density trade-off is central to its role in the energy system. Its lightweight nature makes it an excellent fuel in applications where minimizing mass is essential, but its poor volumetric density requires compression, liquefaction, or advanced material-based storage to achieve practicality. Compared to natural gas and batteries, hydrogen represents both an opportunity and a challenge: it offers superior weight-based efficiency but demands innovative storage strategies to address its volume-based shortcomings. The future of hydrogen as an energy carrier will largely depend on advancements in storage technologies that can reconcile these opposing characteristics.

### 3.2.1 Types of Hydrogen Storage Methods

#### 3.2.1.1 Compressed Hydrogen

Compressed hydrogen is hydrogen gas that is stored under high pressure, typically in the range of 350–700 bar (5,000–10,000 psi), to reduce its volume and facilitate its use as an energy carrier. This method of storage is widely used due to hydrogen's low density at atmospheric pressure, which makes compression essential for practical storage and transportation. Compressed hydrogen is stored in specially designed high-pressure tanks or cylinders made from advanced materials such as carbon–fiber composites to ensure safety and durability. It plays a critical role in hydrogen energy systems, including FCVs, renewable energy storage, and industrial processes. By compressing hydrogen, it can be efficiently transported and stored for later use in generating electricity, powering vehicles, or serving as a feedstock in various chemical applications, thus supporting the transition to cleaner and more sustainable energy systems.

The storage of hydrogen as compressed gas can be modeled using the ideal gas law with considerations for high-pressure conditions:

$$PV = nRT \tag{3.16}$$

where

$P$ = Pressure (Pa) of the compressed hydrogen.
$V$ = Volume ($m^3$) of the storage tank.
$n$ = Number of moles of hydrogen gas stored.
$R$ = Universal gas constant (8.314 J/mol).
$T$ = Temperature (K) inside the tank.

To determine the mass of hydrogen stored ($m$):

$$m = nM$$

where $nM$ is the molar mass of hydrogen ($M$ = 2.016 g/mol).

For higher accuracy, real gas behavior can be accounted for using the compressibility factor $Z$:

$$PV = ZnRT \tag{3.17}$$

where $Z$ corrects deviations from ideal gas behavior under high pressures.

➢ **Chemical Equation for Compressed Hydrogen Storage and Use**

1) **Compression Process**: No chemical reaction occurs during compression; the hydrogen molecules ($H_2$) remain chemically unchanged. It can be represented simply as:

$$H_2\text{(gas at low pressure)} \xrightarrow{\text{Compression}} H_2\text{(gas at high pressure)}$$

2) **Hydrogen Energy Use**: When used in a fuel cell, compressed hydrogen reacts with oxygen to produce water and electricity:

$$2H_2\text{(gas)} + O_2\text{(gas)} \rightarrow 2H_2O\text{(liquid)} + \text{Energy} \tag{3.18}$$

The energy released per mole of hydrogen is calculated based on its enthalpy of combustion:

$$\Delta H = -286\ \text{kJ/mol}$$

This energy is harnessed in fuel cells to generate electricity efficiently.

***Key Features of Compressed Hydrogen Storage***

One of the most defining features of compressed hydrogen storage is the high-pressure levels to which hydrogen gas is subjected in order to achieve usable energy densities. Hydrogen is typically compressed to pressures ranging from 200 to 700 bar (20–70 MPa), depending on the application and storage requirements. At these elevated pressures, the density of hydrogen gas increases significantly, allowing larger quantities of hydrogen to be stored within a limited space. This is particularly important because hydrogen in its natural gaseous state is extremely light and diffuse, making it inefficient to store in uncompressed form. Higher storage pressures enable FCVs and other systems to carry sufficient hydrogen for practical use, thereby bridging the gap between hydrogen's low natural density and the high energy needs of modern applications.

To handle such extreme pressures, storage containers for compressed hydrogen must be designed with advanced engineering and materials. Modern hydrogen storage tanks are usually made from lightweight, high-strength carbon–fiber composites, which are capable of withstanding immense internal pressure while keeping the overall system weight manageable. To enhance durability and prevent hydrogen leakage, these composite tanks are often reinforced with metallic or polymer liners that act as a barrier between the hydrogen gas and the outer structural layers. Such multi-material construction ensures both strength and long-term reliability. Additionally, all compressed hydrogen tanks are manufactured under strict safety standards and certifications, such as those established by ISO and ASME (American Society of Mechanical Engineers). These standards govern not only the design and testing of tanks but also their operation and maintenance, ensuring that they remain safe under demanding conditions.

The applications of compressed hydrogen storage are diverse and rapidly expanding. In the automotive sector, compressed hydrogen is the primary method of storing fuel for FCVs, which are designed to achieve driving ranges comparable to conventional gasoline-powered cars while offering the benefits of zero emissions. The portability, relatively mature technology, and fast refueling times of compressed hydrogen make it a practical choice for passenger cars, buses, and trucks. In industrial applications, compressed hydrogen is used extensively in refineries, fertilizer production, and chemical plants, where hydrogen serves as a critical feedstock for processes such as hydrocracking, ammonia synthesis, and methanol production. Beyond these uses, compressed hydrogen is also playing an increasingly important role in energy storage systems. By converting surplus electricity from renewable sources such as wind and solar into hydrogen through electrolysis and then storing it in compressed form, energy can be effectively stored and later reconverted to electricity or heat when demand rises. This makes compressed hydrogen a key enabler of renewable energy integration and grid stability, supporting the broader transition toward sustainable energy systems. Table 3.4 shows the properties of low-, medium-, high-, and ultra-high-pressure compressed hydrogen. Table 3.5 shows the properties of type-specific storage tanks for compressed hydrogen.

**Table 3.4** Properties of Low-, Medium-, High-, and Ultra-High-Pressure Compressed Hydrogen

| Property | Low Pressure | Medium Pressure | High Pressure | Ultra-High Pressure |
|---|---|---|---|---|
| Pressure range (bar) | <200 | 200–350 | 350–700 | >700 |
| Energy density | Low | Moderate | High | Very high |
| Storage volume | Large | Moderate | Compact | Extremely compact |
| Tank material | Steel or aluminum | Steel with partial reinforcement | Carbon–fiber composites | Advanced carbon–fiber composites |
| Weight | Very heavy | Heavy | Lightweight | Ultra-lightweight |
| Cost | Low | Moderate | High | Very high |
| Safety considerations | Minimal risk at low pressure | Moderate risk | High pressure requires strict safety protocols | Very high risk; advanced safety systems required |
| Applications | Laboratory use, intermediate storage | Industrial processes, backup systems | Fuel cell vehicles, portable hydrogen systems | Aerospace, advanced research |
| Compression energy | Minimal | Moderate | High | Very high |
| Infrastructure | Basic | Moderate | Advanced | Specialized |

***Advantages of Compressed Hydrogen Storage***

One of the greatest strengths of compressed hydrogen storage lies in its status as an established technology. Unlike some newer hydrogen storage methods that are still in the research or pilot stage, compressed hydrogen has been widely studied, tested, and implemented for decades. The technology is mature and well-understood, which makes it a dependable option for a variety of applications. Moreover, infrastructure to support compressed hydrogen, such as HRS, is gradually expanding in certain regions, particularly in countries like Japan, Germany, and the United States, where governments and industries are actively promoting hydrogen-based mobility solutions. This growing infrastructure makes compressed hydrogen more accessible and practical for commercial adoption.

**Table 3.5** Properties of Type-Specific Storage Tanks for Compressed Hydrogen

| Attribute | Type I (All-Metal) | Type II (Hoop-Wrapped Metal) | Type III (Fully Composite with Metal Liner) | Type IV (Fully Composite with Polymer Liner) |
|---|---|---|---|---|
| Material | Steel or aluminum | Steel/ aluminum with fiber wrapping | Metal (e.g., aluminum) with composite wrap | Polymer liner with composite wrap |
| Weight | Very heavy | Heavy | Lightweight | Ultra-lightweight |
| Pressure range (bar) | Up to 200 | 200–350 | 350–700 | >700 |
| Energy density | Low | Moderate | High | Very high |
| Cost | Low | Moderate | High | Very high |
| Durability | High | High | Moderate (metal liner prone to fatigue) | Moderate (polymer liner sensitive to leaks) |
| Safety | Very safe | Safe | Requires advanced safety measures | Requires advanced safety measures |
| Application | Stationary systems, industrial use | Intermediate applications | Fuel cell vehicles, portable hydrogen systems | Aerospace, high-density energy applications |
| Corrosion resistance | Moderate | Moderate | High | Very high |
| Reusability | High | High | Moderate | Moderate |
| Manufacturing complexity | Simple | Moderate | Complex | Very complex |

Another advantage is the ease of use associated with compressed hydrogen systems. Unlike chemical carriers, which require complex processes for hydrogen release, or solid-state methods, which often involve slow kinetics and temperature management, compressed hydrogen storage is relatively straightforward. It involves simply compressing hydrogen gas into high-pressure tanks and dispensing it when needed. This simplicity not only reduces system complexity but also makes it easier to integrate into existing fuel distribution networks, ensuring smooth operation for both industrial and consumer-level applications.

In addition, compressed hydrogen tanks are designed using lightweight composite materials, which improve their performance in mobile applications such as FCVs. Modern storage vessels are typically made of advanced carbon–fiber composites reinforced with liners to ensure durability and leak prevention. By reducing the weight of the tanks while maintaining strength, these innovations help vehicles achieve better efficiency and extended driving ranges. This makes compressed hydrogen particularly suitable for the transportation sector, where every kilogram saved in vehicle weight translates into improved fuel economy and usability.

Another major strength of compressed hydrogen storage is its flexibility across different scales of use. It can be adapted for small-scale applications, such as powering passenger cars, buses, and portable energy units, as well as for large-scale industrial purposes, including supplying refineries, chemical plants, or backup energy storage systems. This versatility allows compressed hydrogen to serve as a bridge technology between mobile, stationary, and industrial energy needs, giving it an edge in terms of adaptability compared to other storage options that may be limited to niche applications.

***Challenges of Compressed Hydrogen Storage***

Despite its benefits, compressed hydrogen storage faces several technical and economic challenges. One of the most significant issues is its energy intensity. Compressing hydrogen to pressures as high as 700 bar requires a substantial amount of energy, often consuming up to 10–15% of the hydrogen's own energy content. This reduces the overall efficiency of hydrogen as an energy carrier and increases the operational costs of storage. For large-scale systems, the energy required for compression can become a significant drawback, limiting the competitiveness of hydrogen compared to other fuels.

Another critical challenge revolves around safety concerns. High-pressure storage systems must be engineered with robust safety measures to prevent leaks or catastrophic failures. Hydrogen itself presents additional challenges because of its small molecular size, which makes it prone to leakage through tiny imperfections or seals, and its high flammability, which creates risks of fire and explosion if leaks occur in confined spaces. These risks require advanced engineering, continuous monitoring, and adherence to strict international safety standards, all of which add complexity and cost to compressed hydrogen storage solutions.

Compressed hydrogen storage also suffers from a low volumetric density compared to conventional liquid fuels. Even when compressed to 700 bar, hydrogen gas occupies a much larger volume than gasoline or diesel for the same energy content. This means that hydrogen-powered vehicles require larger tanks to achieve comparable driving ranges, and stationary storage facilities need significant space to store meaningful amounts of energy. This fundamental limitation of hydrogen gas makes it less practical in applications where space is at a premium, unless combined with other innovations.

Finally, the **cost** of compressed hydrogen storage is another major hurdle. The manufacture of high-strength carbon–fiber composite tanks is both technologically demanding and expensive, contributing significantly to the overall cost of hydrogen systems. While economies of scale and advances in materials science are gradually reducing these costs, compressed hydrogen storage remains less economically competitive than fossil fuels or even some alternative hydrogen storage methods. Until costs decrease further, widespread adoption will remain limited to regions and industries with strong policy support and investment.

***Innovations in Compressed Hydrogen Storage***

To overcome these challenges, several innovations are being explored. One promising direction is the development of improved materials for hydrogen tanks. Research into nano-engineered composites and advanced polymers aims to reduce tank weight and manufacturing costs while simultaneously enhancing durability and resistance to hydrogen leakage. These innovations could make tanks safer, lighter, and more affordable, enabling broader deployment in mobile and stationary applications. Another area of advancement is in optimized compression technologies. Engineers are developing more efficient compression systems that consume less energy while achieving the same or higher pressure levels. For example, advanced multi-stage compressors with heat recovery systems can lower the energy penalty associated with hydrogen compression, improving the overall efficiency of the storage process.

In addition, compressed hydrogen storage is increasingly being integrated with renewable energy systems. On-site hydrogen production using electrolysis powered by solar or wind energy, combined with immediate compression and storage, minimizes the need for transportation and reduces overall system costs. Such localized hydrogen ecosystems improve efficiency, enhance energy security, and reduce the carbon footprint of hydrogen production and storage. By connecting compressed hydrogen directly to renewables, these integrated systems strengthen the role of hydrogen as a clean energy carrier in the global shift toward sustainable energy solutions.

Compressed hydrogen storage represents a critical technology in the hydrogen economy, providing a balance between practicality, efficiency, and safety. Advances in materials, engineering, and infrastructure continue to enhance its feasibility, particularly for applications requiring mobile and flexible energy solutions.

#### 3.2.1.2 Liquid Hydrogen

Hydrogen is cooled to cryogenic temperatures (−253 °C) and stored as a liquid. This method provides high energy density but is energy-intensive due to cooling requirements. $LH_2$ storage involves cooling hydrogen gas to cryogenic temperatures of **−253 °C (−423 °F)**, at which point it becomes a liquid. This method provides a high energy density and is especially suitable for applications where space is limited, such as in aerospace or large-scale transportation systems. Figure 3.5 illustrates a schematic diagram of the process of $LH_2$.

1) ***Density of*** $\boldsymbol{LH_2}$

The density of liquid hydrogen $\rho LH_2$ is approximately 70.85 kg:

$$m = \rho LH_2 \cdot V$$

2) ***Energy Density***

The gravimetric energy density ($E_g$) and volumetric energy density ($E_v$) for $LH_2$ are calculated as:

$$E_g = \frac{\text{Energy per unit mass of hydrogen}}{\text{Mass of hydrogen}} \tag{3.19}$$

$$E_g = \frac{\text{Energy per unit volume of hydrogen}}{\text{Volume of hydrogen}} \tag{3.20}$$

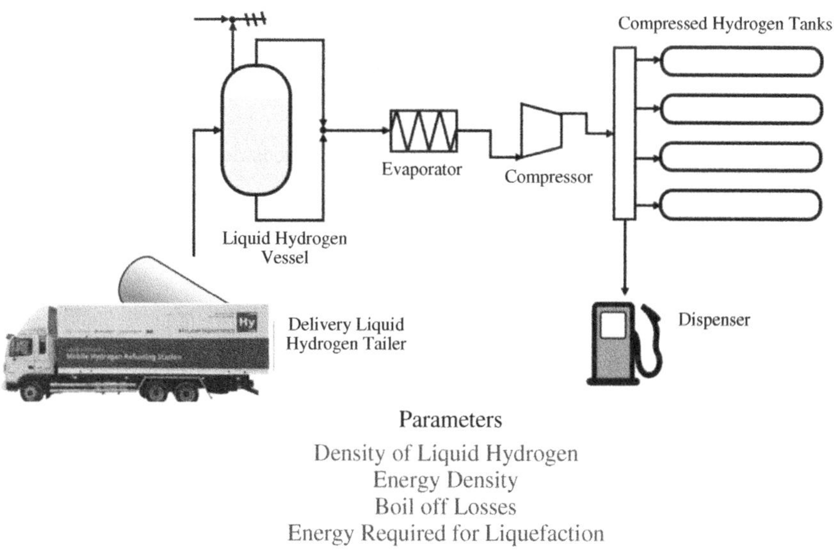


**Figure 3.5** Schematic Diagram of Process of Liquid Hydrogen

For hydrogen:

Lower heating value (LHV): 120 MJ/kg
Gravimetric energy density: 120 MJ/kg
Volumetric energy density: 8.5 MJ/L for $LH_2$.

3) ***Boil-Off Losses***

Cryogenic storage incurs **boil-off losses** due to heat transfer from the surroundings. The rate of boil-off ($R_b$) can be estimated using:

$$R_b = \frac{Q}{L_v.m} \tag{3.21}$$

where

$Q$: Heat influx (J/s) into the storage tank.
$L_v$: Latent heat of vaporization of hydrogen (446 kJ).
$m$: Mass of hydrogen

4) ***Energy Required for Liquefaction***

The energy required to liquefy hydrogen ($E_l$) from gaseous form is significant and is expressed as:

$$E_l = \text{Specific energy of liquefaction} \times \text{Mass of hydrogen}$$

For hydrogen, the specific energy of liquefaction is approximately 10–13 kWh/kg.

5) ***Tank Insulation and Heat Transfer***

The heat transfer rate into a cryogenic storage tank is determined using:

$$Q = \frac{kA(T_{\text{outside}} - T_{\text{inside}})}{d} \tag{3.22}$$

where

- $k$: Thermal conductivity of the insulation material.
- $A$: Surface area of the tank.
- $T_{\text{outside}}$, $T_{\text{inside}}$: Temperatures outside and inside the tank.
- $d$: Thickness of the insulation layer.

***How Liquid Hydrogen Storage Works***

$LH_2$ storage relies on a process known as cryogenic cooling, where hydrogen gas is cooled using advanced refrigeration systems until it reaches its boiling point at atmospheric pressure, which is approximately 20 K (−253 °C). At this ultra-low temperature, hydrogen transitions from a gaseous to a liquid state, significantly increasing its density and making it more practical to store large quantities in compact spaces. To maintain these extreme conditions, cryogenic insulation is employed, which minimizes heat transfer from the external environment and helps preserve the liquid state of hydrogen over extended periods. This step is critical, as even small amounts of heat leakage can lead to vaporization losses, commonly referred to as boil-off.

The storage of $LH_2$ requires specially designed cryogenic tanks, which are generally double-walled and vacuum-insulated to minimize thermal conduction, convection, and radiation. These tanks are often constructed using materials with very low thermal conductivity, such as stainless steel, aluminum alloys, and specialized composite structures. The design of these tanks is optimized to withstand both the low temperatures and the potential pressure fluctuations associated with $LH_2$ storage. The vacuum between the walls of the tank acts as a barrier to heat transfer, while reflective layers and multi-layer insulation further enhance the tank's ability to retain hydrogen in its liquid state. Ensuring safety and minimizing boil-off losses are central challenges in this engineering process.

In terms of applications, $LH_2$ plays a vital role across several industries due to its unique properties. In the aerospace sector, it has long been used as a rocket propellant, valued for its extremely high energy density and lightweight characteristics, which are essential for space exploration missions. The transportation industry is increasingly exploring $LH_2$ as a fuel option for hydrogen-powered ships, trains, and long-haul trucks, where its high volumetric energy density provides a practical solution for extended travel ranges. Moreover, in the field of energy storage, $LH_2$ offers great potential for large-scale storage systems that can integrate renewable energy sources such as solar and wind. By storing excess renewable energy in the form of $LH_2$, it becomes possible to stabilize the power grid, ensure supply during periods of low renewable generation, and advance the development of a clean and sustainable energy economy. Table 3.6 shows the types and attributes of $LH_2$.

### *Advantages of Liquid Hydrogen Storage*

One of the primary advantages of $LH_2$ storage is its high energy density. Compared to compressed hydrogen gas, $LH_2$ offers a much greater volumetric energy density, meaning that significantly more hydrogen can be stored in the same physical space. This characteristic is particularly important for applications where energy-to-volume ratio is critical, such as in space exploration or long-distance transportation. By storing hydrogen in liquid form, it becomes possible to power demanding systems without the need for excessively large storage tanks, which would otherwise be impractical or impossible to integrate.

**Table 3.6** Types and Attributes of Liquid Hydrogen

| Type of Liquid Hydrogen | Attributes |
|---|---|
| Normal liquid hydrogen | – Consists of a 75 : 25 ratio of ortho-hydrogen to para-hydrogen<br>– Boiling point: 20.28 K (−252.87 °C)<br>– Density: ~70.85 $kg/m^3$ at boiling point<br>– Used in cryogenics, rockets, and cooling systems |
| Para-hydrogen | – Contains hydrogen molecules in the lower energy singlet state<br>– Preferred state at very low temperatures (close to absolute zero)<br>– Conversion from ortho to para is exothermic, releasing heat<br>– Essential for efficient storage of liquid hydrogen in cryogenic tanks |
| Ortho-hydrogen | – Contains hydrogen molecules in the higher energy triplet state<br>– More stable at higher temperatures<br>– Spontaneously converts to para-hydrogen over time at cryogenic temperatures, generating heat<br>– Exists in larger amounts in normal hydrogen gas at room temperature |
| Slush hydrogen | – A mix of solid and liquid hydrogen<br>– Created by partially freezing liquid hydrogen.<br>– Denser than normal liquid hydrogen (~81 $kg/m^3$).<br>– Suitable for specific applications requiring increased density and cooling capacity. |
| Metallic hydrogen (hypothetical) | – High-pressure phase of hydrogen, theorized to occur under extreme conditions (e.g., in gas giant cores)<br>– Predicted to have unique properties like superconductivity<br>– Does not exist in liquid state under standard storage conditions but a subject of ongoing research |

Another major benefit is the reduction in storage space that $LH_2$ enables. Because liquefaction allows hydrogen molecules to be packed much more closely together than in gaseous form, the storage tanks required can be much smaller and more compact. This is particularly advantageous in aerospace applications, where space and weight savings are essential, as well as in transportation vehicles like hydrogen-powered trucks, trains, and even ships. The ability to store large amounts of energy in smaller volumes helps reduce the weight burden on vehicles and increases their efficiency, range, and practicality.

$LH_2$ also improves transport efficiency over long distances. In its gaseous form, hydrogen has a very low density, which makes pipeline transport or large-scale storage challenging and costly. By liquefying hydrogen, the volume it occupies is drastically reduced, allowing for more economical transport using specialized cryogenic tankers or ships. This makes $LH_2$ an attractive option for international hydrogen trade, enabling energy-rich countries to export hydrogen in a more compact and cost-effective form to regions with high energy demand.

***Challenges of Liquid Hydrogen Storage***

Despite its benefits, $LH_2$ storage comes with significant challenges, one of the foremost being energy-intensive cooling. The liquefaction process itself is highly energy-demanding, consuming nearly 30–40% of the hydrogen's own energy content. This means that a substantial portion of the potential energy is lost simply in the process of cooling hydrogen gas down to its boiling point of 20 K (−253 °C). This inefficiency represents a major technological and economic barrier to the large-scale adoption of $LH_2$ as a storage and transport medium.

Another challenge is associated with boil-off losses. Even when stored in advanced double-walled, vacuum-insulated tanks, some amount of heat transfer inevitably occurs over time. This causes portions of the $LH_2$ to re-gasify and escape in what is known as "boil-off." While boil-off rates can be minimized using state-of-the-art insulation techniques, they cannot be completely eliminated, leading to long-term energy losses. These losses are especially problematic in applications where hydrogen must be stored for extended periods, such as during shipping or long-term backup energy storage.

Safety concerns also play a critical role in the handling and storage of $LH_2$. At such extremely low cryogenic temperatures, even brief exposure can cause severe cryogenic burns or material embrittlement. Additionally, hydrogen itself is highly flammable and forms explosive mixtures with air even at very low concentrations. Handling cryogenic liquids requires highly trained personnel, robust safety protocols, and specialized equipment to prevent accidents such as leaks, explosions, or material failures. These safety requirements add complexity and cost to $LH_2$ storage systems.

The cost factor represents another barrier to widespread adoption. The infrastructure needed to produce, liquefy, store, and transport hydrogen in its liquid form is highly specialized and expensive. From advanced liquefaction plants and cryogenic tankers to double-walled insulated storage systems, the capital and operational expenses are considerably higher than those associated with compressed hydrogen. Until significant technological improvements and economies of scale are achieved, these costs will continue to limit the commercial feasibility of $LH_2$ in many industries.

***Emerging Innovations in Liquid Hydrogen Storage***

To address these challenges, researchers and industries are exploring several emerging innovations in $LH_2$ storage. One promising area is the development of advanced insulation materials with superior thermal resistance. By minimizing heat transfer into the storage tanks, these innovations significantly reduce boil-off losses, enabling longer storage times and improving efficiency. Advances in nanomaterials, aerogels, and multi-layer reflective barriers are showing potential in enhancing insulation performance while keeping tanks lightweight.

Another area of innovation lies in cryogenic tank design. Traditional $LH_2$ tanks tend to be heavy and expensive, but recent developments in composite materials and lightweight alloys are enabling safer, more durable, and more cost-effective storage solutions. By reducing the overall weight and improving structural resilience, these new designs are particularly attractive for transportation and aerospace applications, where both safety and efficiency are crucial. Researchers are also investigating hybrid storage systems, which combine the advantages of $LH_2$ with other storage techniques such as high-pressure compression. These systems aim to maximize storage flexibility, minimize boil-off, and allow hydrogen to be stored under conditions that balance efficiency with practicality. For instance, cryo-compressed hydrogen ($CcH_2$) storage, which combines cryogenic cooling with moderate compression, is emerging as a promising alternative that offers higher densities and lower losses compared to conventional $LH_2$.

Finally, the push toward green liquefaction is transforming the environmental footprint of $LH_2$. By integrating renewable energy sources such as solar, wind, or hydroelectric power into the hydrogen liquefaction process, it becomes possible to reduce the carbon emissions associated with producing $LH_2$. This innovation not only makes the storage process more sustainable but also strengthens the $LH_2$ role as a clean energy carrier in the global transition toward carbon-neutral energy systems. $LH_2$ storage is critical for enabling hydrogen's role in energy-intensive applications. While technological and economic challenges remain, continued advancements in cryogenics and infrastructure are paving the way for broader adoption.

> ***Key: Clean hydrogen:*** *Hydrogen produced using renewable energy sources (like solar or wind) through water electrolysis, resulting in negligible carbon emissions.*

#### 3.2.1.3 Solid-State Storage

Solid-state hydrogen storage represents one of the most promising and advanced alternatives to conventional compressed and $LH_2$ methods, as it utilizes the inherent ability of certain solid materials to absorb or adsorb hydrogen within their structures in a safe and compact manner. Unlike gaseous storage at high pressures or cryogenic storage at extremely low temperatures, solid-state storage integrates hydrogen directly into the host material through either absorption, where hydrogen atoms penetrate the bulk structure of the material and form stable compounds, or adsorption, where hydrogen molecules attach to the surface of porous materials. This dual mechanism not only reduces safety concerns associated with leaks and explosions but also offers enhanced volumetric storage density, making it particularly suitable for both stationary energy storage and portable applications. Among the most widely researched storage media are metal hydrides, chemical hydrides, and porous materials such as carbon nanotubes or metal–organic frameworks (MOFs), each with unique advantages and challenges. Metal hydrides, such as magnesium hydride or titanium hydride, operate by absorbing hydrogen into the crystal lattice of metals or alloys, forming stable hydride phases that can later release hydrogen upon heating or pressure reduction, thereby offering reversibility and high storage capacity, though often limited by slow kinetics and high desorption temperatures. Chemical hydrides, on the other hand, store hydrogen in chemical bonds within compounds like sodium borohydride or ammonia borane, which liberate hydrogen through chemical reactions triggered by heat or the addition of water, providing high storage density but with challenges in recyclability and regeneration of spent material. Porous materials like zeolites, carbon nanotubes, and MOFs function by adsorbing hydrogen molecules on their extensive surface areas and pore structures, making them lightweight and efficient at low temperatures, though their storage capacity at ambient conditions still requires significant improvement to meet Department of Energy (DOE) targets. The combination of these materials and mechanisms demonstrates the versatility of solid-state storage, balancing factors such as energy density, reversibility, safety, and material cost. Research continues to optimize these systems through nanostructuring, catalyst integration, and hybrid approaches that combine different storage materials to overcome limitations of kinetics, operating conditions, and hydrogen release rates. Overall, solid-state hydrogen storage stands at the forefront of next-generation energy solutions, bridging the gap between high-density storage, safety, and practical applicability for hydrogen-powered transportation, renewable energy buffering, and portable power systems. The general equation for the solid-state hydrogen storage can be represented as:

$$M + x\mathrm{H}_2 \leftrightarrow M\mathrm{H}_{2x} + \Delta H \tag{3.23}$$

where

- $M$: Solid storage material (e.g., metal hydride).
- $\mathrm{H}_2$: Hydrogen gas.
- $M\mathrm{H}_{2x}$: Hydrogen-stored solid compound.
- $\Delta H$: Enthalpy change of the reaction (storage or release process).

The relation between pressure ($P$), temperature ($T$), and hydrogen absorbed ($x$) is described using the van't Hoff equation:

$$\ln P = \frac{\Delta H}{RT} - \frac{\Delta S}{R} \tag{3.24}$$

where

- $P$: Equilibrium pressure of hydrogen gas.
- $\Delta H$: Enthalpy change of absorption or desorption (J/mol).
- $R$: Universal gas constant (8.314 J/(mol·K)).
- $T$: Temperature (K).
- $\Delta S$: Entropy change of the reaction (J/(mol·K)).

The storage capacity ($C$) of a material can be expressed as:

$$C = \frac{m_{H_2}}{m_{material}} \times 100 \tag{3.25}$$

where

- $C$: Storage capacity in weight percentage (wt%).
- $m_{H_2}$: Mass of hydrogen absorbed (kg).
- $m_{material}$: Mass of the storage material (kg).

Table 3.7 shows the types of storage and their attributes.

**Table 3.7** Types of Storage and their Attributes

| Type of Storage | Attributes |
|---|---|
| Metal hydrides | – Composed of metals or alloys that absorb hydrogen to form hydrides<br>– High volumetric hydrogen density<br>– Operates at moderate temperatures and pressures<br>– Examples: $LaNi_5$, TiFe, $MgH_2$<br>– Challenges: Heavy materials, slow kinetics, high desorption temperatures for some hydrides |
| Complex hydrides | – Composed of lightweight hydrides such as borohydrides (e.g., NaBH)<br>– High gravimetric hydrogen density<br>– Requires catalysts for efficient hydrogen release<br>– Challenges: High operating temperatures, regeneration complexity, and cost |

**Table 3.7** (Continued)

| Type of Storage | Attributes |
|---|---|
| Ammonia borane | – High hydrogen storage capacity (~19.6 wt%)<br>– Releases hydrogen through thermal decomposition<br>– Challenges: Complex hydrogen release pathways, material stability, and by-product handling |
| Carbon-based materials | – Includes carbon nanotubes, graphene, and activated carbon<br>– Hydrogen stored through physisorption (weak van der Waals forces)<br>– Operates at low temperatures (e.g., liquid nitrogen temperatures)<br>– Advantages: Lightweight, scalable<br>– Challenges: Low storage capacity under ambient conditions |
| Metal-organic frameworks (MOFs) | – Porous crystalline structures made of metal nodes and organic linkers<br>– High surface area and tunable pore sizes for hydrogen storage<br>– Hydrogen stored through physisorption<br>– Operates at cryogenic temperatures for optimal performance<br>– Challenges: Cost, stability, and limited capacity at room temperature |
| Chemical hydrogen storage | – Includes reversible chemical compounds like formic acid or liquid organic hydrogen carriers (LOHCs)<br>– Hydrogen released through catalytic reactions<br>– Operates at low to moderate temperatures<br>– Advantages: Safe, high energy density<br>– Challenges: Regeneration of spent material can be energy-intensive. |

***Advantages of Solid-State Hydrogen Storage***

One of the most significant advantages of solid-state hydrogen storage lies in safety. Unlike compressed hydrogen gas or cryogenic $LH_2$, solid-state systems store hydrogen in a nongaseous state, eliminating the hazards associated with high pressures or extremely low temperatures. This makes the storage system inherently safer, as the probability of accidental leakage or catastrophic release is much lower compared to conventional methods.

Another benefit is the compact storage capability. Solid-state materials often exhibit high volumetric density, meaning they can store a substantial amount of hydrogen within a relatively small space. This compactness provides an advantage over gaseous hydrogen storage, where large pressurized tanks are required, making solid-state storage suitable for size- and space-sensitive applications. Solid-state hydrogen systems also offer excellent stability. Once hydrogen is absorbed within the solid matrix, it can remain there for extended periods without significant losses, unlike $LH_2$, which suffers from boil-off losses. This long-term stability ensures reliability for stationary storage applications where hydrogen may need to be stored for months at a time.

Lastly, modularity is a key strength. Solid-state storage units can be designed for small-scale and portable applications, making them particularly useful in powering fuel cells for electronics, such as laptops, portable generators, or small vehicles. Their ability to scale down while maintaining efficiency gives them versatility unmatched by other storage methods.

#### *Challenges of Solid-State Hydrogen Storage*

Despite these advantages, several challenges limit widespread adoption. The foremost challenge is material weight. Many solid-state storage materials, such as metal hydrides, are relatively heavy. While they offer good volumetric density, their high weight lowers the gravimetric density (hydrogen stored per unit mass of system), which reduces their attractiveness for lightweight applications like automotive systems. Another drawback is slow kinetics. The processes of hydrogen absorption and desorption in solid-state materials can be sluggish, meaning it takes a longer time to charge or discharge hydrogen compared to gas or liquid systems. This issue necessitates optimization for real-world performance, especially in applications demanding rapid refueling.

The energy requirement is also a concern. To release hydrogen stored within solid matrices, additional heat energy is often necessary. This can reduce overall system efficiency and complicate system design by requiring integrated heating systems. Finally, cost remains a substantial barrier. Advanced solid-state storage materials such as MOFs and high-performance hydrides are expensive to produce. Scaling up these materials to industrial levels is still an ongoing challenge, limiting their commercial viability at present.

#### *Innovations in Solid-State Hydrogen Storage*

Researchers are actively working on overcoming these barriers. One key direction is the development of lightweight alloys. Alloys based on lithium, magnesium, and aluminum are under study for their ability to store hydrogen while maintaining relatively low system weight, thereby improving gravimetric storage capacity. Advances in nanostructured materials are also proving promising. By leveraging nanotechnology, scientists enhance the surface area and diffusion pathways of storage media, improving both capacity and kinetics. Materials like carbon nanotubes and MOFs are being engineered at the nanoscale to allow faster hydrogen uptake and release.

In addition, hybrid systems are emerging, where solid-state materials are combined with other storage technologies such as compressed gas to maximize flexibility and efficiency. This integration allows for systems that can balance gravimetric and volumetric density while ensuring reliable performance. Another significant innovation is the integration of catalysts. Catalysts can accelerate the hydrogen absorption and desorption reactions, addressing one of the key limitations of slow kinetics. With the use of advanced catalytic systems, refueling times and energy costs for hydrogen release can be dramatically reduced.

***Applications of Solid-State Hydrogen Storage***

The applications of solid-state hydrogen storage are broad and impactful. In the realm of portable electronics, compact storage systems paired with miniature fuel cells can provide long-lasting power for devices such as smartphones, laptops, and military equipment, reducing reliance on batteries. In transportation, solid-state hydrogen has potential in specialized vehicles where safety and compactness are paramount. For instance, submarines, aerospace vehicles, and even passenger cars may benefit from the reduced risk of leakage and efficient use of space.

Solid-state systems are also valuable in stationary storage applications, particularly for renewable energy integration. They can store excess electricity from solar or wind power in the form of hydrogen and later release it through fuel cells, thereby providing backup power, grid stabilization, and enhanced energy security for buildings and communities.

➢ **Solid-State Hydrogen Storage Materials**

1) **Interstitial Metal Hydrides (AB_5, AB_2, A–B and BCC Alloys, Mg-Based)**

   These are classic "workhorse" materials where hydrogen occupies interstitial sites in a metallic or intermetallic lattice.

   a) **AB_5 ($LaNi_5$-Type)**
      - Why It's Used: Near-room-temperature operation, flat plateau pressure (easy pressure regulation), excellent cycling stability; long used in NiMH batteries and small $H_2$ buffers.
      - Pros: Good volumetric density, robust kinetics, tolerant to practical handling.
      - Cons: Low gravimetric capacity due to heavy rare-earth and Ni content; material cost.

b) **A–B (TiFe)**
- Why It's Interesting: Similar operating window to AB_5 but cheaper chemistry.
- Challenges: Needs activation (surface cleaning by vacuum/heat or small $O_2$ getters) before it will fully absorb; sensitive to impurities ($H_2O$, $O_2$, CO).

c) **AB_2 (Laves Phases Like Ti–Zr–V–Cr) and BCC Alloys (Ti–V–Cr, V–Ti–Fe)**
- Why They Matter: Higher capacities and tunable plateau pressures via composition. Body-centered cubic (BCC) alloys show rapid kinetics because H diffuses quickly in the open BCC lattice.
- Trade-offs: Hysteresis between absorption/desorption plateaus, sensitivity to impurities, and cost/weight of V-rich chemistries.

d) **Mg-Based Hydrides ($MgH_2$, Mg Alloys)**
- Upside: Very high gravimetric potential (light Mg), excellent volumetric density.
- Headache: Strong Mg—H bonds → high desorption temperature and sluggish kinetics.
- What Helps: Ball-milling/nanostructuring (shorter diffusion paths), catalysts (Ti, $Nb_2O_5$, V, Ni), and destabilization via alloying or reactive hydride composites (e.g., $MgH_2$ + Si/Al or $MgH_2$ + $LiBH_4$) to lower enthalpy and speed release.

2) **Complex/Chemical Hydrides (Alanates, Borohydrides, Amide–Imide Systems, AB)**

These bind hydrogens within **polyatomic anions** or molecular complexes.

a) **Alanates (e.g., $NaAlH_4$)**
- Story: A landmark system once Ti-doped to enable reversible cycling.
- Operation: Multi-step dehydrogenation around ~120–180 °C (composition-dependent) with decent capacities.
- Limits: Thermal management and kinetics; some capacity loss over many cycles.

b) **Borohydrides ($LiBH_4$, $Mg(BH_4)_2$, $NaBH_4$)**
- Upside: Exceptional gravimetric capacity on paper.
- Reality Check: Many require high temperatures for $H_2$ release and/or suffer irreversibility; off-board regeneration (chemically recharging spent fuel) is often needed fine for cartridges or stationary systems, not ideal for onboard closed-loop.

c) **Amide–Imide Systems (Li–Mg–N–H, etc.)**
- Idea: Couple multiple light elements to tailor thermodynamics; some operate at lower temperatures than simple borohydrides.
- Challenge: Reaction complexity, gas purity management (avoid $NH_3$ traces), and long-term cycling.

d) **Ammonia Borane (AB, $NH_3BH_3$)**
- What It Offers: High hydrogen content in a small molecule.
- Caveats: Typically, nonreversible on board; releases by thermolysis or hydrolysis (if water is added). Excellent for portable, single-use cartridges where spent fuel can be centrally regenerated.

3) **Porous Adsorbents (Carbons, Zeolites, MOFs/COFs)—Physisorption**

These materials store $H_2$ on vast internal surfaces; capacity scales with surface area and pore size distribution.

a) **Activated Carbons and Template Carbons**
- Pros: Cheap, conductive (good heat removal), mechanically robust pellets.
- Cons: At room temperature, capacities are modest; they shine at cryogenic (~77 K) "cryo-adsorption" with moderate pressures.

b) **Zeolites**
- Pros: Rigid frameworks, well-defined pores, scalable.
- Cons: Generally lower $H_2$ uptake than high-surface area carbons or MOFs.

c) **MOFs/COFs (Covalent–Organic Frameworks)**
- Why Exciting: Record-high surface areas, tunable open metal sites that strengthen $H_2$ binding, and crystalline porosity that can be engineered for optimal pore size.
- Reality: Best performance is typically at 77 K; at ambient temperature, system-level capacities struggle to meet stringent automotive targets unless combined with precooling or hybridization (e.g., adsorbent + moderate compression).
- Engineering: Densification (pellets/monoliths) to raise volumetric capacity without choking mass transfer; binder choice matters for stability and thermal conduction.

4) **Cross-Cutting Engineering Issues**
- Thermal Management: Hydriding is exothermic; without good heat removal, beds "stall." Dehydriding needs heat input integrate heat exchangers, fins, foams, graphite.
- Kinetics and Catalysts: Surface catalysts (Ni, Pd, Pt, Ti-based) accelerate $H_2$ dissociation; nanostructuring shortens diffusion paths. Ball-milling and core–shell particles are common.
- Hysteresis and Cycling: Many metal hydrides show pressure hysteresis; repeated cycling can cause pulverization/decrepitation (helpful for kinetics but increases pressure drop/dust). Additives and compaction help.
- Impurity Tolerance: Even ppm-level $H_2O$, $O_2$, CO, $NH_3$ can poison or react; gas cleanup and getters are standard.
- Form Factor and Safety: Powders are typically pelletized with conductive, inert binders; some complex hydrides are air/moisture-sensitive and need sealed handling.

5) **How to Choose a Material for an Application?**
   - Portable/One-Way Cartridges: Chemical hydrides (e.g., AB, $NaBH_4$) or adsorbents with small coolers—prioritize energy per device and simplicity; accept off-board regeneration.
   - Stationary Buffers/Renewables: AB_5 / TiFe / AB_2 tanks—durable, easy thermal integration, predictable plateau pressures.
   - Weight-Sensitive Platforms: Push Mg-based or BCC alloys with strong catalysis + nanostructuring, or adsorbents with precooling; consider hybrid tanks (moderate compression + adsorbent) to balance gravimetric/volumetric needs.

At the end, solid-state hydrogen storage represents a promising frontier in the hydrogen economy. Its unique combination of safety, stability, and compactness offers a strong alternative to compressed and $LH_2$ systems. Although challenges such as material weight, slow kinetics, energy requirements, and cost remain, continuous innovations in lightweight alloys, nanostructured materials, hybrid systems, and catalytic integration are pushing the field forward. With ongoing research and technological improvements, solid-state hydrogen storage is poised to play a significant role in enabling a cleaner, safer, and more efficient hydrogen-based energy system across portable, transportation, and stationary domains.

#### 3.2.1.4 Chemical Storage

Hydrogen is stored in chemical compounds (e.g., ammonia, formic acid) and released when needed through chemical reactions. Chemical storage of hydrogen involves storing hydrogen in the form of chemical compounds, which release hydrogen through specific reactions, typically under controlled conditions. This method offers a high-density and safe way to store hydrogen, making it a practical choice for certain applications.

##### *How Chemical Hydrogen Storage Works*

Chemical hydrogen storage relies on compounds that either contain or can produce hydrogen. These compounds can be broadly categorized into:

1) **Chemical Hydrides**:
   - Examples: Ammonia borane ($NH_3BH_3$), sodium borohydride ($NaBH_4$), lithium hydride (LiH).
   - These compounds react with water, acids, or other chemicals to release hydrogen.

2) **Liquid Organic Hydrogen Carriers (LOHCs):**
   - Examples: Toluene-methylcyclohexane system, dibenzyltoluene.
   - LOHCs absorb hydrogen through catalytic hydrogenation and release it through dehydrogenation processes.
3) **Ammonia ($NH_3$):**
   - A hydrogen-rich chemical that can be cracked to release hydrogen and nitrogen using catalysts.
4) **Methanol:**
   - Hydrogen can be extracted from methanol through reforming processes involving heat and catalysts.

Chemical hydrogen storage involves reversible chemical reactions where hydrogen is stored in chemical compounds and released via catalytic or thermal processes. The equations capture the hydrogenation (storage) and dehydrogenation (release) processes. Table 3.8 illustrates attributes of different types of chemical hydrogen storage.

**Table 3.8** Attributes of Different Types of Chemical Hydrogen Storage

| Type of Chemical Hydrogen Storage | Attributes |
|---|---|
| Formic acid (HCOOH) | – Hydrogen stored in liquid form, easily transportable<br>– Releases $H_2$ through catalytic decomposition<br>– Moderate operating conditions (~80–100 °C)<br>– Advantages: High stability, nontoxic, and liquid phase for easy handling<br>– Challenges: Requires efficient catalyst, $CO_2$ by-product must be managed for sustainability |
| Ammonia borane ($NH_3BH_3$) | – Solid compound with high hydrogen content (~19.6 wt%)<br>– Hydrogen released via thermal decomposition in multiple steps<br>– Operates at higher temperatures (~120–200 °C)<br>– Advantages: High hydrogen density, lightweight.<br>– Challenges: Regeneration is energy-intensive, potential ammonia release. |

*(Continued)*

**Table 3.8** (Continued)

| Type of Chemical Hydrogen Storage | Attributes |
|---|---|
| Liquid organic hydrogen carriers (LOHCs) | – Hydrogen stored in reversible liquid compounds (e.g., toluene/methylcyclohexane system).<br>– Hydrogen released through catalytic dehydrogenation.<br>– Operates at temperatures ~200–300 °C<br>– Advantages: Liquid state ensures safe and easy transport and storage.<br>– Challenges: Requires efficient catalyst and significant energy input for regeneration |
| Methanol ($CH_3OH$) | – Hydrogen stored in liquid methanol, released via reforming reaction<br>– Reaction: $CH_3OH + H_2O \rightarrow CO_2 + 3H_2$<br>– Operates at moderate temperatures (~200–250 °C)<br>– Advantages: High energy density, infrastructure compatibility<br>– Challenges: $CO_2$ by-product, complex reaction mechanism |
| Sodium borohydride ($NaBH_4$) | – Hydrogen released upon reaction with water: $NaBH_4 + 2H_2O \rightarrow 4H_2 + NaBO_2$<br>– Operates at ambient conditions with a catalyst<br>– Advantages: High hydrogen yield, stable under dry conditions<br>– Challenges: Regeneration of spent fuel ($NaBO_2$) is energy-intensive. |
| Metal ammonia solutions | – Hydrogen stored in the form of ammonia (NH)<br>– Decomposed to release hydrogen: $NH_3 \rightarrow N_2 + 3H_2$<br>– Operates at ~500–700 °C for decomposition.<br>– Advantages: High storage density, existing transport infrastructure.<br>– Challenges: High temperature required, potential ammonia leakage concerns |

***General Chemical Reaction Equation***

The reversible reaction for chemical hydrogen storage can be expressed as:

$$\text{Compound}_{\text{hydrogen-poor}} + x\text{H}_2 \leftrightarrow \text{Compound}_{\text{hydrogen-rich}} + \Delta H$$

where

$\text{Compound}_{\text{hydrogen-poor}}$: Dehydrogenated compound.
$H_2$: Hydrogen gas.
$\text{Compound}_{\text{hydrogen-rich}}$: Hydrogenated compound.
$\Delta H$: Enthalpy change of the reaction (J/mol).

1) **Formic Acid (HCOOH):**
   - Hydrogen release via catalytic decomposition: $HCOOH \rightarrow H_2 + CO_2$
   - Hydrogenation (regeneration): $CO_2 + H_2 \rightarrow HCOOH$
2) **Ammonia Borane ($NH_3BH_3$):**
   - Dehydrogenation via thermal decomposition: $NH_3BH_3 \rightarrow NH_2BH_2 + H_2$ (first step)
   - $NH_2BH_2 \rightarrow NHBH + H_2$ (second step)
   - Full decomposition releases three moles of $H_2$.
3) **Liquid Organic Hydrogen Carriers (LOHCs):**
   - General reaction for hydrogenation and dehydrogenation:

$$\text{LOHC}_{\text{dehydrogenated}} + x\text{H}_2 \leftrightarrow \text{LOHC}_{\text{hydrogenated}}$$

***Advantages of Chemical Hydrogen Storage***

One of the main advantages of chemical hydrogen storage lies in its high hydrogen density, both gravimetric and volumetric, compared to compressed or $LH_2$ systems. Chemical storage methods, such as using LOHCs, ammonia, formic acid, or ammonia borane, can pack hydrogen in a chemically bonded form, which allows more hydrogen to be stored in the same volume or weight of material. This makes chemical carriers particularly attractive for applications that demand compact storage, such as aviation, shipping, or portable power systems.

Another key benefit is that many chemical hydrogen storage systems enable room-temperature storage, eliminating the need for either extreme compression (200–700 bar) or cryogenic cooling (down to −253 °C). This not only reduces the overall energy requirements of storage but also simplifies the infrastructure needed to handle hydrogen, making the process more cost-effective and practical. Because chemical carriers remain stable under normal ambient conditions, they are safer and easier to integrate into existing supply chains. Safety is another important advantage of chemical storage. Unlike compressed or $LH_2$, which pose risks of leakage and flammability due to hydrogen's small molecular size and volatility, chemical hydrogen storage systems keep hydrogen in a stable molecular form until it is needed. This stability significantly reduces the risks associated with accidental leaks, fires, or explosions, thereby making chemical carriers more appealing in sensitive applications where safety is paramount.

Chemical carriers also offer superior transportability. Substances like ammonia or LOHCs are liquids at ambient conditions and can be transported using existing infrastructure already in place for conventional fuels, such as pipelines, tankers, and fueling stations. This reusability of infrastructure reduces the cost of large-scale adoption and allows hydrogen to be distributed in a manner similar to current fossil fuels, making it easier to integrate into global energy markets.

Finally, chemical storage systems excel in long-term storage. Unlike compressed hydrogen, which can suffer from slow leakage, or $LH_2$, which faces boil-off losses, chemically bound hydrogen can remain stored for extended periods without significant degradation or energy loss. This long-term stability makes chemical hydrogen carriers particularly valuable for seasonal energy storage, emergency reserves, or remote applications where hydrogen may need to be stored for months before use.

***Challenges of Chemical Hydrogen Storage***

Despite these advantages, chemical hydrogen storage faces several challenges, beginning with its energy input requirements. The process of releasing hydrogen, particularly through dehydrogenation reactions, often requires substantial amounts of heat or energy. This reduces the overall efficiency of the hydrogen supply chain and can make chemical storage less attractive unless efficient catalysts or heat recovery systems are employed.

Another issue is by-product management. Extracting hydrogen from chemical carriers usually results in additional compounds or residues that must be handled, recycled, or disposed of properly. For example, using ammonia borane as a carrier releases hydrogen but also leaves behind spent material that needs regeneration. If these by-products are not efficiently managed, the system becomes costly, wasteful, and environmentally challenging. The reliance on catalysts adds another layer of complexity. Many chemical hydrogen storage methods require expensive or complex catalysts often involving precious metals like platinum, ruthenium, or palladium to achieve efficient hydrogen release. The cost, availability, and long-term durability of these catalysts are major hurdles to commercial scalability.

Additionally, regeneration complexity is a significant drawback. Some chemical carriers are difficult to regenerate after hydrogen release, which increases costs and resource consumption. For example, regenerating ammonia borane back into its hydrogen-rich form is a highly energy-intensive process, making it unsuitable for large-scale or repeated use unless regeneration pathways are improved. These factors currently limit the overall efficiency and economic feasibility of chemical hydrogen storage technologies.

***Emerging Innovations in Chemical Hydrogen Storage***

To address these limitations, researchers are focusing on emerging innovations that could make chemical hydrogen storage more practical and sustainable. One area of development is the creation of advanced catalysts that are both cost-effective and highly efficient, reducing the energy requirements of hydrogen release while improving reaction speed and durability. Transition metal catalysts and nanostructured materials are showing promise in this domain. Another innovation is the design of recyclable carriers. For example, advancements in LOHCs are enabling better recyclability and regeneration, allowing these systems to be reused multiple times without losing performance. This greatly enhances the life-cycle economics of chemical hydrogen storage and reduces waste. There is also increasing interest in hybrid systems, which combine chemical storage with other hydrogen storage methods. For instance, pairing chemical carriers with compressed hydrogen or solid-state systems can improve overall efficiency and flexibility, enabling storage solutions tailored to specific applications.

One of the most promising innovations is in ammonia cracking. Ammonia is already a globally traded chemical with well-established infrastructure, and recent progress in catalysts and reactor design is making it more feasible to extract hydrogen from ammonia efficiently and economically. This positions ammonia as a strong candidate for large-scale hydrogen storage and transport, especially for international trade.

***Applications of Chemical Hydrogen Storage***

The potential applications of chemical hydrogen storage are wide-ranging. In transportation, chemical carriers can provide compact, safe, and high-density storage solutions for hydrogen-powered vehicles, especially in areas where refueling infrastructure is still developing. For renewable energy integration, chemical carriers serve as a way to store excess electricity generated by wind or solar power in a stable, transportable form that can be released when demand is higher. In stationary power systems, chemical hydrogen storage can provide reliable backup power for industrial and residential grids, contributing to grid stabilization and energy security. Furthermore, in aerospace and marine industries, where compact and energy-dense storage is crucial, chemical hydrogen carriers offer promising solutions to power hydrogen-based aircraft or ships without the space and safety limitations of gaseous or $LH_2$ storage. Chemical hydrogen storage offers a versatile and scalable solution to hydrogen storage challenges, with ongoing research aimed at improving efficiency, cost, and environmental impact. Its adaptability across diverse applications makes it a key component of the evolving hydrogen economy.

> ***Key: Hydrogen infrastructure:*** *The integrated systems and facilities required for the production, storage, distribution, and utilization of hydrogen, including pipelines and refueling stations.*

#### 3.2.1.5 Material-Based Hydrogen Storage Methods

Material-based storage methods are an important class of hydrogen storage technologies that focus on using various materials capable of absorbing, adsorbing, or chemically binding hydrogen. Unlike conventional gaseous or $LH_2$ storage, which requires either very high pressures or cryogenic temperatures, material-based storage seeks to achieve safer, more compact, and potentially more energy-efficient solutions. By leveraging physical adsorption, chemical reactions, or nanostructured materials, these systems aim to address the challenges of energy density, safety, and scalability in hydrogen storage for energy applications. The key methods within this category include physisorption, chemisorption, chemical hydrogen storage, clathrate-based storage, and nanostructured materials.

***Physisorption-Based Storage (Adsorption)***

Physisorption is a material-based hydrogen storage method in which hydrogen molecules are physically adsorbed onto the surface of porous materials through weak van der Waals forces. Since this interaction does not involve strong chemical bonding, the process is reversible, allowing hydrogen to be released easily when needed. Commonly used materials for this purpose include MOFs and carbon-based materials such as activated carbon, graphene, and carbon nanotubes. MOFs are particularly attractive because of their extremely high surface area and tunable pore sizes, which enable the storage of large amounts of hydrogen under specific conditions. Similarly, carbon nanostructures provide lightweight and highly porous surfaces for adsorption. The advantages of this method include its reversibility and the fact that it can operate at relatively low pressures, making it safer compared to compressed hydrogen systems. However, its primary drawback is that significant hydrogen storage capacity is only achieved at very low temperatures, often close to liquid nitrogen levels (−196 °C), which limits its practicality for large-scale or ambient applications.

***Chemisorption-Based Storage***

Chemisorption-based hydrogen storage involves the chemical reaction of hydrogen with materials to form hydrides. This method differs from physisorption in that hydrogen becomes chemically bound, which not only increases storage density but also requires energy input to release the hydrogen. There are two main classes of materials used in chemisorption storage: metal hydrides and complex hydrides. Metal hydrides, such as magnesium hydride ($MgH_2$) and titanium-based hydrides, form stable metal–hydrogen bonds and are widely studied due to their relatively high hydrogen capacity. Complex hydrides, such as sodium borohydride ($NaBH_4$) and lithium aluminum hydride ($LiAlH_4$), offer even higher hydrogen content but suffer from poor reversibility, meaning that once hydrogen is released, regenerating the original compound is difficult. The advantages of chemisorption-based storage include high hydrogen density and operation under moderate pressure and temperature conditions compared to compressed hydrogen or $LH_2$. However, the main challenges are slow reaction kinetics and the high energy required to release hydrogen, which hinder real-world adoption unless catalysts or advanced thermal management systems are used.

### Chemical Hydrogen Storage

In chemical hydrogen storage, hydrogen is stored in chemical compounds and later released through controlled reactions. This approach focuses on using hydrogen-rich molecules, which serve as carriers. Two prominent examples are ammonia ($NH_3$) and LOHCs. Ammonia contains about 17.6% hydrogen by weight and can be "cracked" to release hydrogen gas using catalysts. LOHCs, such as the toluene/methylcyclohexane system, store hydrogen in organic molecules that can be hydrogenated or dehydrogenated depending on the application. The key advantage of chemical hydrogen storage is its stability and safety. Hydrogen remains bound within the compound, minimizing the risks of leakage or flammability. Additionally, many carriers operate in the liquid phase, making them compatible with existing fuel transport infrastructure like pipelines and tankers. However, challenges include the need for high temperatures to release hydrogen, as well as the requirement for specialized catalysts to ensure efficiency and selectivity. Despite these challenges, chemical hydrogen storage is increasingly seen as one of the most scalable solutions for long-distance hydrogen transport and long-term storage.

### Hydrogen Adsorption in Clathrates

Another interesting material-based storage method is hydrogen adsorption in clathrate hydrates. In this approach, hydrogen molecules are trapped within crystalline water structures, forming cage-like lattices that physically encapsulate the hydrogen. This method is attractive because it operates at moderate pressures and temperatures, avoiding the extremes required for compressed or $LH_2$ storage. Additionally, the materials involved, such as water, are environmentally friendly and abundant, making the process sustainable. The main advantage of clathrate-based storage is its safety and mild operating conditions, which make it suitable for experimental and niche applications. However, the technology faces significant challenges, most notably its low hydrogen storage capacity and the need for precise control over pressure and temperature to maintain stable clathrates. As such, while promising, this method is still in its early stages of development and is not yet competitive with other hydrogen storage technologies.

***Nanostructured and Advanced Materials***

Recent advances in nanotechnology have opened up new possibilities for material-based hydrogen storage using nanostructured and advanced materials. Examples include nonporous materials such as zeolites and aerogels, which provide vast internal surface areas and pore volumes for hydrogen adsorption. Another promising approach involves metal-decorated nanostructures, where nanoparticles of metals like palladium are incorporated into carbon nanomaterials to enhance hydrogen uptake and storage performance. These advanced materials allow researchers to fine-tune storage properties, such as capacity, kinetics, and reversibility, for specific applications ranging from mobile hydrogen storage to large-scale energy storage systems. The advantages of these nanomaterials include the potential for high storage capacities and the ability to tailor material properties for different operating conditions. However, they also face challenges related to high production costs and scalability issues, since synthesizing nanomaterials with the required precision is expensive and often limited to laboratory-scale processes. Despite these challenges, nanostructured materials remain one of the most promising frontiers for hydrogen storage research.

#### 3.2.1.6 Cryo-Compressed and Underground Hydrogen Storage

$CcH_2$ storage combines two storage methodologies: cryogenic $LH_2$ storage and compressed hydrogen gas storage. Hydrogen is cooled to cryogenic temperatures (~20 K) where it exists as a liquid, but is then stored in a pressure vessel capable of sustaining pressures up to 250–350 bar. This hybrid approach maximizes volumetric density while mitigating some of the boil-off issues of pure cryogenic storage.

The density of $LH_2$ at 20 K is approximately 70.8 kg/m$^3$, while compressed hydrogen at 700 bar and 298 K achieves a density of ~42 kg/m$^3$. By combining cryogenic cooling with pressure, $CcH_2$ can achieve densities exceeding 80–85 kg/m$^3$, which translates into significantly higher energy per unit volume.

The stored mass of hydrogen in a cryo-compressed vessel can be expressed as:

$$m = \rho V_t \tag{3.26}$$

where $m$ is stored mass (kg), $\rho$ is the effective hydrogen density (kg/m$^3$), and $V_t$ is the tank volume (m$^3$). For example, consider a 150-L cryo-compressed tank (0.15 m$^3$) operating at ~80 kg/m$^3$ density:

$$m = 80 \times 0.15 = 12\ \text{kg}$$

Given hydrogen's specific energy of 120 MJ/kg, the energy content is:

$$E = m \times 120 = 12 \times 120 = 1440\ \text{MJ}$$

This is sufficient for ~600 km driving range in an FCV with an average consumption of 2.4 MJ/km.

The work required for compression can be estimated using the isothermal compression formula: $W = nRT \ln\left(\frac{P_2}{P_1}\right)$.

For example, compressing hydrogen from 1 to 350 bar at 80 K:

$$W = nRT \ln\left(\frac{350}{1}\right)$$

where $m$ is the mass of hydrogen and $M = 0.002$ kg/mol. Substituting, one can compute the actual energy costs of compression. Additionally, boil-off losses occur due to unavoidable heat influx. The heat transfer rate ($Q$) into the cryogenic vessel is modeled as:

$$Q = UA\Delta T \tag{3.27}$$

where $U$ is the overall heat transfer coefficient, $A$ is the surface area, and $\Delta T$ is the temperature gradient. Minimizing $Q$ requires advanced insulation systems like multi-layer vacuum insulation (MLI).

While hydrogen does not undergo storage-related chemical reactions in $CcH_2$ tanks, one must consider ortho-para conversion. At room temperature, hydrogen exists as ~75% ortho-hydrogen and ~ 25% para-hydrogen. Upon liquefaction, the equilibrium shifts toward para-hydrogen (~99%). The exothermic conversion reaction releases heat:

$$\text{ortho-H}_2 \rightarrow \text{para-H}_2 + \Delta H \tag{3.28}$$

with $\Delta H \approx 703$ J/mol. If not managed, this conversion leads to boil-off losses during storage.

#### *Underground Hydrogen Storage*

Underground hydrogen storage (UHS) involves injecting hydrogen into large-scale geological formations for seasonal or strategic storage. Unlike cryo-compressed storage, which is suited for vehicles and decentralized systems, UHS is used for grid-scale energy storage. Potential formations include:

1) Salt Caverns—Solution-mined salt domes with high permeability.
2) Depleted Gas Fields—Former natural gas reservoirs with existing infrastructure.
3) Aquifers—Deep saline formations capable of storing hydrogen under pressure.

The storage capacity of a geological reservoir is given by:

$$M_{H_2} = \phi V_b \rho H_2 \tag{3.29}$$

where

- $\phi$ = Porosity of formation.
- $V_b$ = Bulk volume of reservoir ($m^3$).
- $\rho H_2$= Density of hydrogen under reservoir conditions.

For instance, a salt cavern of volume 500,000 $m^3$ with an effective density of 20 kg/$m^3$ can store:

$$M_{H_2} = 0.25 \times 500,000 \times 20 = 2.5 \times 10^6 \text{ kg}$$

This corresponds to an energy content of: $E = 2.5 \times 10^6 \times 120 = 3 \times 10^8$ MJ, which is sufficient to power an entire city for weeks.

Suppose a city requires 5 GW of power for 30 days. The total energy demand is:

$$E = 5 \times 10^9 \times 30 \times 24 \times 3600 = 1.296 \times 10^{16} \text{ J}$$

Given hydrogen's specific energy: $m = \dfrac{E}{120 \times 10^6} = 1.08 \times 10^8$ kg.

If stored in a salt cavern at 20 kg/$m^3$ effective density:

$$V = \frac{1.08 \times 10^8}{20} = 5.4 \times 10^6$$

Thus, a cavern of ~5.4 million $m^3$ would be required, which is feasible given existing salt dome capacities.

In underground storage, hydrogen is subjected to thermobaric conditions where both temperature and pressure vary with depth. The governing equations involve Darcy's law for flow:

$$Q = -\frac{kA}{\mu}\frac{dP}{dx} \tag{3.30}$$

where $Q$ is volumetric flow, $k$ is permeability, $A$ is cross-sectional area, $\mu$ is viscosity, and $dP/dx$ is pressure gradient.

Hydrogen's low molecular weight leads to higher diffusivity, which can cause losses due to migration. The retention efficiency is enhanced in salt caverns due to near-zero permeability.

Underground hydrogen is susceptible to several chemical interactions with the geological medium and microbial life:

1) **Methanogenesis (Microbial Reaction)** $4H_2 + CO_2 \rightarrow CH_4 + 2H_2O$

   This leads to hydrogen losses but produces methane.

2) **Sulfate Reduction** $4H_2 + SO_4^{2-} + H^+ \rightarrow HS^- + 4H_2O$
3) **Mineral Reactions**—Hydrogen can reduce certain oxides in reservoir rocks:

$$Fe_2O_3 + 3H_2 \rightarrow 2Fe + 3H_2O$$

These reactions alter storage efficiency and must be accounted for in system modeling. Table 3.9 illustrates a comparative analysis between cryo-compressed versus underground storage.

$CcH_2$ storage and UHS represent two complementary strategies for addressing hydrogen's storage challenges. Cryo-compression offers high-density, mobile storage for transportation applications, leveraging both cryogenic cooling and pressure. It requires careful thermodynamic management to minimize boil-off and address ortho-para conversion. UHS, in contrast, enables seasonal and strategic storage at a gigawatt scale, utilizing geological formations. However, it introduces complexities related to flow dynamics, diffusion, and chemical interactions with the geological environment. Mathematical models ranging from ideal gas law, compressibility factors, and thermodynamic work equations in $CcH_2$, to Darcy's law, porosity models, and reaction kinetics in UHS, enable quantification of storage performance. Chemical equations highlight critical transformations such as methanogenesis and sulfate reduction, which may alter the long-term efficiency of UHS. Ultimately, an integrated approach combining cryogenic-pressurized systems for mobility and geological reservoirs for grid-scale energy balancing will be key in realizing a hydrogen-based energy economy. Table 3.10 illustrates the comparison of the different hydrogen storage techniques. Table 3.11 illustrates the efficiency and efficacy of the different storage methods.

**Table 3.9** Comparative Analysis: Cryo-Compressed Versus Underground Storage

| Parameter | Cryo-Compressed ($CcH_2$) | Underground Storage (UHS) |
|---|---|---|
| Scale | Small-medium (vehicles, portable) | Large-scale (grid, seasonal) |
| Density | 70–85 $kg/m^3$ | 10–30 $kg/m^3$ (effective) |
| Losses | Boil-off due to heat ingress | Diffusion, microbial consumption |
| Chemical stability | Ortho-para conversion | Multiple geochemical reactions |
| Economics | High cost per kg stored | Economical at large scale |
| equation relevance | $PV = nRT$ with cryogenic corrections | Darcy's law, porosity, geochemical kinetics |

**Table 3.10** Comparison of the Different Hydrogen Storage Techniques

| Storage Technique | Typical System Energy Density (Gravimetric) | Typical System Energy Density (Volumetric) | Meets 2025 Target? | Meets Ultimate Target? | Notes (System-Level Context) |
|---|---|---|---|---|---|
| Compressed $H_2$ (700 bar, type IV) | ~1.71 kWh/kg (≈5.1 wt%) | ~0.86 kWh/L (≈0.026 kg/L) | Gravimetric: close; volumetric: no | No | Representative DOE program-record values for state-of-practice 700-bar tanks |
| Liquid $H_2$ ($LH_2$)—Automotive vacuum-jacketed tank (indicative) | ~2.0–2.3 kWh/kg (≈6–7 wt%) | ~1.3–1.5 kWh/L (≈0.039–0.045 kg/L) | Generally yes (both, depending on design) | Volumetric: sometimes short of 1.7; gravimetric: sometimes short | System-level values vary by design; cryogenic losses/dormancy are key challenges. ($CcH_2$ example below gives 2.3/1.5 kWh units) |
| Cryo-compressed $H_2$ ($CcH_2$) | ~2.3 kWh/kg (≈7.1 wt%) | ~1.5 kWh/L (≈0.045 kg/L) | Gravimetric: yes; volumetric: near/at 2025 | Volumetric: below ultimate; gravimetric: near | ANL assessment of a $CcH_2$ system; roadmap notes $CcH_2$ among few options simultaneously near both targets, with dormancy/efficiency challenges |
| Solid-state (metal hydrides) | ~0.3–0.5 kWh/kg (≈1–1.5 wt% system) | Often ≥1.3 kWh/L possible at system level | Gravimetric: no; volumetric: can be yes | Gravimetric/volumetric: generally no | High volumetric density but heavy systems; thermal management and mass are limiting. |
| Material-based sorbents (e.g., MOFs, carbons) | Today: <<1.8 kWh/kg at system (room-temp [RT]) | RT demos up to ~0.4 kWh/L (12 g/L); densified/cryogenic lab results much higher but not yet practical | No (system, RT) | No | DOE analyses show meeting 5.5 wt% and 40 g/L at system requires exceptional RT uptake and packing; current RT systems fall short. |
| Chemical carriers (e.g., ammonia-borane slurry, LOHCs) | Often <1.8 kWh/kg at system | ~1.2–1.3 kWh/L (≈37–40 g/L) demonstrated in DOE studies | Gravimetric: no; volumetric: near 2025 | No | Good volumetric density; but off-board regeneration and system mass/complexity limit onboard suitability. |
| Underground $H_2$ storage (salt caverns, aquifers, depleted reservoirs) | *Not an onboard system* | *Reservoir-scale volumetric energy density*: pure $H_2$ stores ~ 24–28% of the energy per volume of $CH_4$ at similar *P*–*T*; blends scale accordingly | *N/A* | *N/A* | DOE/NETL shows UHS energy-per-volume versus UGS (natural gas); useful for seasonal/storage hubs, not comparable to LDV onboard targets. |

**Table 3.11** Efficiency and Efficacy of the Different Storage Methods

| Storage Method | Energy Efficiency (Indicative Range %) | Key Considerations Affecting Energy Efficacy |
|---|---|---|
| Compressed $H_2$ (700 bar, type IV) | ~55–65% | High compression energy (~10–15% of $H_2$ energy content lost), but widely used in automotive. Efficiency decreases with higher pressure. |
| Liquid $H_2$ ($LH_2$, automotive vacuum-jacketed tank) | ~25–40% | Liquefaction requires very high energy (~30–40% of $H_2$ content), boil-off losses further reduce efficacy. Good gravimetric density but poor efficiency. |
| Cryo-compressed $H_2$ ($CcH_2$) | ~35–50% | Combines compression + cryogenic cooling, higher density but requires cooling energy. Efficiency better than pure $LH_2$, worse than compressed $H_2$. |
| Solid-state (metal hydrides) | ~30–50% | Absorption/desorption requires heat; slow kinetics; heavy storage system reduces system-level efficacy. Efficient at low pressures but limited in mobile use. |
| Material-based sorbents (MOFs, carbons) | ~40–55% (projected) | Adsorption at cryogenic temperatures (~77 K) consumes energy; relatively higher reversibility and lower losses, but still under research for practical use. |
| Chemical carriers (e.g., ammonia-borane slurry, LOHCs) | ~30–45% | Conversion (hydrogenation/dehydrogenation) requires external energy and catalysts; storage stable but low round-trip efficiency. |
| Underground $H_2$ storage (salt caverns, aquifers, depleted reservoirs) | ~70–90% | Large-scale bulk storage; very high energy efficacy since compression is moderate and losses are minimal. Suited for seasonal/grid storage, not mobile. |

### 3.2.2 Emerging Hydrogen Storage Technologies

Emerging hydrogen storage technologies focus on overcoming limitations in efficiency, storage capacity, safety, and cost, making hydrogen a more viable energy carrier for widespread applications. These technologies leverage advancements in materials science, nanotechnology, and engineering. Below is an overview of the most promising emerging hydrogen storage technologies:

### ➢ High-Pressure Hydrogen Storage with Composite Materials

High-pressure hydrogen storage using composite materials represents a groundbreaking advancement in sustainable energy systems. As hydrogen emerges as a clean energy carrier, efficient storage solutions are critical for its widespread adoption in transportation, industry, and power generation. Composite materials, often combining lightweight polymers with high-strength fibers like carbon or glass, offer a superior alternative to traditional metal storage tanks. These materials provide the necessary strength to withstand high pressures, typically up to 700 bar, while being significantly lighter, enhancing fuel efficiency in automotive applications. Moreover, composites exhibit excellent resistance to hydrogen embrittlement, ensuring durability and safety over long-term use. Innovations in manufacturing techniques, such as filament winding and resin transfer molding, are making these storage systems more cost-effective and scalable. The integration of high-pressure composite storage systems with renewable hydrogen production has the potential to revolutionize energy infrastructure, paving the way for a cleaner, more sustainable future.

- **Innovations**:
  - **Carbon–Fiber Reinforced Polymers (CFRPs)**: Lightweight alternatives to traditional steel tanks.
  - **Metal Matrix Composites (MMCs)**: Enhanced durability and hydrogen permeability.
- **Benefits**:
  - Reduced tank weight for mobile applications.
  - Improved safety under high-pressure conditions.
- **Challenges**: High manufacturing costs and energy-intensive production.

### ➢ Liquid Cryogenic Hydrogen Storage

Liquid cryogenic hydrogen storage is a pivotal technology for the efficient handling and utilization of hydrogen as a clean energy carrier. In this method, hydrogen is cooled to extremely low temperatures, approximately −253 °C (−423 °F), to transition into a liquid state, significantly increasing its energy density compared to gaseous storage. This makes it especially suitable for applications requiring compact and lightweight storage, such as aerospace and long-distance transportation. Cryogenic storage tanks are meticulously designed with advanced insulation systems to minimize heat transfer and reduce boil-off losses. Materials used must withstand extreme temperatures and maintain structural integrity under high vacuum conditions. Despite its advantages, challenges remain, including high energy requirements for liquefaction and managing evaporation losses during storage and transportation. Continued advancements in cryogenic technologies, including improved insulation and liquefaction efficiency, are crucial for realizing the full potential of $LH_2$ storage in enabling a global hydrogen economy.

- **Emerging Developments**:
  - Advanced cryogenic insulation materials to reduce boil-off.
  - Integrated cryo-coolers for maintaining low temperatures.
- **Applications**: Aerospace and large-scale hydrogen transportation.
- **Challenges**: Energy-intensive cooling and evaporation losses.

➢ **Chemical Hydrogen Storage with Advanced Carriers**

Chemical hydrogen storage with advanced carriers offers a promising solution for safely storing and transporting hydrogen at ambient conditions. In this approach, hydrogen is chemically bound to carriers such as metal hydrides, ammonia, or organic liquid carriers like formic acid or methylcyclohexane. These carriers enable hydrogen to be stored in a compact and stable form, reducing the risks associated with high-pressure or cryogenic storage. Advanced carriers are designed for high hydrogen density, reversibility, and efficient release of hydrogen through catalytic processes. For instance, ammonia can store hydrogen at high densities and be easily decomposed into nitrogen and hydrogen using specialized catalysts. MOFs and nanostructured materials are also being explored to enhance hydrogen absorption and desorption kinetics. By leveraging advanced carriers, this method addresses logistical challenges and enhances the practicality of hydrogen supply chains for various applications, from fuel cells to industrial processes. Ongoing research into carrier regeneration and scalability is critical to integrating this technology into the hydrogen economy.

- **Innovative Carriers**:
  - **Liquid Organic Hydrogen Carriers (LOHCs)**: New LOHC molecules with lower dehydrogenation temperatures.
  - **Ammonia Derivatives**: Ammonia reforming using advanced catalysts for hydrogen release.
  - **Borohydrides and Formic Acid**: Enhanced catalytic pathways for efficient hydrogen release.
- **Advances**:
  - Reversible and energy-efficient hydrogenation/dehydrogenation cycles.
  - High-purity hydrogen extraction techniques.
- **Challenges**: Catalyst costs and by-product management.

➢ **Hydrogen in Clathrate Hydrates**

Hydrogen in clathrate hydrates presents a novel approach to hydrogen storage, leveraging the ability of water molecules to form crystalline cages under specific temperature and pressure conditions. In this method, hydrogen molecules are trapped within the cavities of the hydrate structure, creating a stable and compact storage medium. Clathrate hydrates can store hydrogen at relatively mild pressures and low temperatures, offering a safer and potentially more cost-effective alternative to high-pressure and cryogenic storage methods. The formation of these hydrates is influenced by factors such as pressure, temperature, and the presence of promoters, which accelerate hydrate formation and improve storage efficiency. One of the significant advantages of hydrogen hydrates is their potential for large-scale applications, such as energy storage in renewable systems or hydrogen transportation over long distances. However, challenges such as slow formation rates, low hydrogen capacity, and the energy required for synthesis and release need to be addressed. Advances in materials science and process optimization hold promise for making hydrogen clathrate hydrates a viable component of the hydrogen economy.

- **Emerging Focus**:
  - Enhanced hydrate formation rates using additives.
  - High-density hydrogen storage hydrate formulations.
- **Applications**: Small-scale storage and niche energy systems.
- **Challenges**: Low storage density and slow kinetics.

➢ **Advanced Nanostructured Materials**

Advanced nanostructured materials are at the forefront of innovation in various fields, from energy storage to biomedical applications, due to their unique properties derived from nanoscale architecture. These materials, which include nanocomposites, nanoparticles, nanowires, and nanotubes, exhibit remarkable characteristics such as high surface area, tunable porosity, and enhanced mechanical, thermal, and electrical properties. In hydrogen storage, nanostructured materials like MOFs, graphene-based materials, and nanoscale metal hydrides are being explored for their ability to absorb and release hydrogen efficiently under practical conditions. In electronics, nanostructures enable the development of faster, smaller, and more energy-efficient devices. Additionally, in medicine, nanostructured materials are revolutionizing drug delivery systems, allowing for targeted therapy and reduced side effects. Ongoing research aims to optimize the synthesis, scalability, and environmental impact of these materials, ensuring their integration into technologies that address global challenges in energy, health, and sustainability.

- **Notable Innovations**:
  - **Graphene-Based Materials**: High surface area and tunable properties for hydrogen adsorption.
  - **Nanotubes and Nanowires**: Improved kinetics for adsorption and release.
  - **Metal–Nanoparticle Hybrids**: Combines the benefits of metals and nanostructures.

- **Advantages**:
  - High storage capacity and faster kinetics.
  - Operates at lower pressures.
- **Challenges**: Expensive synthesis and lack of scalability.

## ➢ Magnetic and Electrochemical Hydrogen Storage

Magnetic and electrochemical hydrogen storage are emerging technologies that offer innovative solutions for hydrogen utilization in energy systems. Magnetic hydrogen storage leverages the interaction between hydrogen molecules and magnetic fields to enhance storage efficiency, particularly in solid-state materials such as magnetic alloys. This approach can improve hydrogen adsorption and desorption kinetics, making it an attractive option for applications requiring rapid and reversible storage. On the other hand, electrochemical hydrogen storage utilizes materials like metal hydrides, graphene, or nanostructured electrodes within battery-like systems to store hydrogen electrochemically. These systems enable the direct conversion of hydrogen into electrical energy or vice versa, seamlessly integrating with renewable energy sources for grid storage or portable power solutions. Both methods address key challenges in traditional hydrogen storage by offering compact, efficient, and safer alternatives. However, further advancements in material design, energy efficiency, and scalability are essential to realize their full potential in the growing hydrogen economy.

- **Key Developments**:
  - **Magnetocaloric Hydrogen Storage**: Magnetic materials that absorb hydrogen upon cooling.
  - **Electrochemical Hydrogen Storage**: Using advanced electrodes for reversible hydrogen storage in batteries or cells.
- **Applications**: Integration into hybrid energy systems.
- **Challenges**: Limited energy density and efficiency.

## ➢ Hybrid Hydrogen Storage Systems

Hybrid hydrogen storage systems combine multiple storage technologies to capitalize on their respective strengths, offering a versatile and efficient solution to meet the diverse demands of hydrogen applications. These systems integrate methods such as high-pressure gaseous storage, cryogenic liquid storage, and solid-state storage using materials like metal hydrides or carbon-based nanostructures. By blending these approaches, hybrid systems address the limitations of individual storage methods, such as the high energy costs of liquefaction, the bulkiness of high-pressure tanks, or the slow kinetics of solid-state storage. For instance, a hybrid system might use compressed hydrogen for immediate energy demands while relying on metal hydrides for long-term storage. Such flexibility makes hybrid systems particularly suitable for applications requiring dynamic hydrogen supply, such as FCVs, renewable energy storage, and industrial processes. Advances in materials science and system design continue to improve the efficiency, safety, and cost-effectiveness of these systems, positioning them as a key enabler for the global hydrogen economy.

- **Examples**:
  - MOF-enhanced high-pressure tanks.
  - LOHC integrated with cryogenic systems.
- **Advantages**:
  - Tailored solutions for specific applications.
  - Enhanced overall efficiency and safety.
- **Challenges**: Complexity in design and operation.

➢ **Additive Manufacturing for Hydrogen Storage**

Additive manufacturing, commonly known as 3D printing, is revolutionizing hydrogen storage by enabling the design and fabrication of highly efficient and customized storage solutions. This technology allows for the creation of complex geometries and porous structures that optimize the weight-to-strength ratio and maximize hydrogen storage capacity. For instance, lattice structures and intricate channels can be 3D-printed to enhance surface area, improving hydrogen adsorption in solid-state storage systems like metal hydrides or carbon-based materials. Additive manufacturing also facilitates the production of lightweight composite tanks for high-pressure or cryogenic hydrogen storage, reducing overall system weight and enhancing transport efficiency. Moreover, this method supports rapid prototyping and scalability, enabling cost-effective production of tailored solutions for specific applications, from portable storage units to large-scale industrial tanks. As research advances in materials like high-strength polymers, ceramics, and metal alloys suitable for hydrogen environments, additive manufacturing holds significant potential to accelerate the development of next-generation hydrogen storage technologies.

- **Advantages**:
  - Customizable designs for better efficiency.
  - Lightweight materials for mobile applications.
- **Applications**: Automotive and portable hydrogen storage devices.
- **Challenges**: Material limitations and durability concerns.

#### 3.2.2.1 Biological Hydrogen Storage

Biological hydrogen storage explores the potential of living organisms and biomaterials to store hydrogen in an environmentally friendly and sustainable manner. This innovative approach leverages the natural processes of microorganisms, such as algae and certain bacteria, which can produce and store hydrogen under specific conditions through metabolic activities. For example, cyanobacteria and green algae generate hydrogen as a by-product of photosynthesis when exposed to light and deprived of sulfur. Additionally, bio-inspired materials, including hydrogenase enzymes and biopolymers, are being studied for their ability to reversibly bind and release hydrogen. These systems offer the advantage of operating under ambient conditions, reducing the need for high pressures or extreme temperatures. Although still in the experimental phase, biological hydrogen storage holds promise for integrating hydrogen production and storage into renewable and circular energy systems. Challenges such as improving storage capacity, stability, and scalability must be addressed to unlock its full potential in contributing to a sustainable hydrogen economy.

- **Innovations**:
  - Hydrogen storage in algae or microbial cells.
  - Bio-based hydrides and storage materials.
- **Applications**: Long-term storage in remote or sustainable setups.
- **Challenges**: Low hydrogen density and technological maturity.

> ***Key: Hydrogen refueling station:*** *Specialized facilities where hydrogen-powered vehicles can refuel, operating similarly to conventional gas stations but using compressed or liquid hydrogen.*

## 3.3 Hydrogen as a Clean Energy Source

**Clean energy** refers to energy sources or systems that produce power with minimal or no GHG emissions, air pollution, or other environmental harm. It is characterized by its sustainability, renewability, and low impact on ecosystems. Clean energy includes sources such as:

- **Solar Energy**: Harnessing sunlight using PV cells or solar thermal systems.
- **Wind Energy**: Generating electricity from wind turbines.
- **Hydropower**: Using flowing or falling water to generate power.
- **Geothermal Energy**: Utilizing heat from beneath the Earth's surface.
- **Biomass Energy**: Organic material used for energy production in a sustainable manner.

Clean energy aims to reduce reliance on fossil fuels, combat climate change, and promote a transition to sustainable energy systems.

Hydrogen qualifies as a form of clean energy when it is produced and used in environmentally sustainable ways. Its clean energy potential lies in its ability to act as a zero-emission fuel, with water as the only by-product when used in fuel cells or combustion. Hydrogen is poised to play a pivotal role in the global clean energy transition, especially for sectors difficult to decarbonize, such as heavy industry and long-haul transportation. Advances in production, storage, and distribution technologies are expected to accelerate its adoption as a cornerstone of clean energy systems. Clean energy, reimagined through the lens of hydrogen energy, represents a transformative force that harnesses the most abundant element in the universe to power our future sustainably. It is the art of converting hydrogen into a universal energy carrier that bridges the gap between renewable sources and practical applications, enabling a zero-emission energy cycle. Table 3.12 shows the comparison of different clean energy sources with key attributes.

### 3.3.1 Production of Clean Hydrogen

The production of clean hydrogen refers to methods of producing hydrogen with minimal environmental impact, primarily using renewable energy sources or low-carbon processes. Clean hydrogen plays a key role in achieving global sustainability goals by enabling a transition to carbon-neutral energy systems. Following are the primary methods of producing clean hydrogen:

1) **Green Hydrogen (Electrolysis of Water)**
   - **Process**: Green hydrogen is produced using electricity from renewable energy sources (such as wind, solar, or hydroelectric power) to split water ($H_2O$) into hydrogen ($H_2$) and oxygen ($O_2$) through a process called electrolysis.
   - **Technology**:
     - **AELs**: Use an alkaline solution as the electrolyte.
     - **PEM Electrolyzers**: Use a solid polymer membrane as the electrolyte, offering higher efficiency and faster response times.
     - **AEM Electrolyzers**: A newer technology combining advantages of AEL and PEM electrolyzers.
   - **Advantages**:
     - Zero emissions when powered by renewable energy sources.
     - Highly sustainable, as it only requires water and renewable electricity.
   - **Challenges**:
     - High costs of electrolyzers and renewable energy infrastructure.
     - Low efficiency compared to some other hydrogen production methods.

**Table 3.12** Comparison of Different Clean Energy Sources with Key Attributes

| Attribute | Solar Energy | Wind Energy | Hydropower | Geothermal Energy | Biomass Energy |
|---|---|---|---|---|---|
| Energy source | Sunlight | Wind | Flow of water | Heat from Earth's interior | Organic materials (plants, waste) |
| Renewability | Renewable | Renewable | Renewable | Renewable | Renewable |
| Environmental impact | Low environmental impact | Low environmental impact | Low, but can disrupt ecosystems | Low environmental impact | Can cause deforestation, pollution |
| Efficiency | 15–20% (varies with location) | 35–45% (depends on location) | 60–90% | 10–20% (depends on resources) | 20–25% |
| Land/space requirements | High (depends on location) | High (land area required for turbines) | Medium (for dam construction) | Low to medium | Medium (for crops, land management) |
| Initial cost | High (installation cost) | Moderate (turbine setup cost) | High (dam, infrastructure) | High (drilling, plant setup) | Low to moderate (depending on setup) |
| Operational cost | Low | Low | Moderate | Low | Moderate to high |
| Intermittency | Intermittent (daytime only) | Intermittent (depends on wind) | Continuous (if water flow is stable) | Continuous (if geothermal source is stable) | Intermittent (depends on feedstock) |
| Storage requirements | High (battery storage needed) | High (battery or backup systems) | Low (usually not required) | Low (if geothermal resources are stable) | High (requires biomass storage systems) |
| Geographic dependency | High (sunlight availability) | High (wind speed and consistency) | Moderate (depends on water resources) | High (geothermal hot spots) | Low to moderate (depends on agricultural conditions) |

(Continued)

**Table 3.12** (Continued)

| Attribute | Solar Energy | Wind Energy | Hydropower | Geothermal Energy | Biomass Energy |
|---|---|---|---|---|---|
| Lifespan | 20–30 years | 20–25 years | 30–50 years | 25–40 years | 10–20 years |
| By-products/ emissions | None (no emissions) | None (no emissions) | Low emissions during construction | Low emissions (minor gases from geothermal plants) | Low emissions (from biomass processing) |
| Scale | Small- to large-scale (solar farms) | Small- to large-scale (wind farms) | Large-scale (hydroelectric dams) | Small- to medium-scale | Small- to large-scale (depending on resources) |
| Flexibility in application | Highly flexible (residential to industrial) | Flexible (land-based) | Limited (needs rivers, large sites) | Limited (site-specific) | Flexible (can use waste materials) |
| Examples of use | Solar panels, solar farms | Wind turbines, wind farms | Dams, hydroelectric plants | Geothermal plants, hot springs | Bioenergy plants, biomass power plants |

2) **Blue Hydrogen (Natural Gas Reforming with CCS)**
   - **Process**: Blue hydrogen is produced through **SMR**, where natural gas (methane, $CH_4$) reacts with steam to produce hydrogen and carbon monoxide (CO). The CO is then reacted with more steam to produce additional hydrogen and carbon dioxide ($CO_2$), which is captured and stored using **CCS** technology.
   - **Technology**:
     - **SMR with CCS**: Captures $CO_2$ emissions from the reforming process and stores it underground to prevent its release into the atmosphere.
   - **Advantages**:
     - Lower emissions than conventional hydrogen production (gray hydrogen).
     - Utilizes existing natural gas infrastructure.
   - **Challenges**:
     - CCS technology is still in development and has scalability issues.
     - Requires ongoing investment in capturing and storing large quantities of $CO_2$.
3) **Pink Hydrogen (Electrolysis Powered by Nuclear Energy)**
   - **Process**: Pink hydrogen is produced using **nuclear-powered electrolysis**, where nuclear energy is used to generate electricity to split water into hydrogen and oxygen.
   - **Technology**: Electrolysis cells powered by the heat or electricity generated from nuclear reactors.
   - **Advantages**:
     - Zero direct emissions, as nuclear energy produces minimal $CO_2$.
     - Potential for continuous, large-scale hydrogen production.
   - **Challenges**:
     - Public concerns over nuclear safety and waste disposal.
     - High initial infrastructure costs.
4) **Turquoise Hydrogen (Pyrolysis of Methane)**
   - **Process**: Turquoise hydrogen is produced through **methane pyrolysis**, a process where methane is heated to high temperatures in the absence of oxygen, breaking it down into hydrogen and solid carbon (carbon black).
   - **Technology**: Pyrolysis reactors that operate at high temperatures (typically 800–1,100 °C) to decompose methane into hydrogen and solid carbon.
   - **Advantages**:
     - No $CO_2$ emissions—solid carbon can be used in various industries (e.g., manufacturing, batteries).
     - Potential for high hydrogen purity.
   - **Challenges**:
     - High energy input required to maintain the necessary temperature.
     - Market and infrastructure development for solid carbon utilization.

5) **Biomass-Based Hydrogen (Biological Processes)**
   - **Process**: Biomass-derived hydrogen is produced from organic materials (like agricultural waste, wood, or municipal solid waste) through various biological processes such as biomass gasification or biological hydrogen production.
     - **Biomass Gasification**: Biomass is heated in the presence of limited oxygen, producing a gas mixture that includes hydrogen.
     - **Biological Hydrogen Production**: Certain microorganisms (e.g., algae or bacteria) generate hydrogen through metabolic processes.
   - **Advantages**:
     - Utilizes renewable waste materials.
     - Offers potential for low-cost hydrogen production.
   - **Challenges**:
     - Efficiency is still relatively low compared to other methods.
     - Difficulty in scaling up biological hydrogen production processes.
6) **Solar Hydrogen (Photoelectrochemical [PEC] Water Splitting)**
   - **Process**: Solar hydrogen is produced by directly using sunlight to drive water splitting through a **PEC cell.**
   - **Technology**: PEC cells integrate solar energy into the water electrolysis process, using semiconductor materials to directly convert solar energy into chemical energy to produce hydrogen.
   - **Advantages**:
     - Directly converts sunlight into hydrogen, reducing the need for complex infrastructure.
     - Highly sustainable and environmentally friendly.
   - **Challenges**:
     - Low efficiency and current technological limitations.
     - High cost of developing efficient PEC cells.
7) **Wind Hydrogen (Using Wind Power for Electrolysis)**
   - **Process**: Wind hydrogen is produced using wind energy to power water electrolysis, where the electricity generated by wind turbines splits water into hydrogen and oxygen.
   - **Advantages**:
     - Harnesses abundant, renewable wind energy.
     - Can be deployed in locations with high wind potential, such as offshore or remote areas.
   - **Challenges**:
     - Intermittency of wind power affects hydrogen production.
     - Requires substantial investment in both wind energy infrastructure and electrolysis technology. Table 3.13 shows the summary of clean hydrogen production methods.

**Table 3.13** Summary of Clean Hydrogen Production Methods

| Method | Key Feature | Emissions | Challenges |
|---|---|---|---|
| Green hydrogen | Electrolysis powered by renewable energy | Zero emissions | High cost, scalability issues |
| Blue hydrogen | SMR with carbon capture and storage (CCS) | Low emissions | High cost of CCS, limited scalability |
| Pink hydrogen | Electrolysis powered by nuclear energy | Zero emissions | Public safety concerns, high infrastructure cost |
| Turquoise hydrogen | Methane pyrolysis (produces solid carbon) | No $CO_2$ emissions | High energy input, market for carbon products |
| Biomass hydrogen | Biomass gasification or biological processes | Low emissions | Low efficiency, limited scalability |
| Solar hydrogen | PEC cells for direct solar-to-hydrogen conversion | Zero emissions | Low efficiency, technology limitations |
| Wind hydrogen | Wind-powered electrolysis | Zero emissions | Intermittent energy supply, high infrastructure cost |

Hydrogen energy offers significant environmental benefits, positioning it as a cornerstone in the global transition to a sustainable, low-carbon energy future. Following are some of the key environmental advantages of hydrogen energy:

1) **Zero Emissions at Point of Use**

   When hydrogen is used in fuel cells or combustion engines, the only by-product is water vapor ($H_2O$). This makes hydrogen an ideal energy carrier for applications where reducing air pollution and GHG emissions is a priority, such as in transportation and power generation. In hydrogen fuel cells, hydrogen reacts with oxygen from the air, producing electricity, heat, and water as the sole emissions. If hydrogen is used in internal combustion engines (ICEs) or turbines, the only emissions are water vapor and trace amounts of nitrogen oxides ($NO_x$), which can be minimized with appropriate technologies.

2) **Reduces GHG Emissions**

   Hydrogen has the potential to significantly reduce global GHG emissions, particularly if it is produced via clean methods like electrolysis powered by renewable energy. Hydrogen's ability to replace fossil fuels in industries, transport, and power generation helps cut down emissions from sectors that are difficult to decarbonize. Hydrogen-powered vehicles produce zero tailpipe emissions, which can help reduce carbon footprints from traditional gasoline or diesel-powered vehicles. Hydrogen can replace carbon-intensive processes in industries such as steel manufacturing, cement production, and refining, where alternative decarbonization solutions are challenging to implement.

3) **Integration with Renewable Energy Sources**

   Hydrogen enables the integration of renewable energy (e.g., solar, wind, hydro) into energy systems by providing an efficient energy storage solution. Hydrogen can act as a long-term energy storage medium for excess electricity generated by intermittent renewable sources. Excess renewable electricity can be converted into hydrogen through electrolysis, storing it for later use during periods of high demand or low renewable generation. This helps balance the grid and ensures stable energy supply, reducing the need for fossil fuel-based backup power.

4) **Reduces Dependence on Fossil Fuels**

   Hydrogen, especially when produced from renewable sources, reduces the reliance on fossil fuels for energy production, leading to significant environmental benefits such as reducing the burning of coal, oil, and natural gas lowers the emissions of particulate matter (PM), sulfur oxides ($SO_x$), nitrogen oxides ($NO_x$), and volatile organic compounds (VOCs), leading to better air quality and public health. By reducing reliance on fossil fuels, hydrogen energy helps mitigate the effects of climate change by lowering the overall carbon intensity of the energy system.

5) **Versatile Application for Decarbonizing Hard-to-Abate Sectors**

   Hydrogen offers a pathway to decarbonize industries and sectors that are difficult to electrify directly, including hydrogen-powered trucks, trains, ships, and airplanes can significantly reduce emissions from the transport sector, particularly for long-distance or heavy-duty applications where BEVs are less feasible. Hydrogen can be used as a feedstock or fuel in industries like steel production, cement manufacturing, and chemical production, which currently rely on fossil fuels and emit large amounts of $CO_2$. Hydrogen can also replace natural gas in residential, commercial, and industrial heating, reducing emissions from fossil fuel-based heating systems.

6) **Supports a Circular Economy**

   Hydrogen, particularly green hydrogen, contributes to a circular economy by enabling closed-loop systems where waste products are minimized or repurposed. For instance, hydrogen fuel cells produce only water vapor, which can be safely returned to the environment, and hydrogen itself can be stored and reused, promoting sustainability in the energy cycle.

7) **Reduces Water Pollution and Conserves Resources**

   Hydrogen production through electrolysis (using renewable energy) and biomass gasification avoids many of the environmental harms associated with fossil fuel extraction, such as Traditional energy sources like coal mining and oil extraction can lead to significant water pollution through runoff, chemical contamination, and oil spills. Hydrogen production, when done sustainably, avoids these risks. Hydrogen allows for the more efficient use of renewable resources, such as wind and solar, which can be harnessed without depleting natural resources. This leads to less environmental disruption compared to resource extraction processes used in fossil fuel industries.

8) **Promotes Global Energy Equity**

Hydrogen energy has the potential to be produced in nearly any region, especially in areas with abundant renewable energy resources, such as deserts (solar power) or coastal areas (wind power). Hydrogen energy can help supply clean, affordable energy to regions that currently lack access to electricity or have limited access to fossil fuels, contributing to sustainable development and energy equity. The ability to produce hydrogen locally reduces the need for extensive energy transportation infrastructure, decreasing the environmental impact associated with energy transmission.

9) **Minimal Land Use Impact**

Compared to large-scale renewable energy generation facilities, such as wind farms or solar parks, hydrogen production (especially from water electrolysis) requires minimal land space. It can be produced in small-scale, decentralized facilities, reducing the need for large land areas, which can impact ecosystems and biodiversity. Table 3.14 shows the summary of environmental benefits of hydrogen energy.

**Table 3.14** Summary of Environmental Benefits of Hydrogen Energy

| Environmental Benefit | Impact |
|---|---|
| Zero emissions at point of use | Produces only water vapor as a by-product |
| Reduces greenhouse gas emissions | Significant reduction in $CO_2$ emissions across sectors |
| Integration with renewable energy | Balances intermittent renewable energy and provides storage |
| Reduces fossil fuel dependence | Lowers air pollution, improves climate resilience |
| Decarbonizes hard-to-abate sectors | Cuts emissions in heavy transport and industries |
| Supports circular economy | Promotes sustainability with closed-loop systems |
| Reduces water pollution | Avoids environmental harms from fossil fuel extraction |
| Conserves resources | Decreases resource depletion and environmental disruption |

> ***Key: Carbon pricing:*** *An economic strategy to reduce carbon emissions by assigning a cost (or tax) to the release of $CO_2$, incentivizing cleaner energy alternatives like green hydrogen.*

### 3.3.2 Applications in Clean Transportation

Hydrogen energy has a wide range of applications in clean transportation, offering a sustainable alternative to conventional fossil fuels. As a zero-emission fuel, hydrogen plays a key role in decarbonizing the transport sector, especially in areas where electrification may be challenging. Here are some of the primary applications of hydrogen energy in clean transportation:

1) **Hydrogen FCVs**

   Hydrogen FCVs use hydrogen as a fuel to produce electricity, which powers an electric motor. The fuel cell combines hydrogen with oxygen from the air to generate electricity, emitting only water vapor and heat as by-products. This makes them a clean alternative to conventional ICE vehicles.

   - **Passenger Vehicles**:
     - Hydrogen-powered cars, such as the Toyota Mirai, Hyundai Nexo, and Honda Clarity, offer long driving ranges (up to 400 miles or more) and quick refueling times, making them ideal for urban and highway driving.
     - Hydrogen FCVs can serve as an alternative for EVs, especially in regions with limited charging infrastructure or for consumers who need long-range, fast refueling options.
   - **Advantages**:
     - Zero tailpipe emissions (only water vapor).
     - Quick refueling (typically 3–5 minutes).
     - Longer range compared to many BEVs.
   - **Challenges**:
     - High production cost of fuel cells and hydrogen storage.
     - Limited HRS.

2) **Hydrogen-Powered Buses**

   Hydrogen fuel cell buses are a popular solution for public transportation systems aiming to reduce their carbon footprints. They offer zero emissions, long range, and quick refueling, making them an ideal choice for city buses and long-distance intercity routes.

   - **Public Transit**: Several cities worldwide, such as London, Los Angeles, and Beijing, have deployed hydrogen-powered buses in their fleets to replace diesel-powered buses and reduce air pollution.

- **Long-Distance Coaches**: Hydrogen-powered long-distance coaches can reduce emissions in the intercity bus transportation sector, where EVs may face challenges due to their larger battery requirements.
- **Advantages**:
  - Zero emissions, improving urban air quality.
  - High passenger capacity and long operating ranges.
  - Fast refueling compared to battery electric buses.
- **Challenges**:
  - High infrastructure costs for refueling stations.
  - Initial investment and operational cost challenges.

3) **Hydrogen-Powered Trucks (Heavy-Duty Transport)**

**Hydrogen fuel cell trucks** are being developed to replace **diesel-powered heavy-duty trucks**, providing a cleaner alternative for freight transportation. Hydrogen-powered trucks can be used for both long-haul and short-haul applications, reducing carbon emissions in the freight sector

- **Long-Haul Trucks**: Hydrogen-powered long-haul trucks, such as those developed by **Nikola Motors** and **Hyundai**, are designed to reduce emissions in freight transportation, where battery electric trucks may struggle with range and fast refueling needs.
- **Advantages**:
  - Long range, ideal for long-haul freight where battery electric trucks may not be feasible due to energy density limitations.
  - Zero tailpipe emissions (water vapor).
  - Fast refueling for continuous freight operations.
- **Challenges**:
  - Expensive fuel cell technology and hydrogen storage solutions.
  - Development of refueling infrastructure along major transport routes.

4) **Hydrogen Trains (Rail Transport)**

**Hydrogen fuel cell trains** are being explored as an alternative to diesel-powered trains, especially in regions without extensive electrification of rail networks. Hydrogen trains can operate on nonelectrified tracks, providing a clean alternative for regional and intercity rail transport.

- **Regional and Intercity Rail**: Hydrogen-powered trains, such as the **Alstom Coradia iLint**, are already in operation in Germany, with plans for deployment in other countries.
- **Advantages**:
  - Zero emissions, contributing to cleaner air along rail corridors.
  - Suitable for nonelectrified tracks, reducing the need for extensive electrification infrastructure.
  - Quiet operation compared to diesel trains.

- **Challenges**:
  - High initial costs for hydrogen fuel cells and train infrastructure.
  - Developing a hydrogen refueling network along rail routes.

5) **Hydrogen-Powered Ships (Maritime Transport)**

   **Hydrogen-powered ships** are emerging as a solution for decarbonizing the maritime sector, which is responsible for significant GHG emissions. Hydrogen can be used in fuel cells or combustion engines to power ships, especially for short-distance or ferry operations.

   - **Ferries and Short-Distance Shipping**: Several hydrogen-powered ferries are being developed or are in operation in countries like Norway, with the **HySeas III** ferry being a notable example.
   - **Large Cargo Ships**: Although more challenging, hydrogen is being explored for use in larger cargo ships through fuel cell or ammonia-based solutions.
   - **Advantages**:
     - Zero emissions, helping to reduce maritime air pollution.
     - Hydrogen can be stored as a liquid or ammonia for easier transportation and storage on ships.
   - **Challenges**:
     - Storing and transporting hydrogen in a safe and efficient manner on ships.
     - Developing the required refueling infrastructure for maritime operations.

6) **Hydrogen-Powered Aircraft (Aviation)**

   The aviation industry is exploring **hydrogen fuel cells** and **hydrogen combustion** as a means to decarbonize air travel. Hydrogen has the potential to replace jet fuel in commercial aircraft, contributing to a reduction in aviation's carbon emissions.

   - **Short-Haul Aircraft**: Hydrogen-powered small planes for regional and short-haul flights are in development by companies like **ZeroAvia** and **Airbus**.
   - **Long-Haul Aircraft**: Hydrogen-powered solutions for long-distance flights are still in early stages of development, but there are significant research efforts into hydrogen combustion and hybrid systems that could be viable in the future.
   - **Advantages**:
     - Zero-carbon emissions from hydrogen-powered aircraft.
     - Potential to significantly reduce aviation's contribution to global warming.
   - **Challenges**:
     - The need for lightweight and high-efficiency fuel storage.
     - Development of new infrastructure for hydrogen refueling at airports.

7) **Hydrogen in Rail and Bus Stations, Ports, and Airports**

Hydrogen can also be used to provide **clean energy** for supporting infrastructure in transport hubs, such as:

- **Hydrogen-Powered Depots and Stations**: Refueling stations for hydrogen FCVs, buses, trucks, and trains can reduce the environmental footprint of public transportation and freight logistics.
- **Hydrogen-Powered Cranes and Forklifts**: In ports and airports, hydrogen can power equipment like cranes and cargo handlers, further reducing emissions in logistics operations.

## 3.4 Hydrogen Infrastructure Development

Hydrogen energy infrastructure involves the production, storage, distribution, and utilization of hydrogen as a clean energy source. Hydrogen is a versatile and sustainable fuel that plays a critical role in transitioning to a low-carbon economy, especially in sectors that are difficult to decarbonize, such as heavy industry and long-haul transportation. Figure 3.6 shows the key components of the hydrogen infrastructure.

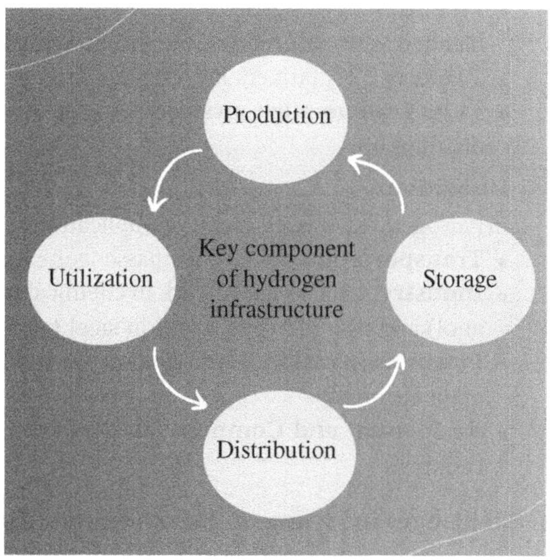


**Figure 3.6** Key Components of Hydrogen Infrastructure

### Key Components of Hydrogen Infrastructure

1) **Production**

   Hydrogen can be produced using various methods, which are categorized based on their carbon footprint:
   - **Green Hydrogen**: Produced via electrolysis of water using renewable energy (e.g., solar, wind).
   - **Blue Hydrogen**: Produced from natural gas with CCS.
   - **Gray Hydrogen**: Produced from natural gas without CCS, resulting in $CO_2$ emissions.
   - **Pink Hydrogen**: Produced via electrolysis using nuclear energy.

2) **Storage**

   Hydrogen storage is essential to balance supply and demand due to its intermittent production (from renewables):
   - **Compressed Gas Storage**: Hydrogen is stored under high pressure in tanks.
   - **$LH_2$ Storage**: Hydrogen is cooled to cryogenic temperatures for compact storage.
   - **Chemical Storage**: Using materials like ammonia, metal hydrides, or LOHCs.

3) **Distribution**

   Efficient distribution networks are required to transport hydrogen from production sites to end-users:
   - **Pipelines**: Hydrogen can be transported through dedicated pipelines or blended with natural gas in existing networks.
   - **Tankers**: $LH_2$ tankers for long-distance transport.
   - **Tube Trailers**: High-pressure cylinders mounted on trucks for smaller-scale distribution.

4) **Utilization**

   Hydrogen has a wide range of applications across various sectors:
   - **Transportation**: FCEVs for passenger cars, buses, trains, and ships.
   - **Industry**: Used as a feedstock in chemical production (e.g., ammonia, methanol) and as a reducing agent in steel manufacturing.
   - **Power Generation**: Hydrogen can be used in fuel cells or turbines to generate electricity.
   - **Residential and Commercial**: Blended with natural gas for heating and cooking.

### Challenges in Hydrogen Infrastructure Development

1) **High Costs**: Electrolyzers, fuel cells, and CCS technologies are expensive.
2) **Storage and Transportation**: Hydrogen's low energy density poses challenges for storage and transport.

3) **Safety Concerns**: Hydrogen is highly flammable and requires robust safety standards.
4) **Infrastructure Gaps**: Insufficient refueling stations and pipelines limit scalability.
5) **Regulatory Framework**: Standardized regulations and policies are needed to promote adoption.

### Future Directions and Innovations

- **Scaling Green Hydrogen**: Increased investment in renewable energy-driven hydrogen production.
- **Hydrogen Hubs**: Developing centralized clusters for production, storage, and utilization.
- **Advanced Materials**: Research into better catalysts, membranes, and storage materials.
- **Global Collaboration**: Countries forming alliances to build a global hydrogen economy.
- **Hydrogen Blending**: Gradual blending of hydrogen into natural gas pipelines as a transition strategy.

Hydrogen energy infrastructure is poised to revolutionize the global energy landscape, but its success depends on collaborative efforts between governments, industry, and research institutions to address current challenges and unlock its full potential.

The efficient distribution of hydrogen is a critical aspect of building a sustainable hydrogen economy. Distribution networks are essential to transport hydrogen from production sites to end-users across industries, transportation, and residential sectors. These networks include pipelines, road and rail transport, and marine shipping solutions, depending on the scale and distance of distribution. Figure 3.7 shows the hydrogen distribution network.

### Types of Hydrogen Distribution Networks

1) **Pipelines**
   - **Description**: Hydrogen is transported as a gas through pipelines, which can be dedicated hydrogen pipelines or modified natural gas pipelines.
   - **Applications**: Suitable for large-scale and long-term hydrogen distribution between production sites and industrial users.
   - **Advantages**:
     - Cost-effective for large volumes over short to medium distances.
     - Continuous supply reduces the need for storage.

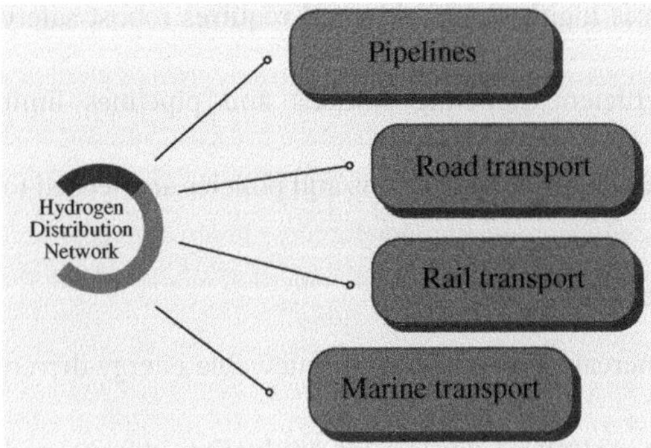


**Figure 3.7** Hydrogen Distribution Network

- **Challenges**:
  - Requires high-purity hydrogen to avoid pipeline degradation.
  - Initial infrastructure development is capital-intensive.
- **Examples**:
  - The European Hydrogen Backbone initiative aims to create a pan-European pipeline network.
  - U.S. Gulf Coast hydrogen pipeline system.

2) **Road Transport (Tube Trailers)**
   - **Description**: Hydrogen is transported in high-pressure cylinders (compressed gas) or cryogenic tanks ($LH_2$) mounted on trucks.
   - **Applications**: Ideal for small- to medium-scale distribution, such as to HRS.
   - **Advantages**:
     - Flexible and scalable.
     - Suitable for regions without pipeline infrastructure.
   - **Challenges**:
     - High costs per unit of hydrogen transported.
     - Safety concerns with pressurized tanks.
   - **Use Cases**:
     - Supplying remote HRS.
     - Pilot projects for localized hydrogen distribution.
3) **Rail Transport**
   - **Description**: Hydrogen is transported via rail in tankers, either as compressed gas or as liquid.
   - **Applications**: Large-scale transport for industrial applications or across regions with extensive rail networks.
   - **Advantages**:
     - Cost-effective for bulk transport over long distances.
     - Reduced carbon footprint compared to road transport.

- **Challenges**:
  - Requires specialized railcars and loading/unloading facilities.
  - Infrastructure adaptation for hydrogen transport.

4) **Marine Transport**
   - **Description**: Hydrogen is shipped over long distances in liquid form or as derivatives like ammonia or methanol.
   - **Applications**: International transport of hydrogen from production-rich regions (e.g., Australia, the Middle East) to consumption-heavy regions (e.g., Europe, Japan).
   - **Advantages**:
     - Enables large-scale hydrogen trade across continents.
     - Derivatives like ammonia are easier to handle and transport.
   - **Challenges**:
     - Energy-intensive liquefaction process.
     - Development of specialized shipping vessels.
   - **Examples**:
     - Japan-Australia Hydrogen Energy Supply Chain (HESC).

**Innovations in Hydrogen Distribution**

1) **Liquid Organic Hydrogen Carriers**:
   - Hydrogen is chemically bonded to a liquid carrier for safe and efficient transport.
   - The carrier is reusable, making it a sustainable option.
2) **Cryogenic Compression**:
   - Combining compression and cooling to store and transport hydrogen efficiently.
3) **Ammonia as a Carrier**:
   - Ammonia can be easily liquefied and transported, then converted back to hydrogen at the destination.
4) **Hydrogen Drones**:
   - Small-scale distribution using drones for remote areas or specific applications.

**Challenges in Hydrogen Distribution**

- **Cost**: High costs of infrastructure development and maintenance.
- **Energy Loss**: Conversion and transportation processes can be energy-intensive.
- **Safety**: Hydrogen's flammability requires stringent safety standards.
- **Standardization**: Lack of uniform standards and regulations across regions.

**Future Directions**

- Development of hydrogen corridors to connect high-demand regions.
- Investment in dedicated hydrogen infrastructure for long-term scalability.
- Integration of digital technologies like IoT and AI for efficient monitoring and control of distribution networks.
- International collaborations to establish a global hydrogen supply chain.

Hydrogen distribution networks are pivotal in realizing the hydrogen economy, and continued innovation and investment will enhance their efficiency, scalability, and safety.

### 3.4.1 Hydrogen Refueling Stations

HRS are critical components of the hydrogen economy, enabling the use of hydrogen as a fuel for vehicles, particularly FCEVs. These stations facilitate the transfer of hydrogen from storage to vehicle tanks, ensuring safe, efficient, and timely refueling.

**Components of a HRS**

1) **Hydrogen Supply**
   - **On-Site Generation**: Hydrogen is produced at the station using electrolysis or reforming.
   - **Delivery**: Hydrogen is transported to the station as compressed gas or liquid via trailers or pipelines.
   - **Storage**: Hydrogen is stored on-site in high-pressure tanks or cryogenic containers.
2) **Compression System**
   - Hydrogen is compressed to high pressures (typically 350–700 bar) for storage and transfer into vehicle tanks.
3) **Dispensers**
   - Equipped with specialized nozzles to transfer hydrogen safely into FCEVs.
   - Automated systems monitor pressure, temperature, and flow rate to ensure precise fueling.
4) **Cooling System**
   - Pre-cools hydrogen to −40 °C to reduce heat generated during refueling and ensure compatibility with vehicle storage systems.
5) **Safety Systems**
   - Includes hydrogen leak detectors, automatic shut-off valves, and robust protocols to manage high pressures and flammability risks.

**Types of HRS**

1) **Public Refueling Stations**
   - Accessible to all hydrogen-powered vehicles.
   - Common in urban areas and along highways for consumer convenience.
2) **Private Refueling Stations**
   - Restricted to specific fleets or industrial users, such as buses, trucks, or forklifts.
3) **Mobile Refueling Units**
   - Portable stations used for temporary refueling needs or in areas with low demand.
4) **Integrated Stations**
   - Co-located with conventional refueling options (e.g., gasoline, electric charging).

**Refueling Process**

1) **Vehicle Arrival**: The vehicle connects to the dispenser using a standardized nozzle.
2) **Authentication**: Payment or identification through digital systems.
3) **Pre-Check**: Safety checks for pressure and compatibility.
4) **Fueling**: Hydrogen flows into the vehicle tank, monitored for temperature and pressure.
5) **Completion**: The system automatically stops when the tank is full, and the user disconnects the nozzle.

Refueling typically takes 3–5 minutes, comparable to conventional fuel.

**Global Deployment of HRS**

- **Japan**: A leader with a national strategy to establish widespread HRS for FCEVs.
- **Europe**: Countries like Germany, France, and the United Kingdom are building hydrogen corridors.
- **The United States**: California leads with the most HRS, supporting its zero-emission vehicle goals.
- **South Korea**: Rapid expansion driven by government and industry initiatives.

**Challenges in HRS Deployment**

1) **High Initial Costs**: Building an HRS can cost \$1–\$2 million.
2) **Infrastructure Gaps**: Limited availability compared to traditional refueling stations.

3) **Safety Concerns**: Hydrogen's flammability requires stringent standards.
4) **Demand Uncertainty**: Adoption of FCEVs is still in the early stages.

**Future Trends and Innovations**

1) **Modular HRS**: Smaller, scalable units to reduce costs and enable gradual expansion.
2) **On-Site Green Hydrogen Production**: Solar or wind-powered electrolysis at the station.
3) **High-Capacity Stations**: Designed for heavy-duty vehicles like trucks and buses.
4) **Digital Integration**: IoT and AI-enabled monitoring systems for efficiency and safety.
5) **Expansion of Hydrogen Highways**: Networked stations along major transport routes to support long-distance travel.

HRS are essential for advancing hydrogen mobility and achieving decarbonization goals in the transportation sector. Investments in infrastructure, innovation, and supportive policies will be crucial for their widespread adoption.

## 3.5 Conclusion

The transition to a sustainable energy future is imperative in addressing global climate change and reducing dependence on fossil fuels. Hydrogen has emerged as a key enabler in this transformation, offering a versatile and clean energy solution with applications across multiple sectors. This chapter has provided a comprehensive analysis of hydrogen's role in the energy transition by examining its fundamental characteristics, technological advancements, economic feasibility, and policy landscape.

Beginning with an introduction to energy transition, we explored the growing need for a shift toward renewable and low-carbon energy sources. Hydrogen, as discussed in understanding hydrogen as an energy carrier, possesses unique properties that make it an efficient medium for energy storage, transport, and utilization. Its ability to complement renewable energy sources, such as wind and solar power, enhances grid stability and facilitates large-scale energy distribution.

As a clean energy source, hydrogen presents a viable solution for decarbonizing hard-to-abate sectors like heavy industry, aviation, and long-haul transportation. However, the realization of a hydrogen-based economy requires substantial investments in infrastructure development, including production facilities, efficient storage solutions, and distribution networks. Developing cost-effective and scalable hydrogen infrastructure remains a challenge but is crucial for its widespread adoption.

From an economic perspective, hydrogen production and utilization must become more cost-competitive with conventional energy sources. Advances in technology, economies of scale, and declining renewable energy costs are gradually making hydrogen a more attractive option. However, financial support, incentives, and public–private partnerships are essential to accelerate market adoption.

The success of hydrogen in the energy transition largely depends on the policy and regulatory frameworks that govern its development. Governments worldwide are implementing hydrogen strategies, setting carbon reduction targets, and offering incentives to drive investment in hydrogen technologies. International collaboration and standardization will further enhance hydrogen's role in achieving a decarbonized global economy.

In conclusion, hydrogen is a promising and indispensable component of the clean energy transition. While challenges remain in infrastructure, costs, and policy alignment, continued innovation, supportive regulations, and strategic investments will unlock hydrogen's full potential. A concerted effort from governments, industries, and research institutions will be essential in realizing a sustainable, hydrogen-powered future.

## 3.6 Case Study

- **Hydrogen's Role in Japan's Energy Transition**

  **Introduction to Energy Transition**

  Japan, a nation heavily reliant on imported fossil fuels, has been at the forefront of the global energy transition. With limited domestic energy resources and a strong commitment to reducing carbon emissions, Japan has embraced hydrogen as a key component of its clean energy strategy. This case study explores how Japan is integrating hydrogen into its energy system to achieve sustainability, energy security, and economic growth.

  **Understanding Hydrogen as an Energy Carrier**

  Recognizing hydrogen's potential as a clean and flexible energy carrier, Japan has invested in diverse hydrogen production methods, including blue hydrogen (derived from natural gas with carbon capture) and green hydrogen (produced using renewable energy). Hydrogen is being utilized in multiple applications, from electricity generation and industrial processes to FCVs and residential energy supply.

**Hydrogen as a Clean Energy Source**

Japan's commitment to hydrogen is evident in projects such as the Fukushima Hydrogen Energy Research Field (FH2R), one of the world's largest green hydrogen production facilities. Powered by renewable energy, FH2R produces hydrogen via electrolysis, demonstrating its potential to support grid stability and energy storage. Additionally, hydrogen FCVs, such as Toyota's Mirai, are being promoted as an alternative to conventional gasoline-powered cars, reducing emissions in the transportation sector.

**Hydrogen Infrastructure Development**

To facilitate the adoption of hydrogen, Japan has made significant strides in infrastructure development. The country has established a network of HRS, aiming for over 1,000 stations by 2030. Moreover, Japan is pioneering the development of hydrogen supply chains, including the importation of liquefied hydrogen from Australia through the HESC project.

**Economic Considerations**

Despite its potential, hydrogen adoption faces economic challenges, including high production and transportation costs. Japan has focused on cost reduction through technological advancements and economies of scale. Government subsidies and incentives have encouraged private sector investments, making hydrogen-based solutions more financially viable.

**Policy and Regulatory Frameworks**

Japan's government has played a crucial role in hydrogen deployment through strategic policies such as the Basic Hydrogen Strategy (2017) and the Green Growth Strategy (2020). These policies outline long-term goals, funding mechanisms, and international collaboration efforts to position Japan as a global leader in hydrogen innovation.

**Conclusion**

Japan's hydrogen initiatives serve as a model for other nations seeking a sustainable energy transition. By leveraging hydrogen as a clean energy source, investing in infrastructure, addressing economic barriers, and implementing supportive policies, Japan is paving the way for a hydrogen-powered future.

- **Germany's Hydrogen Strategy for a Sustainable Energy Transition**

**Introduction to Energy Transition**

Germany has long been a global leader in renewable energy and decarbonization efforts. As part of its transition away from fossil fuels and nuclear energy, the country has identified hydrogen as a crucial component in achieving climate neutrality by 2045. This case study examines Germany's approach to integrating hydrogen into its energy system, highlighting key developments in production, infrastructure, economic considerations, and policy support.

**Understanding Hydrogen as an Energy Carrier**
Germany views hydrogen as an essential energy carrier that can complement renewable energy sources like wind and solar. Since excess renewable energy can be used to produce hydrogen via electrolysis, hydrogen serves as an effective means of energy storage and grid balancing. It also enables sector coupling by providing clean energy for industries, heating, and transportation.

**Hydrogen as a Clean Energy Source**
Germany is heavily investing in green hydrogen, which is produced using renewable electricity. The country has launched several pilot projects, including the REFHYNE project, Europe's largest PEM electrolyzer, to produce green hydrogen for industrial use. Additionally, hydrogen is being explored for decarbonizing steel production, with companies like Thyssenkrupp and Salzgitter AG replacing coal-based processes with hydrogen in steel manufacturing.

**Hydrogen Infrastructure Development**
To ensure large-scale hydrogen adoption, Germany is building an extensive hydrogen pipeline network and expanding its HRS for FCVs. The $H_2$ Mobility initiative aims to establish a nationwide hydrogen fueling network to support the growing fleet of hydrogen-powered buses, trucks, and trains. Moreover, Germany is working on international partnerships to import hydrogen from countries with abundant renewable energy resources, such as Australia and North Africa.

**Economic Considerations**
One of the key challenges in Germany's hydrogen transition is the high cost of green hydrogen production. The government is addressing this through the National Hydrogen Strategy (NWS), which includes €9 billion in funding to develop hydrogen infrastructure, support research and innovation, and subsidize early-stage hydrogen projects to enhance competitiveness.

**Policy and Regulatory Frameworks**
Germany's hydrogen strategy is backed by strong regulatory frameworks and policy initiatives. The NWS (2020) sets ambitious targets, aiming for 5 GW of electrolysis capacity by 2030 and positioning Germany as a global leader in hydrogen technology. The government is also aligning its hydrogen policies with the European Union's Hydrogen Strategy, ensuring coordinated efforts across member states.

## 3.7 Exercise

1. What are the key drivers behind the global energy transition, and why is hydrogen considered a vital component of this shift?

2. How does hydrogen compare to other renewable energy sources in terms of efficiency, storage, and applicability?

3. What are the major challenges in transitioning from a fossil fuel-based economy to a hydrogen-based energy system?

4. What are the different types of hydrogen (green, blue, gray, etc.), and how are they produced?

5. How does hydrogen function as an energy carrier, and what are its advantages over traditional fuels?

6. What role does hydrogen play in energy storage and grid balancing for renewable energy sources?

7. How can hydrogen contribute to decarbonizing hard-to-abate sectors such as transportation, industry, and power generation?

8. What are the environmental benefits of using hydrogen as a clean energy source compared to fossil fuels?

9. What are the potential risks or challenges associated with hydrogen production, storage, and usage in terms of safety and efficiency?

10. What are the key components of a hydrogen infrastructure, and how do they enable large-scale hydrogen adoption?

11. How does the development of hydrogen refueling stations and pipeline networks impact the commercialization of hydrogen technologies?

12. What are the technological advancements needed to enhance hydrogen production, transportation, and storage?

13. What are the main economic barriers to widespread hydrogen adoption, and how can they be overcome?

14 How do economies of scale and government incentives influence the cost-competitiveness of hydrogen in the energy market?

15 What are the key policy measures and regulatory frameworks that support hydrogen development at the national and global levels?

## 3.8 Objective Type Questions

1 Which of the following best defines the term "energy transition"?
**a)** Shift from nuclear to coal energy
**b)** Shift from fossil fuels to renewable and clean energy sources
**c)** Shift from hydro to wind power only
**d)** Increase in global energy consumption

2 One of the key drivers of energy transition is:
**a)** Decline in energy usage
**b)** Technological stagnation
**c)** Climate change and sustainability goals
**d)** Abundance of fossil fuels

3 The historical context of energy transition involves moving from:
**a)** Wind to solar energy
**b)** Renewable to fossil fuels
**c)** Biomass and coal to oil and gas, and now to renewables
**d)** Nuclear energy to coal

4 Hydrogen storage using high-pressure tanks is an example of:
**a)** Material-based storage
**b)** Chemical storage
**c)** Physical storage
**d)** Biological storage

5 Metal hydrides are used in which hydrogen storage method?
**a)** Liquid hydrogen storage
**b)** High-pressure gas storage
**c)** Material-based storage
**d)** Compressed air storage

6 Which of the following is NOT a physical storage method for hydrogen?
   a) High-pressure gas
   b) Cryogenic liquid hydrogen
   c) Metal–organic frameworks
   d) Composite pressure tanks

7 One of the key benefits of hydrogen as a clean energy source is:
   a) High levels of carbon emissions
   b) Zero emissions at the point of use
   c) High noise pollution
   d) Low energy density

8 Which method is primarily used to produce clean hydrogen?
   a) Coal gasification
   b) Water electrolysis using renewable energy
   c) Natural gas reforming
   d) Biomass combustion

9 What is the main environmental benefit of green hydrogen?
   a) Low water usage
   b) Reduced land use
   c) No greenhouse gas emissions during use
   d) High waste generation

10 Hydrogen fuel cell vehicles contribute to clean transportation by:
   a) Using diesel backup engines
   b) Releasing $CO_2$ into the air
   c) Producing only water as a by-product
   d) Using fossil fuel blends

11 Which of the following is considered part of hydrogen infrastructure development?
   a) Wind turbine farms
   b) Hydrogen refueling stations
   c) Coal mines
   d) Electricity billing systems

12 Hydrogen distribution networks are designed for:
   a) Water supply systems
   b) Delivering hydrogen to end-users
   c) Solar energy distribution
   d) Fueling nuclear reactors

**13** A challenge in establishing hydrogen infrastructure is:
**a)** Excess demand for coal
**b)** Lack of sunlight
**c)** High cost of pipeline construction
**d)** Water pollution

**14** Economies of scale in hydrogen production mean:
**a)** Higher costs with larger volumes
**b)** Lower costs as production scales up
**c)** Constant costs regardless of production
**d)** Reduced efficiency with larger plants

**15** Which factor directly affects the cost of hydrogen production?
**a)** Number of gas stations
**b)** Type of electrolyzer and electricity cost
**c)** Age of the user
**d)** Wind direction

**16** Hydrogen infrastructure cost includes expenses related to:
**a)** PowerPoint presentations
**b)** Office electricity bills
**c)** Storage, transport, and refueling networks
**d)** Data management systems only

**17** Material-based hydrogen storage is still evolving due to:
**a)** Safety concerns in water usage
**b)** Lack of interest from the industry
**c)** Complex chemistry and high cost
**d)** Government regulations banning it

**18** Which of the following is NOT an application of clean hydrogen?
**a)** Powering fuel cell electric vehicles
**b)** Backup power in remote areas
**c)** Coal-powered thermal plants
**d)** Grid energy storage systems

**19** Supply chain economics for hydrogen includes:
**a)** Market price volatility and logistics
**b)** Solar cell installation
**c)** Internet pricing models
**d)** Carbon trading algorithms

**20** A major determinant of hydrogen market demand is:
- **a)** Color of the hydrogen
- **b)** Public awareness and policy incentives
- **c)** Number of fuel stations
- **d)** Size of the hydrogen molecule

## 3.9 Glossary

1) **Energy Transition**: The global shift from fossil fuel-based energy systems to renewable and low-carbon alternatives, aimed at reducing greenhouse gas emissions and promoting sustainability.
2) **Hydrogen**: A versatile and abundant element that can act as a clean energy carrier, with applications in power generation, transportation, and industry.
3) **Energy Carrier**: A substance or system (like hydrogen, electricity, or biofuels) that stores and transports energy, enabling its use across different sectors and applications.
4) **Clean Energy**: Energy produced without significant greenhouse gas emissions, often sourced from renewables such as solar, wind, hydro, or green hydrogen.
5) **Hydrogen as an Energy Carrier**: The use of hydrogen to store and deliver energy across sectors due to its high energy density, flexibility, and ability to decarbonize multiple industries.
6) **Electrolysis**: A process that uses electricity, ideally from renewable sources, to split water into hydrogen and oxygen, producing clean hydrogen without carbon emissions.
7) **Decarbonization**: The process of reducing or eliminating carbon dioxide emissions from energy systems, industries, and transportation to mitigate climate change.
8) **Infrastructure Development (Hydrogen)**: The establishment of facilities and systems required for hydrogen production, storage, transportation, and distribution to enable large-scale adoption.
9) **Hydrogen Storage**: Technologies and methods used to safely store hydrogen in gaseous, liquid, or material-based forms for later use in energy applications.
10) **Distribution Networks**: The pipelines, transport systems, and refueling stations necessary to deliver hydrogen from production sites to end-users across industries and sectors.
11) **Economic Considerations**: The financial aspects of hydrogen adoption, including cost competitiveness, investment requirements, market demand, and long-term economic opportunities.

12) **Economic Analysis (Hydrogen Economy)**: Evaluation of the costs, benefits, and potential profitability of hydrogen technologies in comparison with conventional energy systems.
13) **Policy and Regulatory Frameworks**: Government strategies, laws, and regulations that guide and support the development, deployment, and safe usage of hydrogen technologies.
14) **Governance in Energy Transition**: The role of institutions, policies, and international cooperation in managing the adoption of hydrogen and ensuring a coordinated energy transition.
15) **Market Opportunities**: Potential business and investment avenues created by the growth of the hydrogen economy, including technology development, infrastructure, and global trade.

## References


1 Chen, L., Wang, Y., Jiang, Y., Zhang, C., Liao, Q., Li, J., Wu, J., & Gao, X. (2024). Life cycle assessment of liquid hydrogen fuel for vehicles with different production routes in China. *Energy*, 299, 131472 https://doi.org/10.1016/J.ENERGY.2024.131472.

2 Filina, O. A., Malozyomov, B. V., & Shchurov, N. I. (2024). Generation of hydrogen fuel on board vehicles with internal combustion engines. *International Journal of Hydrogen Energy*, 93, 320–327 https://doi.org/10.1016/J.IJHYDENE.2024.08.042.

3 Hu, D., Huang, J., Lu, D., & Wang, J. (2024). Hydrogen consumption estimation of fuel cell vehicle based on vehicle energy transfer. *Sustainable Energy Technologies and Assessments*, 68, 103854 https://doi.org/10.1016/J.SETA.2024.103854.

4 Shchurov, N. I., Dedov, S. I., & Shtang, A. A. (2024). Modelling and simulation hybrid electric vehicle with hydrogen fuel cells. *International Journal of Hydrogen Energy*, 95, 481–488 https://doi.org/10.1016/J.IJHYDENE.2024.08.044.

5 Dai, J., Yang, S., Yang, Y., & Fang, Q. (2024). Research on the consequences of hydrogen leakage and explosion accidents of fuel cell vehicles in underground parking garages. *International Journal of Hydrogen Energy*, 77, 1296–1306 https://doi.org/10.1016/J.IJHYDENE.2024.06.278.

6 El-Sayed, W. T., Awad, A. S. A., Al-Abri, R., Alawasa, K., Onen, A., & Ahshan, R. (2024). Integrated planning of hydrogen supply chain and reinforcement of power distribution network for accommodating fuel cell electric vehicles. *International Journal of Hydrogen Energy*, 81, 865–877 https://doi.org/10.1016/J.IJHYDENE.2024.07.316.

7 Fandi, G., Novák, J., Chyský, J., & Šrom, J. (2024). Review and modeling on hydrogen fuel cells electric vehicle (HFCEV), in comparison with battery electrical vehicle (BEV) using MATLAB environment. Case study: Postal car. *Energy*

*Conversion and Management: X*, 24, 100684 https://doi.org/10.1016/J.ECMX.2024.100684.

**8** Gao, Y., Liu, H., & Hou, Y. (2024). Effects of leakage location and ventilation condition on hydrogen leakage during shipping of fuel cell vehicles. *International Journal of Hydrogen Energy*, 54, 1532–1543 https://doi.org/10.1016/J.IJHYDENE.2023.12.095.

**9** Gong, K., Zheng, W., & Shu, Y. (2024). Battery leasing business for hydrogen fuel cell vehicles: Motorists' costs, adoption, and manufacturers' profits. *Energy*, 293, 130657 https://doi.org/10.1016/J.ENERGY.2024.130657.

**10** Halder, P., Babaie, M., Salek, F., Shah, K., Stevanovic, S., Bodisco, T. A., & Zare, A. (2024). Performance, emissions and economic analyses of hydrogen fuel cell vehicles. *Renewable and Sustainable Energy Reviews*, 199, 114543 https://doi.org/10.1016/J.RSER.2024.114543.

**11** Hu, D., Hou, W., Cheng, Z., Feng, C., Lu, D., Yi, F., Yang, Q., Li, J., & Wang, J. (2024). A temperature fluctuation suppression control method of fuel cell vehicles to reduce hydrogen consumption. *Energy*, 305, 132378 https://doi.org/10.1016/J.ENERGY.2024.132378.

**12** Hu, D., Hu, Z., Wang, J., Li, J., Lu, M., Ding, H., Wei, W., Zhang, X., & Wang, C. (2025). CL-Kansformer model for SOC prediction of hydrogen refueling process in fuel cell vehicles. *Journal of Power Sources*, 626, 235772 https://doi.org/10.1016/J.JPOWSOUR.2024.235772.

# 4

# The Urgency of Decarbonization: Environmental and Climate Imperatives

## 4.1 Introduction to Decarbonization

The accelerating impacts of climate change have underscored the urgent need for decarbonization—the systematic reduction of carbon emissions across industries and energy systems. As global temperatures continue to rise due to excessive GHG emissions, particularly carbon dioxide ($CO_2$), the transition to a low-carbon economy has become an environmental and economic necessity. The reliance on fossil fuels for energy generation, industrial processes, and transportation has significantly contributed to climate disruptions, necessitating a paradigm shift toward sustainable, low-emission alternatives such as hydrogen technologies.

This chapter explores the multifaceted urgency of decarbonization, beginning with an overview of its principles and objectives. The environmental consequences of carbon emissions—including extreme weather events, rising sea levels, biodiversity loss, and deteriorating air quality—are examined to highlight the necessity of immediate action. In addition to environmental degradation, carbon emissions impose significant economic costs, affecting global productivity, increasing healthcare burdens, and exacerbating disaster recovery expenses. The financial implications of inaction make decarbonization not only a climate imperative but also an economic one [1, 2].

Technological innovation has emerged as a key driver in the decarbonization transition. Breakthroughs in renewable energy, CCS, electrification, and hydrogen-based solutions offer promising pathways to reduce emissions across critical sectors. Additionally, the role of business industries is increasingly pivotal, as corporate sustainability strategies, carbon pricing mechanisms, and green investments shape the trajectory of industrial decarbonization.

As industries and policymakers seek viable strategies for a sustainable future, hydrogen technologies hold immense potential in revolutionizing energy systems and decarbonizing high-emission industries. This chapter sets the stage for a

*Hydrogen Technologies in Decarbonizing Industries and Energy Systems*, First Edition.
Vikas Khare, Sanjeet K. Dwivedi, Monica Bhatia, and K. K. Pant.

Published 2026 by John Wiley & Sons, Inc.

deeper exploration of decarbonization pathways, emphasizing the intersection of policy, technology, and industry leadership in addressing the global climate crisis.

> ***Key: Decarbonization:*** *The process of reducing carbon dioxide emissions from energy production, industrial activities, and transportation to combat climate change.*

Decarbonization refers to the process of reducing or eliminating carbon dioxide ($CO_2$) emissions, primarily from energy sources like fossil fuels (coal, oil, and natural gas), which are significant contributors to climate change. It involves transitioning to low-carbon or carbon-neutral alternatives, such as renewable energy (solar, wind, hydroelectric), nuclear energy, and the adoption of energy-efficient technologies, alongside practices like CCS and reforestation. Decarbonization in the context of green hydrogen refers to the process of reducing carbon emissions by producing hydrogen through environmentally friendly means, specifically using renewable energy sources like wind, solar, or hydroelectric power to split water ($H_2O$) into hydrogen ($H_2$) and oxygen ($O_2$) via electrolysis. Unlike conventional hydrogen production methods that rely on natural gas (producing significant $CO_2$ emissions), green hydrogen is entirely carbon-free, making it a key element in the transition to a low-carbon economy [3, 4].

The decarbonization through green hydrogen can be modeled using a mathematical equation based on the key aspects of energy consumption, carbon emissions reduction, and hydrogen production. Following is a simplified equation:

**Carbon Emission Function:**

The total carbon emissions at time "$t$" can be expressed as:

$$E(t) = \sum_{i=1}^{N} \alpha_i S_i(t) \tag{4.1}$$

where

$E(t)$ = Total carbon emissions at time "'$t$."
$S_i(t)$ = Energy produced source $i$.
$\alpha_i$ = Emission factor (amount of $CO_2$ emitted per unit of energy from source "$i$").

For fossil fuels, $\alpha_i$ is high, and for renewables, $\alpha_i = 0$. The goal is to reduce $E(t)$ by increasing the share of low-carbon sources.

**Decarbonization Rate:**

Decarbonization can be measured as the rate of decline in carbon intensity per unit of energy over time:

$$\lambda(t) = -\frac{d}{dt}\left(\frac{E(t)}{S(t)}\right) \tag{4.2}$$

where

$\lambda(t)$ = Decarbonization rate.
$S(t)$ = Total energy supply.
$E(t)/S(t)$ = Carbon intensity (emissions per unit of energy).

A higher $\lambda(t)$ indicates a faster shift to clean energy.

**Economic Cost of Decarbonization**
The cost of decarbonization includes investment in renewables, carbon taxes, and operational costs.

$$C_{total} = \int_0^T (C_r(S_r) + (C_0(S_r) + PE(t))dt) \tag{4.3}$$

where

$C_r(S_r)$ = Investment in renewable energy.
$C_0(S_r)$ = Operational costs.
$P$ = Carbon tax per unit of emission.
$PE(t)$ = Cost penalty for carbon emissions.

Minimizing $C_{total}$ while achieving net-zero emission is an optimization problem.

**CCS Model**
Carbon removal through CCS can be modeled as:

$$E_{net}(t) = E(t) - C_{CCS}(t) \tag{4.4}$$

where
$C_{CCS}(t)$ = Amount of carbon captured and stored at time $t$.

If $E_{net}(t) \leq 0$, the economic achieves carbon neutrality.

*Based on the above-mentioned mathematical interpretation, following are the significance of the decarbonization:*

**Zero-Carbon Emissions:** Green hydrogen offers a clean alternative for sectors that are hard to decarbonize, such as heavy industry (e.g., steel and cement production), aviation, and shipping. When used as fuel, green hydrogen produces only water vapor, contributing to the reduction of global $CO_2$ emissions. Figure 4.1 shows the significance of decarbonization through green hydrogen.

**Energy Storage and Grid Stabilization:** Green hydrogen can act as an energy storage medium, storing excess energy generated from renewable sources during times of low demand. This helps stabilize power grids, particularly those relying heavily on intermittent renewable sources like solar and wind.

**Decarbonization of Transport:** Hydrogen fuel cells powered by green hydrogen can replace fossil fuels in vehicles, buses, trucks, and trains, significantly lowering emissions in the transportation sector. It also offers a solution for long-haul and heavy-duty transport, where BEVs might not be feasible due to weight and range limitations [5, 6].

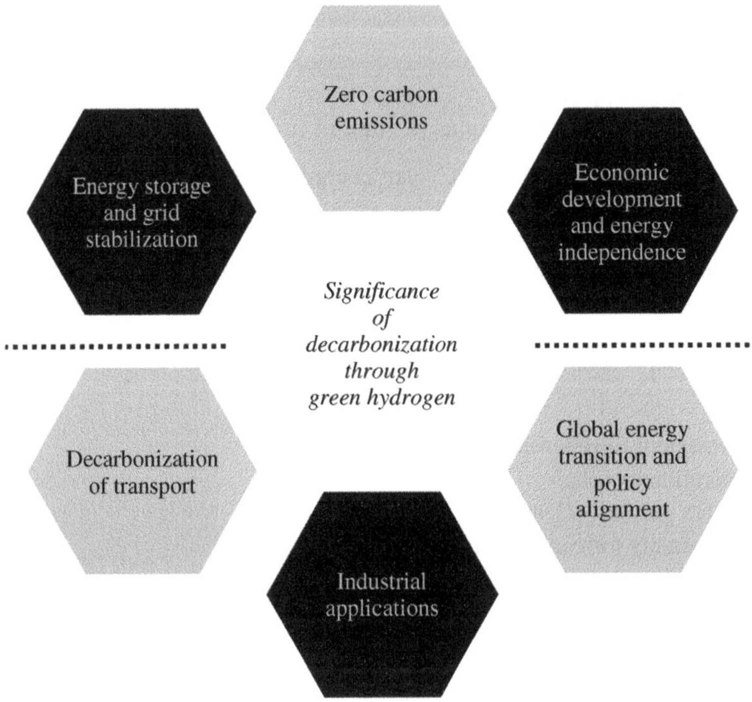


**Figure 4.1** Significance of Decarbonization Through Green Hydrogen

**Industrial Applications:** Green hydrogen is critical for decarbonizing industries like chemicals, refining, and fertilizers, where it is already used but traditionally produced using carbon-intensive methods (gray hydrogen). Green hydrogen can replace gray hydrogen, cutting emissions in these industries.

**Global Energy Transition and Policy Alignment:** Many countries are incorporating green hydrogen into their national energy strategies to meet climate targets set by the Paris Agreement. It helps bridge the gap between current energy demands and future sustainability goals, making it essential for long-term global decarbonization efforts.

**Economic Development and Energy Independence:** Investing in green hydrogen technology can foster economic growth through new industries and job creation in the renewable energy sector. It also reduces reliance on imported fossil fuels, enhancing energy independence and security for countries adopting the technology.

The significance of decarbonization lies in its potential to mitigate the adverse effects of climate change, improve public health, and drive sustainable economic growth. These far-reaching benefits serve as the foundation upon which the objectives and goals of decarbonization efforts are built. By recognizing the critical need to reduce GHG emissions, global initiatives are increasingly focused on transitioning to renewable energy sources, enhancing energy efficiency, and fostering low-carbon innovations. These objectives aim not only to protect the environment but also to ensure energy security, promote green jobs, and build resilient communities. Thus, the significance of decarbonization and its strategic goals are intrinsically linked, each reinforcing the other in the pursuit of a cleaner, more sustainable future.

➢ **Following are the objectives and goals of decarbonization efforts:**

**Reduction of GHG Emissions**

Objective: Use green hydrogen to cut emissions by 30–50% in heavy industries (steel, cement, chemicals) and transportation by 2030.

Goal: Achieve significant reductions in GHG emissions by replacing fossil fuels with green hydrogen in key sectors such as industry, transportation, and energy.

**Energy Transition and Renewable Energy Integration**

Objective: Achieve 100% renewable energy-based hydrogen production, contributing to grid stability and the broader transition to clean energy.

Goal: Promote the use of renewable energy sources (e.g., wind and solar) to power electrolysis for green hydrogen production.

**Industrial Decarbonization**

Objective: Replace gray hydrogen in industrial processes with green hydrogen, targeting full decarbonization of these sectors by 2050.

Goal: Decarbonize hard-to-abate sectors such as steel, cement, and chemicals.

**Green Hydrogen Production Scale-Up**

Objective: Support the deployment of large-scale electrolyzers and promote investments in hydrogen production infrastructure.

Goal: Scale up green hydrogen production to 10 million tons per annum globally by 2030.

**Energy Storage and Grid Flexibility**

Objective: Develop hydrogen storage facilities and integrate hydrogen fuel cells into power grids to balance intermittent renewable energy sources.

Goal: Utilize green hydrogen as an energy storage medium to stabilize grids with high renewable penetration.

**Decarbonization of Transportation**

Objective: Achieve a 30% penetration of hydrogen-powered vehicles in the global fleet by 2040.

Goal: Promote the use of green hydrogen in heavy transportation sectors like shipping, aviation, and long-haul trucking.

**Economic Growth and Job Creation**

Objective: Create over 1 million jobs in the green hydrogen sector by 2030 through R&D, manufacturing, and infrastructure development.

Goal: Foster economic opportunities through the creation of new jobs and industries centered around green hydrogen technologies.

**Policy and Regulatory Support**
Objective: Establish carbon pricing, green hydrogen subsidies, and mandates for its use in key industries by 2025.

Goal: Implement supportive policies and regulations to accelerate green hydrogen adoption.

**International Cooperation and Market Development**
Objective: Develop international standards for hydrogen production and transportation and establish global hydrogen trade routes by 2030.

Goal: Promote global collaboration for green hydrogen trade and technology sharing.

**Cost Competitiveness**
Objective: Invest in R&D to reduce the costs of electrolyzers and renewable energy inputs for hydrogen production.

Goal: Lower the cost of green hydrogen production to less than $2 per kilogram by 2030.

> ***Key: Carbon footprint:*** *The total amount of GHGs, particularly $CO_2$, emitted directly or indirectly by individuals, organizations, or products throughout their lifecycle.*

### 4.1.1 Historical Context and Evolution of Decarbonization Initiatives

Decarbonization refers to the process of reducing carbon dioxide ($CO_2$) emissions associated with human activities, primarily from the combustion of fossil fuels. Over the past few decades, addressing climate change and its related impacts has emerged as one of the most significant global challenges. Nations, organizations, and industries around the world have increasingly pursued initiatives aimed at curbing GHG emissions. In this context, green hydrogen has come to be recognized as a potential game-changer, capable of supporting the global transition to a low-carbon future. Figure 4.2 and Table 4.1 show the historical context and evolution of decarbonization initiatives.

Hydrogen has been touted as a versatile energy carrier since the 1970s, primarily because of its potential to decarbonize hard-to-abate sectors like heavy industry, aviation, and long-haul transportation. However, hydrogen production has traditionally relied on fossil fuels, especially natural gas, leading to substantial $CO_2$ emissions. This is known as "gray hydrogen." Approximately 95% of hydrogen produced today is gray, meaning that while hydrogen as a fuel burns cleanly, its production process is highly carbon-intensive.

The next step in hydrogen's evolution came with the development of "blue hydrogen." This involves producing hydrogen from natural gas, similar to gray hydrogen, but with the added step of capturing and storing the $CO_2$ emissions produced during the process using CCUS technologies. While this method is less

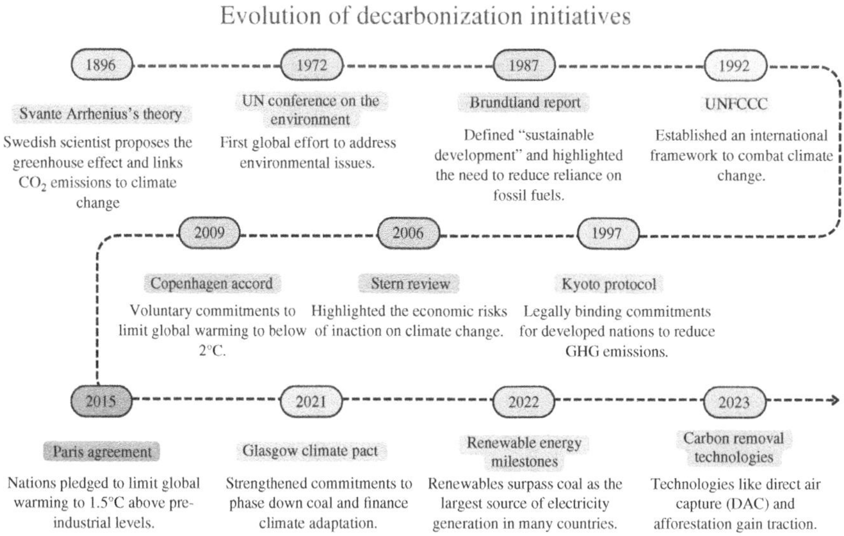


**Figure 4.2** Evolution of Decarbonization Initiatives

carbon-intensive than gray hydrogen, it still relies on fossil fuels, which limits its long-term sustainability in a decarbonized world [7, 8].

The idea of "green hydrogen" revolutionized decarbonization efforts. Green hydrogen is produced by splitting water ($H_2O$) into hydrogen and oxygen using electrolysis, powered by renewable energy sources such as wind, solar, or hydropower. The key advantage of green hydrogen is that the entire process is emissions-free, making it a critical component in achieving net-zero emissions targets.

Green hydrogen is particularly important because it can address sectors that are difficult to decarbonize through electrification alone. While renewable electricity from solar and wind has made significant strides in replacing fossil fuels in the power generation sector, industries like steelmaking, cement production, and aviation still rely heavily on carbon-intensive processes. Green hydrogen offers a solution by serving as a clean fuel for these sectors, either by being burned directly or through the development of hydrogen-derived fuels like ammonia or synthetic methane. Figure 4.3 illustrates key stakeholders in decarbonization.

The EU has been a global leader in green hydrogen development. The European Green Deal, launched in 2019, set the ambitious target of making the continent climate-neutral by 2050. As part of this strategy, the EU unveiled a Hydrogen Strategy in 2020 that aims to scale up the production and use of green hydrogen, with a target of installing at least 40 GW of renewable hydrogen electrolyzers by 2030. Similarly, countries like Germany and the Netherlands have invested heavily in building hydrogen infrastructures, such as hydrogen pipelines and storage facilities.

**Table 4.1** Historical Context and Evolution of Decarbonization Initiatives

| Time Period | Key Events/Initiatives | Description | Impact/Significance |
|---|---|---|---|
| Preindustrial era | — | Low-carbon emissions due to limited industrial activity | Natural ecosystems maintained a balance of carbon cycles |
| 18th century | Industrial Revolution | Rapid industrialization and reliance on fossil fuels (coal, oil) | Significant increase in carbon emissions; awareness of environmental degradation begins |
| 1896 | Svante Arrhenius's Theory | Swedish scientist proposes the greenhouse effect and links $CO_2$ emissions to climate change | Early scientific recognition of the relationship between human activity and climate change |
| 1950s–1960s | Post-War Industrial Boom | Accelerated carbon emissions due to increased industrial production and energy consumption | Rise in global $CO_2$ levels observed; environmental movements begin to emerge |
| 1972 | UN Conference on the Human Environment (Stockholm) | First global effort to address environmental issues | Planted the seeds for international cooperation on sustainability and climate change |
| 1987 | Brundtland Report | Defined "sustainable development" and highlighted the need to reduce reliance on fossil fuels | Set the stage for integrating environmental concerns into economic policies |
| 1992 | United Nations Framework Convention on Climate Change (UNFCCC) | Established an international framework to combat climate change | Laid the foundation for future agreements like the Kyoto Protocol and Paris Agreement |
| 1997 | Kyoto Protocol | Legally binding commitments for developed nations to reduce GHG emissions | First international treaty focused on reducing carbon emissions; not universally ratified |
| 2005 | Kyoto Protocol Comes into Effect | Global carbon markets and emission trading systems introduced | Encouraged market-driven approaches to reduce emissions |
| 2006 | Stern Review on the Economics of Climate Change | Highlighted the economic risks of inaction on climate change | Elevated the urgency of decarbonization from an economic perspective |
| 2009 | Copenhagen Accord | Voluntary commitments to limit global warming to below 2 °C | Lacked binding commitments but emphasized collective global responsibility |

| | | | |
|---|---|---|---|
| 2015 | Paris Agreement | Nations pledged to limit global warming to 1.5 °C above preindustrial levels | Shifted focus to long-term decarbonization targets and net-zero goals |
| 2016 | Carbon Pricing Initiatives Expand | Many countries adopt carbon taxes and cap-and-trade systems | Encouraged businesses to transition to low-carbon technologies |
| 2020 | Net-Zero Commitments Proliferate | Countries and companies begin announcing net-zero emission targets by 2050–2060 | Accelerated global focus on renewable energy, electrification, and green technologies |
| 2021 | COP26 (Glasgow Climate Pact) | Strengthened commitments to phase down coal and finance climate adaptation | Emphasis on transitioning away from fossil fuels and supporting developing nations |
| 2022 | Renewable Energy Milestones | Renewables surpass coal as the largest source of electricity generation in many countries | Marked a significant shift toward cleaner energy sources globally |
| 2023–2024 | Emergence of Carbon Removal Technologies | Technologies like direct air capture (DAC) and afforestation gain traction | Complemented emission reductions with active carbon removal efforts |

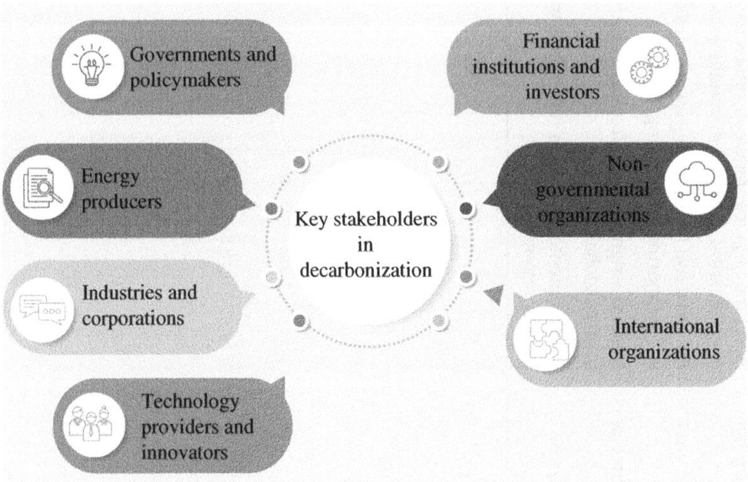


**Figure 4.3** Key Stakeholders in Decarbonization

> ***Key: Climate imperatives:*** *Urgent and necessary actions required to stabilize global temperatures, reduce extreme weather events, and prevent irreversible damage to ecosystems.*

In the United States, the Biden administration's 2021 infrastructure bill included significant funding for clean energy projects, including green hydrogen. The U.S. DOE has launched a "Hydrogen Energy Earthshot" initiative with the goal of reducing the cost of green hydrogen to $1 per kilogram within a decade, making it cost competitive with fossil fuels [9, 10].

China, too, has recognized the potential of green hydrogen in decarbonizing its industrial base and transportation sector. The country has set ambitious goals for hydrogen fuel cell vehicle deployment and is investing in large-scale hydrogen projects, particularly in its industrial heartland. Table 4.2 shows the summary of key stakeholders involved in decarbonization.

### 4.1.2 Challenges and Barriers to Decarbonization

Decarbonization through green hydrogen presents promising opportunities for reducing global carbon emissions, particularly in sectors like transportation, heavy industry, and energy. However, several challenges and barriers exist in the widespread adoption and scaling of green hydrogen technology:

1) **High Production Costs:** The process of producing green hydrogen via electrolysis is still expensive, largely due to the cost of renewable electricity needed to split water into hydrogen and oxygen. Green hydrogen production requires

**Table 4.2** Summary of Key Stakeholders Involved in Decarbonization

| Stakeholder | Role in Decarbonization | Key Responsibilities | Key Attributes |
|---|---|---|---|
| Governments and policymakers | Set regulations, policies, and incentives for decarbonization | Implement carbon pricing, subsidies, emission regulations | Policy-driven, regulatory authority, long-term planning |
| Energy producers | Transition to renewable and low-carbon energy sources | Invest in renewable energy (solar, wind, hydrogen), phase out fossil fuels | Technological adoption, infrastructure investment |
| Industries and corporations | Reduce carbon footprint through innovation and sustainable practices | Implement energy-efficient processes, adopt green technologies, report emissions | Profit-driven yet sustainability-focused, corporate social responsibility (CSR) |
| Technology providers and innovators | Develop and commercialize low-carbon technologies | Research and development (R&D) hydrogen, CCS, electrification solutions | Innovation-driven, research-intensive |
| Financial institutions and investors | Fund and support green energy projects and decarbonization initiatives | Provide green financing, issue climate bonds, invest in low-carbon projects | Risk assessment, impact-driven, ESG (environmental, social, governance) aligned |
| Nongovernmental organizations (NGOs) and advocacy groups | Raise awareness, push for policies, and monitor climate actions | Advocate for stronger climate policies, conduct research, ensure corporate accountability | Public engagement, independent oversight |
| Academia and research institutions | Conduct research and develop decarbonization solutions | Develop new technologies, analyze climate data, educate workforce | Knowledge-driven, research-focused |
| Consumers and society | Drive demand for sustainable products and lifestyles | Support green businesses, adopt energy-efficient behaviors | Awareness-based, behavior-influencing |
| International organizations (e.g., UN, IPCC, and IEA) | Facilitate global climate agreements and collaborations | Set emission targets, coordinate global efforts, provide climate reports | Global influence, data-driven policymaking |

a significant amount of electricity from renewable sources (solar, wind, etc.), which can be costly and not yet universally available.

2) **Infrastructure Development:** Currently, there is inadequate infrastructure for hydrogen production, storage, and transportation. Building pipelines, storage facilities, and refueling stations would require a massive investment. Existing natural gas pipelines and other energy infrastructures are not easily adaptable for hydrogen, raising the cost and time required for conversion or new construction [11, 12].
3) **Energy Efficiency:** Green hydrogen production and utilization, including storage and reconversion back to electricity, suffer from energy losses at various stages. This reduces the overall efficiency of the hydrogen economy compared to alternatives like direct electrification.
4) **Technological Maturity:** Electrolyzers, which are essential for green hydrogen production, are still developing. Current technologies are less efficient than needed for mass adoption, with ongoing research aiming to improve efficiency and reduce costs. Hydrogen is difficult to store because it is the smallest and lightest molecule, making containment challenging. It either requires high-pressure tanks or cryogenic temperatures, both of which are technologically complex and costly.
5) **Economic and Market Barriers:** Green hydrogen is currently not cost competitive with gray hydrogen (produced from natural gas) or other fossil fuels. Without strong financial incentives or carbon pricing, industries may find it economically unfavorable to switch. The intermittent nature of renewable energy sources like solar and wind impacts the reliability of green hydrogen production, which needs a stable and constant electricity supply to be cost-effective.
6) **Policy and Regulatory Framework:** Governments globally have yet to establish clear, consistent policies and regulations for the development of a green hydrogen economy. This uncertainty inhibits investments and industrial scale-up. Continuing subsidies for fossil fuel industries can slow down the shift toward green hydrogen, as conventional energy remains artificially cheaper.
7) **Water Availability and Environmental Impact:** Electrolysis requires large amounts of water to produce hydrogen. In regions where water is scarce, the production of green hydrogen could exacerbate water shortages, leading to environmental concerns. Scaling up renewable energy capacity for green hydrogen will require large areas of land for wind farms, solar panels, or hydroelectric plants, which may conflict with other land use needs like agriculture or conservation.
8) **Public Acceptance and Awareness:** Hydrogen is highly flammable, and the risks associated with its storage and transportation can lead to public concerns

over safety, which could hinder its social acceptance. Limited public knowledge about green hydrogen and its benefits creates a barrier to widespread adoption, as people and businesses are often unaware of how hydrogen can contribute to decarbonization.

9) **International Collaboration and Trade:** Many countries may not have sufficient renewable resources to produce hydrogen domestically, necessitating international trade in hydrogen. Developing a global market with standardized regulations and efficient logistics is a major challenge. Dependence on certain regions for renewable energy or critical materials required for electrolysis (like platinum group metals) could create new geopolitical dependencies, similar to the current oil and gas market [13, 14].
10) **Material Constraints:** The production of electrolyzers often relies on scarce and expensive materials such as platinum and iridium, leading to concerns about supply constraints as demand for green hydrogen increases.

***Key: Net-zero emissions:*** *Achieving a balance between the GHGs emitted into the atmosphere and those removed or offset, ideally by mid-century to limit global warming.*

## 4.2 Environmental Impacts of Carbon Emissions

Carbon emissions have profound environmental impacts, contributing significantly to global climate change and ecological degradation. The excessive release of carbon dioxide ($CO_2$) and other GHGs from industrial activities, transportation, and deforestation traps heat in the Earth's atmosphere, leading to rising global temperatures. This phenomenon, known as global warming, triggers severe consequences such as melting glaciers, rising sea levels, and more frequent and intense extreme weather events, including hurricanes, droughts, and wildfires. Additionally, increased $CO_2$ levels cause ocean acidification, which disrupts marine ecosystems and threatens biodiversity. Air pollution from carbon emissions also contributes to respiratory diseases, reducing public health and quality of life. Mitigating these impacts requires a shift toward renewable energy sources, improved energy efficiency, afforestation, and sustainable industrial practices to reduce carbon footprints and promote environmental conservation. Carbon emissions refer to the release of carbon dioxide ($CO_2$) and other carbon-containing compounds into the atmosphere, predominantly from human activities. These emissions are the leading contributors to global climate change due to their role in the greenhouse effect, where gases trap heat in the Earth's atmosphere, causing global temperatures to rise. The primary sources of carbon emissions are fossil fuel combustion, deforestation, industrial activities, and agricultural practices [15, 16].

As the world grapples with climate change, reducing carbon emissions has become a critical goal. Many strategies are being explored, and one of the most promising solutions is the development of green hydrogen. Unlike fossil fuels, which release $CO_2$ when burned, green hydrogen can be produced and used without emitting carbon, positioning it as a key component in the transition to a low-carbon economy. Table 4.3 shows the different attributes of carbon emission and their sources.

**Table 4.3** Different Attributes of Carbon Emissions and Their Sources

| Attribute | Description | Major Sources | Examples |
|---|---|---|---|
| Energy production | Emissions from burning fossil fuels for electricity and heat generation | Coal-fired power plants, natural gas plants, and oil-based electricity generation | Coal power plants, industrial boilers, natural gas turbines |
| Transportation | Emissions from the combustion of fuels in vehicles, ships, planes, and trains | Road transport (cars, trucks), aviation, shipping, railways | Gasoline-powered cars, diesel trucks, commercial airplanes |
| Industry | Emissions from industrial processes and energy used in manufacturing | Cement production, steelmaking, chemical manufacturing, and industrial boilers | $CO_2$ from limestone calcination in cement production, emissions from blast furnaces in steelmaking |
| Agriculture | Emissions from farming practices and land-use changes | Enteric fermentation (livestock), manure management, rice cultivation, and deforestation for farmland | Methane from cattle digestion, nitrous oxide from fertilizer use, $CO_2$ from forest clearing for croplands |
| Residential and commercial | Emissions from energy use in homes and buildings | Space heating, cooking, cooling, and lighting powered by fossil fuels | Emissions from natural gas heating, kerosene heaters, and propane stoves |
| Land use changes | Emissions from deforestation, urbanization, and other land conversions | Logging, slash-and-burn agriculture, construction of urban infrastructure | $CO_2$ released from cutting tropical forests, emissions from drained wetlands for agriculture |

**Table 4.3** (Continued)

| Attribute | Description | Major Sources | Examples |
|---|---|---|---|
| Waste management | Emissions from waste decomposition and treatment | Landfills, wastewater treatment plants, incineration | Methane emissions from anaerobic decomposition in landfills, $CO_2$ from burning solid waste |
| Energy extraction | Emissions during fossil fuel extraction and processing | Oil drilling, coal mining, natural gas flaring, and shale gas fracking | Methane leaks during fracking, $CO_2$ from oil refining |
| Shipping and aviation | Specialized subcategories of transportation with high emissions intensity | Long-distance flights, container ships, and cruise liners | Jet fuel combustion in aviation, heavy fuel oil burned by cargo ships |
| Fugitive emissions | Unintentional releases of greenhouse gases during industrial activities | Leakage from pipelines, storage tanks, and gas processing facilities | Methane leaks from natural gas infrastructure, $CO_2$ from aging equipment |
| Energy inefficiency | Excess emissions due to outdated or poorly maintained systems | Old vehicles, inefficient HVAC systems, and poorly insulated buildings | High emissions from older cars without catalytic converters, inefficient coal-fired boilers |
| Chemical processes | Emissions from chemical reactions in industrial operations | Production of ammonia, hydrogen, and chlorofluorocarbons (CFCs) | $CO_2$ from ammonia synthesis for fertilizers, greenhouse gases from hydrofluorocarbons (HFCs) used as refrigerants |
| Military activities | Emissions from fuel use and logistics in defense operations | Jet fighters, armored vehicles, naval ships | High carbon footprint from military operations and exercises |

Green hydrogen, heralded as the fuel of the future, holds tremendous potential for combating climate change, particularly when it comes to mitigating carbon emissions. As the world confronts the escalating threats posed by climate change, reducing carbon emissions has emerged as a pivotal goal in global climate strategies. Traditional energy sources, mainly fossil fuels, have been major contributors to GHG emissions, trapping heat in the atmosphere and driving global warming. The transition to green hydrogen, a form of hydrogen produced using renewable energy—presents a promising alternative, poised to revolutionize how we power industries, transportation, and homes. This transformation can drastically reduce the carbon footprint of sectors that have traditionally been difficult to decarbonize. In this context, green hydrogen can be a game-changer, but its deployment comes with both opportunities and challenges related to carbon emissions and broader climate impacts [17].

➢ **Need and Potential Climate Benefits of Green Hydrogen**

The primary driver of climate change is the accumulation of carbon dioxide ($CO_2$) in the atmosphere, primarily from the burning of fossil fuels such as coal, oil, and natural gas. Carbon emissions, alongside other GHGs like methane ($CH_4$) and nitrous oxide ($N_2O$), form a layer in the atmosphere that traps heat, leading to the warming of the planet—a phenomenon known as the greenhouse effect. This warming effect has a range of consequences, including more frequent and severe weather events (such as hurricanes, droughts, and floods), rising sea levels due to the melting of polar ice caps, and disruptions to ecosystems and biodiversity.

Currently, global energy consumption is heavily reliant on carbon-intensive sources. According to the International Energy Agency (IEA), nearly 75% of global $CO_2$ emissions stem from energy-related activities, including electricity generation, transportation, and industrial processes. The Paris Agreement, an international treaty on climate change, aims to limit global warming to below 2 °C, with efforts to cap it at 1.5 °C above preindustrial levels. Achieving this ambitious target will require a drastic reduction in carbon emissions, particularly from the energy sector. This is where green hydrogen comes in.

Green hydrogen is produced through the process of electrolysis, where water ($H_2O$) is split into hydrogen ($H_2$) and oxygen ($O_2$) using electricity generated from renewable sources like wind, solar, and hydropower. Unlike traditional "gray" hydrogen, which is derived from natural gas and results in significant $CO_2$ emissions, green hydrogen is virtually emissions-free. When green hydrogen is burned or used in fuel cells, the only by-product is water vapor, making it a clean and sustainable alternative to fossil fuels [18].

The adoption of green hydrogen could offer a multitude of climate benefits, particularly in sectors where direct electrification is challenging. Some of the key areas where green hydrogen can make a significant impact are as follows:

1) **Decarbonizing Heavy Industries**

   Industries such as steel, cement, and chemicals are among the largest emitters of $CO_2$ globally. These industries often require high-temperature heat and use fossil fuels both as a source of energy and as raw materials in production processes. Green hydrogen can replace natural gas and coal in these industries, providing a low-carbon alternative for generating heat and serving as a feedstock for chemical reactions. For example, in steel production, hydrogen can be used in direct reduction processes to eliminate the need for carbon-based inputs, drastically cutting down emissions.

2) **Reducing Emissions from Transportation**

   The transportation sector is another major contributor to global $CO_2$ emissions, accounting for about 25% of total emissions. While BEVs are gaining traction in the passenger vehicle market, heavy-duty transport modes such as trucks, ships, and airplanes are harder to electrify due to the high energy demands and long distances involved. Green hydrogen, when used in fuel cells, can power these vehicles without emitting $CO_2$, providing a clean energy solution for long-haul transport. Hydrogen-powered FCEVs have the potential to replace diesel trucks, ships, and even planes, significantly reducing the carbon footprint of the transportation sector.

3) **Storing Renewable Energy**

   One of the challenges associated with renewable energy sources like solar and wind is their intermittency—they do not generate electricity consistently, as they depend on weather conditions. Green hydrogen can act as a storage solution for excess renewable energy. During periods of high renewable energy generation, surplus electricity can be used to produce hydrogen, which can then be stored and used later when energy demand exceeds supply. This capability not only reduces reliance on fossil fuel backup generators but also helps stabilize the grid, promoting a more resilient and reliable energy system [19, 20].

4) **Decarbonizing Power Generation**

   In addition to being used directly as a fuel or feedstock, green hydrogen can also contribute to decarbonizing electricity generation. Hydrogen can be used in fuel cells to generate electricity with zero emissions, providing a cleaner alternative to natural gas-fired power plants. In some cases, hydrogen can be blended with natural gas and used in existing gas turbines, gradually reducing the carbon intensity of power generation while taking advantage of existing infrastructure [21].

> ***Key: Environmental sustainability:*** *The responsible interaction with the environment to avoid depletion of natural resources, ensuring long-term planetary health.*

### 4.2.1 Environmental Degradation and Ecosystem Impacts of Carbon Emissions

Carbon emissions, primarily in the form of carbon dioxide ($CO_2$) and methane ($CH_4$), are major contributors to environmental degradation and climate change. These emissions, driven by human activities such as industrial processes, transportation, deforestation, and agriculture, disrupt natural ecosystems, accelerate global warming, and pose serious risks to biodiversity and human well-being.

1) **Global Warming and Climate Change:** Excessive $CO_2$ traps heat in the atmosphere, causing global temperatures to rise. This phenomenon, known as the greenhouse effect, leads to changes in weather patterns, prolonged heatwaves, and increased frequency of extreme weather events. Warmer temperatures accelerate the melting of polar ice caps and glaciers, contributing to rising sea levels that threaten coastal communities and habitats. The absorption of $CO_2$ by oceans reduces pH levels, harming marine life such as coral reefs, shellfish, and plankton.
2) **Impacts on Biodiversity:** Altered climates and deforestation for agriculture and urban development destroy natural habitats, leaving many species homeless and vulnerable to extinction. Changes in temperature and precipitation patterns affect the migration, reproduction, and survival of species, disrupting ecological balance. Warmer temperatures may allow invasive species to thrive in new regions, outcompeting native flora and fauna.
3) **Soil and Water Degradation:** The clearing of forests for agriculture or industrial purposes releases stored carbon and reduces the soil's ability to retain nutrients and water. Changes in precipitation due to climate change affect the availability and quality of freshwater resources, leading to droughts, floods, and reduced agricultural productivity. Figure 4.4 shows the environment degradation and ecosystem impacts of carbon emission.

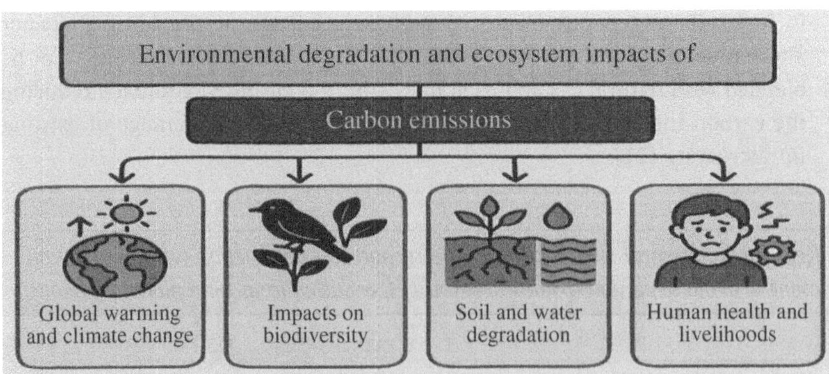


**Figure 4.4** Environment Degradation and Ecosystem Impacts of Carbon Emission

4) **Human Health and Livelihoods:** Carbon emissions are often accompanied by PM and other pollutants, causing respiratory and cardiovascular diseases. Shifts in growing seasons and reduced agricultural yields threaten food supply chains. Rising temperatures and extreme weather events incur significant costs for disaster recovery, infrastructure repair, and mitigation efforts.

The environmental degradation and ecosystem impacts of carbon emissions demand urgent and collaborative action. By addressing the root causes and implementing sustainable practices, humanity can mitigate these effects, preserve biodiversity, and ensure a healthier planet for future generations.

### 4.2.2 Regional and Global Disparities in Carbon Emissions and Their Environmental Impacts

Carbon emissions and their environmental impacts are not evenly distributed across the globe, as they reflect significant disparities in industrialization, energy consumption, population density, and economic development. These inequalities highlight the global challenge of addressing climate change in a manner that is fair and just, while also balancing environmental sustainability with pressing social and economic needs.

**Regional Disparities in Carbon Emissions**: Industrialized nations such as the United States, Canada, and many European countries have historically contributed the highest per capita carbon emissions due to their early and intensive industrialization, reliance on fossil fuels, and energy-intensive lifestyles. While many of these countries have now stabilized or even reduced their emissions through transitions to cleaner energy and stricter environmental policies, their historical contributions remain a major driver of climate change. On the other hand, developing nations, including China and India, are experiencing rapidly rising emissions. This growth is primarily fueled by industrialization, urbanization, and population expansion. Despite their lower per capita emissions compared to industrialized nations, the sheer scale of their economic growth and dependence on coal and other fossil fuels is pushing global emissions upward, though they are also expanding renewable energy use. Meanwhile, least developed countries (LDCs), particularly those in Sub-Saharan Africa, contribute minimally to global carbon emissions due to their low levels of industrialization and energy use. However, these nations are disproportionately vulnerable to climate change impacts such as extreme weather, rising sea levels, and food insecurity, despite playing almost no role in causing the crisis.

**Global Disparities in Environmental Impacts**: The environmental consequences of carbon emissions are also unevenly distributed across the world. Extreme weather events, such as hurricanes and wildfires, affect both wealthy and poor nations, but their outcomes differ drastically. Wealthier nations, though they face economic losses, often recover faster due to stronger infrastructure and resources. In contrast, developing regions in the Global South experience more severe impacts, including widespread displacement, loss of life, and economic collapse, with little adaptive capacity. Rising sea levels represent another stark disparity. Small island nations like the Maldives and Kiribati face existential threats, risking complete submersion and loss of territory, while densely populated coastal cities in Asia, Africa, and Latin America are also highly vulnerable to flooding and damage to critical infrastructure. Food and water security are similarly threatened, with regions such as Africa and the Middle East enduring prolonged droughts and desertification, while global shifts in climate reduce agricultural productivity, disproportionately affecting smallholder farmers in developing nations. Health impacts compound these vulnerabilities: urban centers in countries like India and China suffer from severe air pollution due to emissions, resulting in respiratory and cardiovascular diseases. Additionally, tropical and subtropical regions experience increasingly frequent and intense heatwaves, which drive mortality rates higher and reduce labor productivity.

**Drivers of Disparities**: The disparities in emissions and environmental impacts are driven by several underlying factors. Economic development plays a crucial role, as industrialized nations' historical emissions contrast sharply with the growth-driven emissions of developing countries. Access to energy is another factor, with many regions still heavily dependent on fossil fuels or lacking infrastructure to adopt renewable alternatives. Technological capacity further widens the gap, as wealthier nations can invest in advanced mitigation and adaptation technologies, while poorer nations lack such resources. Finally, disparities in climate policy implementation and international cooperation hinder collective action, leaving vulnerable nations exposed to severe climate risks.

**Bridging the Gap**: Addressing these inequities requires coordinated global efforts. Equitable climate policies must recognize differentiated responsibilities, holding industrialized nations accountable for their historical contributions while supporting developing nations in achieving sustainable growth. Mechanisms such as carbon markets and credits can provide incentives for emissions reduction, particularly benefitting LDCs. Technology transfer is another vital strategy, involving the facilitation of renewable energy adoption and investment in resilient infrastructure across vulnerable regions. Financial support plays an equally critical role, as developed nations must honor commitments to provide climate finance and integrate green development aid into international assistance. Global cooperation, anchored in the Paris Agreement, remains essential to limiting global warming to 1.5 °C. This cooperation must go beyond agreements to include knowledge sharing, research collaboration, and the spread of best practices for climate resilience.

The disparities in carbon emissions and their resulting environmental impacts underline the necessity of a just and balanced approach to climate action. Industrialized nations must lead in emissions reduction and provide financial, technological, and policy support to vulnerable regions. At the same time, developing nations must be empowered with resources to transition toward sustainable pathways. Only through such inclusive and equitable measures can the world strengthen resilience, mitigate climate impacts, and achieve true climate justice.

The disparities in carbon emissions and their environmental impacts underscore the need for a just and balanced approach to climate action. Industrialized nations must lead in emissions reduction and support, while developing regions should be empowered with resources and technology to pursue sustainable growth. Addressing these inequalities is essential for fostering global resilience and achieving climate justice.

## 4.3 Economic Costs Assessment of Carbon Emissions with Risk and Uncertainty

Climate change, driven by carbon emissions, poses profound challenges to global economies. Its impacts disrupt infrastructure, productivity, and resource availability, increasing costs and creating new vulnerabilities for businesses, governments, and individuals. Figure 4.5 illustrates carbon emissions with risk and uncertainty. Following is an overview of the major economic consequences of climate change:

1) **Damage to Infrastructure and Property:** Hurricanes, floods, and wildfires cause significant destruction to homes, businesses, and public infrastructure. For example, the cost of hurricane damage in the United States has risen steadily, with events like Hurricane Katrina costing over $125 billion. Coastal infrastructure faces increased risks of flooding and erosion, requiring costly investments in sea defenses or relocation efforts. The growing frequency and severity of climate-related disasters are driving up insurance premiums, making coverage unaffordable for many.
2) **Decline in Agricultural Productivity:** Higher temperatures, altered precipitation patterns, and extreme weather events disrupt farming, leading to reduced yields of staple crops like wheat, rice, and maize. Changes in growing seasons and increased pest outbreaks create financial instability for farmers, especially in developing nations. Reduced agricultural productivity leads to higher food prices, increasing the cost of living and disproportionately affecting low-income populations.
3) **Impact on Labor Productivity:** Rising temperatures reduce workers' physical ability to perform outdoor tasks, especially in agriculture and construction.

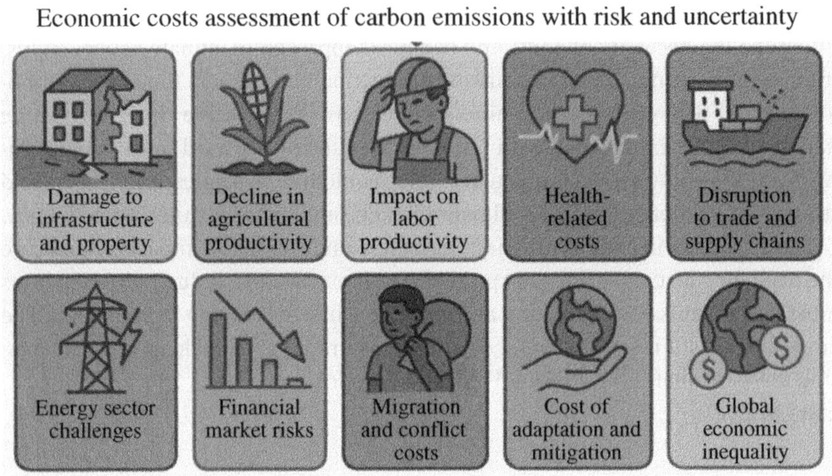


**Figure 4.5** Carbon Emissions with Risk and Uncertainty

Lower productivity translates to economic losses, with global gross domestic product (GDP) projected to decrease by 11–14% by 2050 under high-emission scenarios.

4) **Health-Related Costs:** Air pollution, heatwaves, and the spread of diseases like malaria and dengue lead to higher healthcare expenditures and lost workdays. Loss of life and long-term illnesses impose economic costs on families and reduce the workforce's overall capacity.
5) **Disruption to Trade and Supply Chains:** Droughts and natural disasters disrupt the supply of critical resources like water and raw materials, leading to production slowdowns. Globalized supply chains are sensitive to disruptions from climate events, such as flooding in manufacturing hubs or shipping route blockages. Developing countries reliant on exports of climate-sensitive goods (e.g., agriculture) face significant economic losses.
6) **Energy Sector Challenges:** Rising temperatures lead to higher electricity demand for air conditioning, straining energy grids and increasing costs. Extreme weather events damage infrastructure like power plants and oil rigs, causing interruptions in energy supply. Economies reliant on fossil fuel production face transition risks as global efforts shift toward renewable energy sources.
7) **Financial Market Risks:** Physical risks from climate change can reduce the value of real estate, businesses, and other assets. Transitioning to a low-carbon economy could render fossil fuel reserves and related infrastructure obsolete, leading to financial losses. Climate risks introduce uncertainty in financial markets, deterring investment and increasing volatility.

8) **Migration and Conflict Costs:** Rising sea levels, droughts, and natural disasters force populations to migrate, increasing pressure on urban centers and social services. Competition for scarce resources like water and arable land exacerbates tensions, leading to social instability and economic costs.
9) **Cost of Adaptation and Mitigation:** Building resilient infrastructure, such as flood defenses and heat-resistant buildings, requires substantial upfront costs. Shifting to renewable energy and implementing carbon-neutral technologies entail significant economic investments. Funds directed toward climate adaptation may divert resources from other critical areas, such as education or healthcare.
10) **Global Economic Inequality:** Developing nations, despite contributing the least to emissions, face the highest economic costs from climate change due to limited resources for adaptation. Richer nations may need to provide financial aid to poorer regions, altering global economic dynamics.

The economic consequences of climate change are far-reaching, affecting every sector and region. Addressing these challenges requires global cooperation, strategic investments in resilience, and a commitment to transitioning toward a sustainable, low-carbon economy. By doing so, we can mitigate economic losses while seizing opportunities for innovation and growth.

### 4.3.1 Cost–Benefit Analysis of Mitigation Strategies to Reduce Carbon Emissions

A cost–benefit analysis (CBA) for carbon mitigation strategies evaluates the economic costs of implementing these strategies against the benefits derived from reducing carbon emissions. The primary objective is to assess the economic feasibility and prioritize actions that yield the highest net benefits.

**Costs of Mitigation Strategies:** The costs associated with mitigation strategies encompass direct and indirect expenditures. Direct costs include investments in renewable energy, energy efficiency, electrification of transportation, and CCS. For instance, the installation of utility-scale solar plants can cost between $1,000 and $1,500 per kW, while retrofitting buildings for energy efficiency may range from $25 to $150 per square meter. The shift to EVs involves manufacturing costs and infrastructure development, with EV production costs currently higher but declining due to advancements in battery technology. Similarly, CCS technology can cost between $50 and $100 per ton of $CO_2$ captured. Indirect costs involve broader economic restructuring, policy implementation such as carbon taxes and subsidies, and investments in R&D of low-carbon technologies.

**Benefits of Mitigation Strategies:** The benefits span environmental, social, and economic dimensions. Direct benefits include avoided climate damages and health improvements. For example, avoiding a 2 °C rise in global temperature could prevent up to $54 trillion in damages by 2,100, while air pollution reduction could save the United States up to $160 billion annually in healthcare costs. Economic benefits include job creation in renewable and energy efficiency sectors, energy cost savings, and increased innovation and global competitiveness. Environmental benefits include preserved ecosystems, enhanced biodiversity, and improved agricultural yields due to stable climate conditions.

**Net Present Value (NPV) of Mitigation:** The NPV metric compares costs and benefits over time, adjusting for discount rates. For instance, a strategy costing $1 trillion over 20 years and providing $2 trillion in benefits, with a 3% discount rate, yields a positive net benefit—confirming economic viability.

**Examples of Mitigation Strategies:** Carbon pricing involves administrative and market setup costs but promotes innovation, emissions reduction, and revenue generation. Afforestation and reforestation, costing $500–$1,000 per hectare, contribute to long-term carbon sequestration and biodiversity. Renewable energy deployment, though costing $2,000–$5,000 per installed kW initially, leads to reduced fossil fuel dependency and enhanced energy security. Implementing energy efficiency standards, costing $25–$100 per unit, offers immediate savings and emission reductions.

**Challenges in CBA:** Several challenges exist in conducting CBAs, including the uncertainty of long-term climate impacts and the effectiveness of mitigation actions. Equity considerations also arise, as costs and benefits are unevenly distributed across regions, industries, and income groups. Moreover, debates over discount rates can significantly influence outcomes, as higher rates tend to undervalue future benefits, potentially discouraging investment in mitigation.

**Global Examples:** The European Green Deal exemplifies a comprehensive mitigation plan, with an estimated cost of €1 trillion over 10 years and benefits including job creation, energy independence, and reduced climate risks. China's investment of $350 billion in renewable energy from 2017 to 2020 helped avoid emissions equivalent to 450 million tons of $CO_2$ annually.

In conclusion, while carbon mitigation strategies often require substantial upfront investments, they deliver extensive long-term economic, environmental, and social benefits. Comprehensive CBAs consistently demonstrate that the benefits of climate action far exceed the costs, underscoring the importance of proactive and globally coordinated mitigation efforts.

## 4.4 Technological Innovations for Decarbonization

Decarbonization, the process of reducing carbon dioxide ($CO_2$) and other GHG emissions, is a vital strategy in mitigating the impacts of climate change. Technological advancements across various sectors are accelerating the transition toward a low-carbon future. In the energy sector, the largest contributor to global emissions, renewable energy technologies are leading the transformation. Innovations in solar PVs (such as perovskite solar cells), wind energy (like offshore and floating wind farms), and hydropower (through small modular systems) are making energy generation cleaner and more efficient. Complementing these are advances in energy storage, including solid-state, sodium-ion, and flow batteries, along with hydrogen storage solutions that address the challenge of seasonal energy variability. Smart grid technologies using real-time monitoring and AI are optimizing energy distribution and integrating various renewable sources, while CCS methods—such as DAC and enhanced mineralization—help in removing $CO_2$ from the atmosphere and industrial processes.

The transportation sector is undergoing a massive shift toward zero emission vehicles and alternative fuels. EVs are becoming more viable due to solid-state batteries, fast-charging capabilities, and even dynamic wireless charging technologies. Hydrogen fuel cells, especially suited for heavy-duty transport like buses and trains, offer extended ranges and rapid refueling. SAFs, derived from biomass or captured $CO_2$, serve as drop-in replacements for conventional jet fuels, cutting aviation-related emissions. Meanwhile, autonomous and shared mobility solutions, powered by AI, optimize route efficiency, reduce fuel consumption, and minimize emissions. In the industrial sector, traditionally hard to decarbonize due to its energy intensity, major strides are being made with alternative binders like fly ash and slag in cement production, and green hydrogen in steelmaking. Electrifying high-temperature industrial processes using renewable energy, embracing circular economy principles through recycling and reuse, and applying carbon capture technologies like amine scrubbing and membrane separation, are critical steps toward reducing industrial emissions.

Agriculture and land use also significantly contribute to GHG emissions, particularly methane ($CH_4$) and nitrous oxide ($N_2O$). Precision agriculture, enabled by AI and IoT, optimizes the use of water, fertilizers, and pesticides, thereby increasing efficiency and reducing emissions. The emergence of lab-grown meat and plant-based proteins offers alternatives to traditional livestock farming, which is a major emitter of methane. Techniques like no-till farming and the application of biochar help sequester carbon in soil, while large-scale reforestation and afforestation efforts, supported by satellite and drone technologies, contribute to carbon absorption and biodiversity. The buildings and construction sector is embracing energy-efficient designs such as smart HVAC systems, light emitting diode (LED) lighting, passive solar techniques, and net-zero buildings powered by on-site renewables. Sustainable materials like cross-laminated timber (CLT) and green cement help reduce the embodied emissions of construction, while retrofitting older buildings with better insulation, windows, and systems significantly reduces operational energy use.

Information technology is a powerful enabler across all sectors. AI is optimizing energy use in industries, transportation, and smart grids, while blockchain technology enhances transparency in carbon markets and offset systems. Digital twins—virtual models of physical systems—allow simulation and optimization of energy consumption. In waste management, innovations like waste-to-energy (WTE) systems convert organic waste into bioenergy, reducing dependence on fossil fuels. Advanced chemical recycling methods break plastics down into reusable components, and methane capture technologies at landfills prevent the release of potent GHGs.

Cross-sector innovations are playing a vital role in scaling decarbonization. Green hydrogen, generated via electrolysis using renewable electricity, functions as a clean energy carrier for use in industries, transport, and power. Carbon markets, including cap-and-trade systems, provide economic incentives for emission reductions across sectors. Nature-based solutions (NBS), such as ecosystem restoration and conservation, not only contribute to carbon sequestration but also address biodiversity loss and improve ecosystem resilience. Despite these promising developments, challenges remain, such as the high upfront costs of new technologies, policy and regulatory hurdles, and unequal access to financial and technical resources. However, these are counterbalanced by numerous opportunities: strategic collaboration among governments, industries, and academia; increasing investments to drive down costs and increase adoption; and growing consumer demand for environmentally responsible products and solutions.

Technological innovations are at the heart of global efforts to achieve a sustainable, low-carbon future. From energy and transportation to agriculture and information technology, every sector is witnessing transformative changes. While challenges persist, sustained innovation, bolstered by comprehensive policy support and financial frameworks, will be instrumental in steering the world toward net-zero emissions.

Technological innovations for decarbonization are deeply intertwined with breakthroughs in renewable energy technologies, as the transition to a low-carbon economy hinges on the ability to generate, store, and distribute clean energy efficiently and sustainably. Advances in solar PVs, such as the development of perovskite solar cells, and innovations in wind energy, including floating offshore turbines, are not only revolutionizing energy generation but also serving as the backbone of global decarbonization strategies. These renewable technologies are complemented by improvements in energy storage systems like solid-state and flow batteries, which address the intermittent nature of solar and wind power, enabling reliable integration into smart grids. Furthermore, the synergy between renewable energy and emerging technologies like green hydrogen production and carbon capture amplifies the potential to decarbonize hard-to-abate sectors such as heavy industry and long-haul transport. Thus, the ongoing breakthroughs in renewable energy technologies are not isolated advancements but essential drivers in the broader movement toward decarbonization, shaping a resilient and sustainable energy future.

The renewable energy sector has seen significant technological breakthroughs in recent years, which have improved the efficiency, scalability, and affordability of energy generation from renewable sources. These advancements are crucial to meeting global climate targets and transitioning to a sustainable, low-carbon future. Following are some key breakthroughs in various renewable energy technologies:

1) **Solar Energy**

   **Perovskite Solar Cells:** Perovskite solar cells are an emerging technology that offers higher efficiency and lower production costs compared to traditional silicon-based solar cells. These cells are made from a unique crystalline structure that enhances light absorption and electron movement. Perovskite cells have the potential to exceed the efficiency of silicon solar cells, making solar energy more accessible and affordable.

   **Bifacial Solar Panels:** Bifacial solar panels can capture sunlight on both the front and rear sides of the panel. By using the ground to reflect sunlight onto the rear side, they can generate more electricity. This increases the energy output of solar farms without significantly increasing costs, particularly in areas with high albedo (reflective surfaces like snow or sand).

   **Floating Solar Farms:** Floating solar panels are installed on water bodies like lakes, reservoirs, or oceans, allowing for the efficient use of water surface area. Floating solar reduces land use and can increase energy efficiency due to the cooling effect of water. These systems are especially beneficial in densely populated or land-scarce regions.

   **Solar Windows and Transparent Solar Panels:** Transparent solar cells can be integrated into windows or building facades, generating electricity while still allowing light to pass through. This breakthrough allows buildings to generate their own solar power without compromising aesthetics or functionality.

2) **Wind Energy**

   **Offshore Wind Turbines:** Offshore wind farms, particularly those located in deeper waters, utilize larger and more efficient turbines that are capable of generating substantial amounts of power. Offshore wind has higher and more consistent wind speeds, resulting in greater energy output. Innovations in floating turbine platforms allow wind farms to be deployed in deeper waters, expanding potential wind farm locations.

**Larger and More Efficient Wind Turbines:** Turbine manufacturers are designing larger turbines with longer blades and more efficient designs, capable of generating more electricity at lower costs. Larger turbines with higher capacity factors (more consistent energy production) help lower the cost of wind energy, making it more competitive with fossil fuels.

**Vertical-Axis Wind Turbines (VAWTs):** VAWTs differ from traditional horizontal-axis turbines in that their axis of rotation is vertical. This design can make them more efficient in low-wind conditions and easier to maintain. VAWTs are particularly promising for urban and small-scale applications where space is limited.

3) **Hydropower**

**Small Modular Hydropower:** Small modular hydropower (SMHP) systems are scalable and designed for smaller rivers, dams, and streams where large-scale hydropower systems are not feasible. These systems provide renewable energy to remote or off-grid areas without the environmental impacts associated with large dams, making them more sustainable and adaptable.

4) **Bioenergy**

**Second-Generation Biofuels:** Second-generation biofuels are made from non-food biomass, such as agricultural waste, wood, or algae, rather than food crops like corn or sugarcane. These biofuels avoid the "food versus fuel" debate and can be produced with lower environmental impacts. They are being developed to replace fossil fuels in transportation, power generation, and industrial processes.

**Algae-Based Biofuels:** Algae-based biofuels are derived from microalgae, which can produce oils that can be converted into biodiesel or jet fuel. Algae can grow in nonarable land, consume $CO_2$, and produce biofuels more efficiently than land-based crops. This could significantly reduce emissions and enhance energy security.

5) **Geothermal Energy**

**Enhanced Geothermal Systems (EGSs):** EGS technology involves creating artificial geothermal reservoirs by injecting water into hot, dry rock formations, which increases the amount of energy that can be extracted from geothermal sources. This breakthrough makes geothermal energy more scalable and accessible in regions where natural geothermal resources are limited, expanding its global potential.

**Low-Temperature Geothermal:** Low-temperature geothermal systems operate at temperatures much lower than traditional geothermal plants, which makes them suitable for more widespread applications. These systems could be used for district heating, cooling, or even power generation in areas with moderate geothermal potential.

6) **Energy Storage and Grid Integration**

   **Solid-State Batteries:** Solid-state batteries use a solid electrolyte instead of the liquid electrolyte found in traditional lithium-ion batteries, which increases energy density and safety. These batteries have the potential to revolutionize energy storage, providing more efficient storage for solar and wind energy, and supporting EVs with longer range and faster charging.

   **Grid-Scale Energy Storage**

   Innovation: Technologies such as pumped hydro storage, flow batteries, and compressed air energy storage (CAES) enable large-scale storage of renewable energy, addressing the intermittency issue of solar and wind power.

   Impact: Energy storage is essential for balancing supply and demand and integrating renewables into the grid more effectively.

The breakthroughs in renewable energy technologies are key to reducing global dependence on fossil fuels and mitigating climate change. These advancements in solar, wind, hydropower, bioenergy, geothermal energy, and energy storage are paving the way for a cleaner, more sustainable energy future. As these technologies continue to mature and scale, they will help drive down costs, improve efficiency, and make renewable energy more accessible worldwide. The combination of innovation, investment, and policy support is critical to accelerating the adoption of these technologies in the global energy mix.

### 4.4.1 Carbon Capture and Storage (CCS) Technologies for Mitigating Emissions from Fossil Fuel Combustion

CCS technologies play a pivotal role in the global effort to mitigate climate change by significantly reducing carbon dioxide ($CO_2$) emissions, particularly from industries and power plants that are heavily reliant on fossil fuels. CCS works by capturing $CO_2$ produced during fossil fuel combustion, transporting it, and securely storing it in geological formations, thereby preventing its release into the atmosphere. The process involves three main stages: capture, transport, and storage. During the capture phase, $CO_2$ is separated from other gases emitted during industrial processes, such as those in power generation or manufacturing. Once captured, the $CO_2$ is transported typically via pipelines, ships, or trucks—to designated storage sites. These sites often include deep underground geological formations like depleted oil and gas fields, deep saline aquifers, or unminable coal seams, where the $CO_2$ is stored long-term. CCS is especially valuable for industries like cement, steel, and chemical manufacturing, which face significant challenges in shifting to cleaner energy sources. By capturing emissions at their origin, CCS helps reduce the carbon footprint of these hard-to-abate sectors, supporting

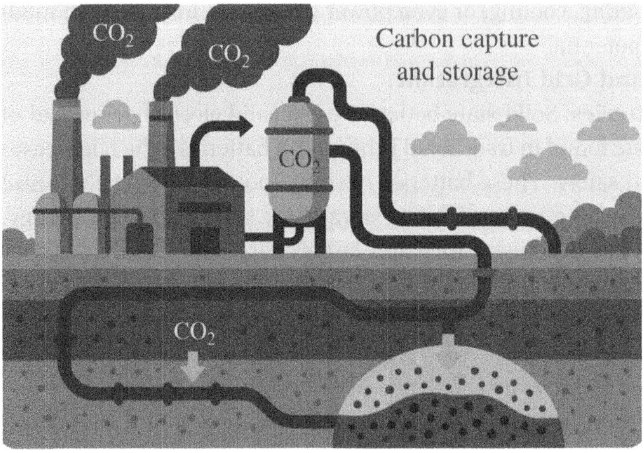


**Figure 4.6** Schematic Diagram of the Carbon Capture and Storage Technologies

broader decarbonization goals and facilitating the transition to a more sustainable, low-carbon economy. Figure 4.6 illustrates a schematic diagram of the CCS technologies.

➢ **Key CCS Technologies**

A) **Capture Technologies**

1) **Pre-Combustion Capture**
   - How it works: Fossil fuels are gasified to produce a mixture of hydrogen and carbon dioxide. The $CO_2$ is separated before combustion, leaving primarily hydrogen as the fuel for power generation.
   - Applications: Pre-combustion capture is commonly used in integrated gasification combined cycle (IGCC) power plants, as well as in industrial applications like fertilizer production.
2) **Post-Combustion Capture**
   - **How it works**: $CO_2$ is captured from the flue gas after fossil fuels are burned. This is typically done using solvents (amines) or other absorbent materials that chemically bond with $CO_2$, removing it from the gas stream.
   - **Applications**: This technology is most applicable to existing power plants and industries that burn fossil fuels, as it can be retrofitted onto existing infrastructure.
3) **Oxyfuel Combustion**
   - How it works: Fossil fuels are burned in a mixture of oxygen and recycled flue gas instead of air. This process results in a flue gas that is primarily $CO_2$ and water vapor, making $CO_2$ easier to capture.

- Applications: This is still a developing technology but is being tested in both power plants and some industrial processes.

B) **$CO_2$ Transport**

- **Pipelines**: The most common method for transporting captured $CO_2$ over long distances is through pipelines, which are capable of moving large volumes of gas. This method is already in use in some regions, such as North America and Europe, where $CO_2$ is transported for storage in underground reservoirs.
- **Ships**: $CO_2$ can also be transported via specially designed ships, similar to the transportation of LNG. This method is useful for regions where pipeline infrastructure is not feasible.
- **Trucks**: For shorter distances or small-scale projects, $CO_2$ can be transported by trucks in liquid form.

C) **$CO_2$ Storage**

1) **Geological Sequestration**
   - **How it works**: $CO_2$ is injected into deep geological formations such as depleted oil and gas fields, deep saline aquifers, or unminable coal seams. These formations are considered stable and impermeable, trapping $CO_2$ underground for thousands of years.
   - **Applications**: This is the most commonly used storage method and has been proven effective in several large-scale projects. It is particularly suited for regions with abundant geological formations.
2) **Enhanced Oil Recovery (EOR)**
   - **How it works**: $CO_2$ is injected into mature oil fields to help increase oil production by reducing the viscosity of the oil, allowing it to flow more easily. The $CO_2$ is then stored underground in the oil field.
   - **Applications**: This method has been used for decades in the oil industry, and the $CO_2$ captured from power plants can be used to enhance oil recovery while simultaneously storing $CO_2$.
3) **Mineral Carbonation**
   - **How it works**: $CO_2$ reacts with naturally occurring minerals to form stable carbonates, effectively "locking" the $CO_2$ in solid form. This process mimics the natural weathering of rocks but on an accelerated scale.
   - **Applications**: While still in the early stages of research, mineral carbonation has the potential to offer a permanent solution for $CO_2$ storage, especially in areas with abundant mineral resources.
4) **DAC Storage**
   - **How it works**: DAC technologies capture $CO_2$ directly from the ambient air, and the captured $CO_2$ is then stored in geological formations or used in various applications.

- **Applications**: DAC is seen as a potential way to achieve negative emissions, as it allows for the capture and storage of $CO_2$ that has already been released into the atmosphere.

➢ **Challenges and Limitations of CCS**

Despite its potential, there are several challenges and limitations to the widespread deployment of CCS:

- **Cost**: The infrastructure for capturing, transporting, and storing $CO_2$ is expensive. This includes the cost of building capture facilities, pipelines, and storage sites. Without financial incentives or regulatory frameworks (e.g., carbon pricing), it is challenging for companies to adopt CCS at scale.
- **Storage Capacity**: While large geological formations offer significant potential for $CO_2$ storage, concerns about the long-term safety and integrity of storage sites remain. For example, the risk of $CO_2$ leakage from storage sites could undermine the effectiveness of CCS.
- **Public Acceptance**: Public concerns about the safety of $CO_2$ storage, especially regarding the potential for leaks or seismic events triggered by injection, can be a barrier to the development of CCS projects.
- **Scalability**: Scaling CCS to the level required to meet global climate targets is a major challenge. The technology must be deployed across a wide range of industries and power generation facilities worldwide to make a significant impact on global emissions.

➢ **Future of CCS Technologies**

The future of CCS depends on continued research, technological innovation, and investment. Some key areas of focus for advancing CCS technologies include:

- **Cost Reduction**: Research into new capture materials and more efficient processes could lower the cost of CCS, making it more economically viable.
- **Integration with Renewable Energy**: Combining CCS with renewable energy sources like biomass (BECCS) can achieve negative emissions, potentially offsetting hard-to-abate sectors.
- **Policy Support**: Government policies and carbon pricing mechanisms, such as carbon taxes or cap-and-trade systems, are critical in incentivizing the adoption of CCS and helping offset its high costs.
- **Global Cooperation**: CCS projects need to be scaled globally, and international collaboration can help ensure that knowledge, technology, and best practices are shared, accelerating the deployment of CCS worldwide.

CCS technologies are an essential part of the global strategy for reducing GHG emissions, particularly for industries that rely on fossil fuels and are difficult to decarbonize. Despite challenges such as high costs and public concerns, ongoing innovations and policy support are expected to make CCS a key tool in mitigating the impacts of climate change. Successful implementation of CCS, alongside other decarbonization strategies, can play a significant role in achieving global climate goals and securing a sustainable future.

### 4.4.2 Energy Efficiency and Their Integration with Digital Technologies

Energy efficiency technologies and practices play a critical role in reducing energy consumption, minimizing GHG emissions, and supporting sustainable development. Advancements in this field are transforming how energy is produced, distributed, and consumed, making it possible to achieve more with fewer resources. These innovations span across various sectors—including residential, industrial, commercial, and transportation—enabling significant economic and environmental benefits.

Energy efficiency refers to the ability to use less energy to achieve the same level of service or output. It involves enhancing the performance of energy conversion, transmission, and end-use systems. Improving energy efficiency not only reduces operational costs but also contributes significantly to climate goals by minimizing dependence on fossil fuels. Following are the key advancements in energy efficiency systems:

**Smart Energy Management Systems:** Smart meters and sensors provide real-time data on energy usage, allowing users to monitor and optimize consumption patterns effectively. Home energy management systems (HEMS) connect with appliances and smart devices, managing energy use based on user preferences, time-of-use rates, and renewable energy availability. Similarly, BEMS are employed in commercial settings to oversee and optimize energy use in heating, cooling, lighting, and more.

**Industrial Energy Efficiency:** In the industrial sector, energy-efficient motors, such as permanent magnet and synchronous reluctance motors, help reduce operational energy losses. Waste heat recovery systems, including CHP and heat exchangers, repurpose energy that would otherwise be lost. AI and machine learning are being increasingly applied to optimize industrial processes, reducing energy use without compromising productivity.

**Advancements in Building Design:** Modern buildings utilize passive design techniques to maximize natural ventilation, daylight, and thermal regulation, minimizing reliance on artificial systems. High-performance insulation materials—such as aerogels and phase-change materials—enhance thermal resistance. Net-zero energy buildings (NZEBs) combine renewable energy systems, efficient appliances, and advanced materials to produce as much energy as they consume.

**Transportation Sector Innovations:** EVs have seen rapid advancements in battery technology, regenerative braking, and lightweight design, dramatically improving energy efficiency. Aerodynamic improvements in vehicle design also contribute to reduced fuel consumption. Intelligent transportation systems (ITS) enhance traffic flow and minimize energy use through smart traffic lights, real-time navigation, and vehicle-to-vehicle communication.

Following are the practices that enhance energy efficiency:

**Energy Auditing:** Conducting regular energy audits helps identify inefficiencies in existing systems and processes, enabling focused improvements that enhance overall energy performance.

**Demand Response Programs:** These programs incentivize consumers to shift or reduce their energy consumption during peak demand periods, ensuring a more balanced and efficient energy grid.

**Behavioral Energy Efficiency:** Encouraging energy-conscious behavior—such as turning off unused appliances and adjusting thermostat settings—can lead to substantial energy savings. Awareness campaigns and incentives play a crucial role in promoting such practices.

**Decentralized Energy Systems:** Community-based microgrids and localized renewable energy systems reduce transmission losses and support greater energy efficiency at the local level. Figure 4.7 illustrates energy efficiency and integration with technologies.

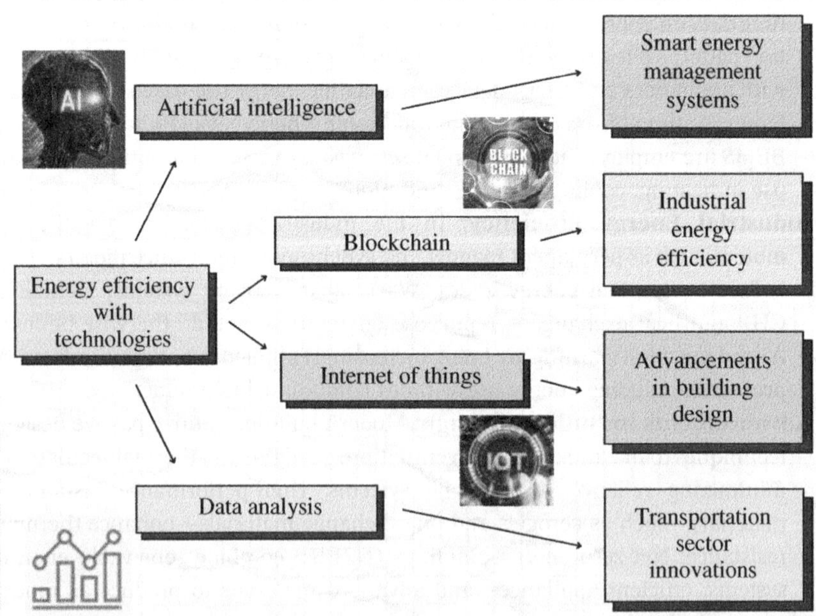


**Figure 4.7** Energy Efficiency and Integration with Technologies. *Source:* Vladgrin / Adobe Stock Photos, photon_photo / Adobe Stock Photos.

The integration of digital technologies such as AI, the IoT, and blockchain has emerged as a transformative strategy in optimizing decarbonization efforts across multiple sectors. These technologies not only enhance operational efficiency but also offer real-time insights and enable innovative approaches to reduce GHG emissions, thus contributing significantly to global sustainability goals. Digital technologies are crucial in managing complex energy systems, optimizing resource usage, and mitigating emissions. Through data-driven insights, they allow real-time monitoring of energy consumption, emissions, and system inefficiencies. Predictive and prescriptive analytics further aid in identifying trends and formulating actionable strategies. Automation and intelligent controls streamline processes, boosting overall performance while minimizing energy waste. Following are the key digital technologies and their applications to enhance the decarbonization efforts:

**Artificial Intelligence (AI):** AI plays a vital role in enhancing decarbonization efforts by supporting intelligent decision-making. It enables accurate energy demand forecasting, ensuring efficient grid operations and minimizing peak-time fossil fuel usage. Process optimization through AI algorithms reduces energy waste in industrial operations, while predictive maintenance helps preempt equipment failures, avoiding unnecessary downtime and emissions. In smart agriculture, AI optimizes inputs like water and fertilizer, cutting emissions from farming activities.

**Internet of Things (IoT):** IoT connects physical devices, enabling seamless communication and control across energy systems. Smart grids leverage IoT-enabled meters and sensors to balance supply and demand effectively. Building management systems (BMS) use IoT to manage lighting, heating, and cooling, significantly improving building energy efficiency. In the transportation sector, IoT monitors vehicle performance and emissions, supporting eco-friendly driving practices. Within industrial environments, IoT sensors continuously track energy usage and emissions, enabling real-time adjustments.

**Blockchain Technology:** Blockchain enhances transparency, traceability, and trust in decarbonization initiatives. It is particularly impactful in carbon credit trading, offering secure platforms for compliance and exchange. It ensures accountability in renewable energy certificates (RECs) by tracking energy origins. Blockchain also supports peer-to-peer energy trading, empowering consumers to share excess renewable energy locally. In supply chains, blockchain tracks the carbon footprint of products, promoting greener practices throughout the lifecycle.

**Advanced Data Analytics:** Data analytics helps in emission mapping, identifying geographic emission hotspots for focused intervention. Scenario analysis uses simulations to evaluate the impact of policy changes and new technologies on emissions. Carbon footprint calculators, tailored for individuals or organizations, empower users to monitor and actively reduce their environmental impact.

## 4.5 Corporate Sustainability and Industrial Strategies for Decarbonization

Businesses and industries play a pivotal role in advancing decarbonization efforts, given their significant contribution to global GHG emissions and their immense capacity for innovation, investment, and influence. As major players in the global economy, they hold the power to drive meaningful environmental change through the integration of sustainable practices, technologies, and strategies that not only mitigate their own environmental footprints but also shape broader societal transitions toward a low-carbon future. Industrial sectors—particularly energy, manufacturing, construction, and transportation—are among the largest sources of $CO_2$ emissions due to their resource-intensive operations, high energy consumption, and dependence on fossil fuels. Moreover, the environmental impact of businesses often extends beyond their direct operations, as emissions embedded in global supply chains typically outweigh those from internal processes. Recognizing their central role, businesses are now becoming catalysts for decarbonization through economic influence, technological innovation, market power, and policy engagement. Economically, businesses drive growth and employment, and by investing in green technologies and sustainable business models, they create new market opportunities while supporting climate resilience. Technologically, industries lead the development and deployment of energy-efficient systems, renewable energy solutions, and circular economy innovations, accelerating the transition away from carbon-intensive practices. Their market power allows them to shift consumer behavior by offering sustainable products, while their involvement in policy advocacy helps shape regulatory landscapes that support emission reduction goals.

In practice, many businesses are transitioning to renewable energy through PPAs and direct investments in clean energy infrastructure. They are also implementing energy-efficient production methods, collaborating with suppliers to reduce Scope 3 emissions, and redesigning products and packaging to align with circular economy principles. Carbon offsetting and removal initiatives such as reforestation, CCS, and DAC are gaining momentum as businesses seek to balance unavoidable emissions. Furthermore, innovations like EVs, sustainable building materials, and smart appliances are making it easier for consumers to embrace low-carbon lifestyles. These efforts not only contribute to climate goals but also offer substantial business benefits, including operational cost savings, enhanced brand reputation, regulatory compliance, improved investor confidence, and competitive advantage in sustainability-driven markets. However, challenges persist. High initial costs, technology gaps, market barriers, and inconsistent policy environments can hinder the pace of decarbonization. To overcome these, businesses

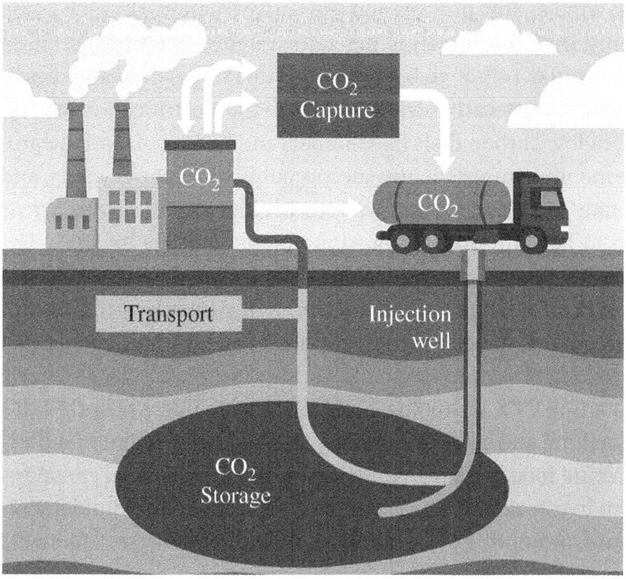


**Figure 4.8** Schematic Diagram of Industrial Strategies for Decarbonization

must commit to science-based emission reduction targets, invest in green innovations, engage in multi-stakeholder collaborations, ensure transparency through robust emissions reporting, and promote consumer awareness around sustainable choices. Ultimately, businesses and industries are indispensable to the global decarbonization agenda. Their leadership in adopting sustainable practices, developing innovative technologies, and influencing both markets and policies will be crucial to achieving a resilient, low-carbon economy. By embracing their role as agents of change, they can ensure long-term profitability while safeguarding the planet for future generations. Figure 4.8 illustrates a schematic diagram of industrial strategies for decarbonization.

Industry sectors play a pivotal role in global decarbonization efforts, as they collectively account for a substantial share of GHG emissions. Through the adoption of sustainable practices, innovative technologies, and a shift toward cleaner energy sources, these sectors can significantly curtail their carbon footprint and support global climate goals. The energy sector, being the largest contributor to carbon emissions due to its reliance on fossil fuel combustion for power generation, is central to these efforts. Decarbonization strategies here include transitioning to renewable energy sources like wind, solar, and hydro, modernizing energy grids with storage and demand-response capabilities, and phasing out coal-fired plants in favor of transitional fuels like natural gas. This has been evidenced by the rapid expansion of solar and wind energy projects and legislative frameworks such as the EU's Renewable Energy Directive targeting net-zero systems. The transportation sector, responsible for around 14% of global emissions due to gasoline and diesel-powered vehicles, is advancing decarbonization through the electrification of vehicles, the development of low-carbon fuels like biofuels and hydrogen, and the promotion of sustainable mobility infrastructure such as public transport, cycling, and walking. Companies like Tesla lead EV innovation, while policy measures across several countries aim to ban internal combustion engine vehicles by 2035. The industrial sector, including heavy industries such as steel, cement, and chemicals, contributes to about 30% of $CO_2$ emissions due to energy-intensive operations. Strategies to decarbonize this sector include the use of low-carbon production techniques, such as green hydrogen in steelmaking, improving material efficiency and recycling, and adopting CCS technologies. Examples include ArcelorMittal's hydrogen-based steelmaking and circular economy initiatives in cement manufacturing. The agriculture and food industry, contributing 10–12% of global emissions through methane from livestock and nitrous oxide from fertilizers, is implementing precision agriculture, promoting plant-based diets, and employing sustainable land management to sequester carbon. Notable players in this shift include Beyond Meat and Impossible Foods, as well as large-scale regenerative farming projects. The building and construction sector, responsible for 39% of energy-related emissions, is reducing its carbon footprint through energy-efficient designs and retrofits, the use of sustainable materials such as bamboo and recycled steel, and the integration of renewable energy systems like rooftop solar. Certifications such as Leadership in Energy and Environmental Design (LEED) and design models like passive houses exemplify this transformation. While the technology and digital sector is not a primary emitter, its energy demand is growing rapidly due to the proliferation of data centers, blockchain, and AI applications. Decarbonization here involves powering data infrastructure with renewables, enhancing energy efficiency, and using digital tools to optimize energy use across sectors, as seen in Google's carbon-neutral data centers and blockchain-based carbon credit systems. The financial sector also plays a crucial role by influencing investment patterns and supporting green economic activity. Financial institutions are shifting capital away from fossil fuels, developing green bonds and sustainability-linked loans, and promoting ESG disclosures and carbon pricing mechanisms, with firms like BlackRock leading the way. The healthcare sector, often overlooked, contributes to emissions through energy use, medical waste, and supply chain operations. Measures such as energy-efficient hospital designs, sustainable procurement, and improved waste management are key strategies, with institutions like the NHS in the UK setting ambitious net-zero targets. Cross-sectoral collaboration further enhances decarbonization efforts through PPPs for R&D, regulatory frameworks that enforce standards, innovation hubs fostering sustainable technologies, and global alliances like the Cement Sustainability Initiative and Clean Skies for Tomorrow Coalition. However, the path to industry-wide decarbonization is not without challenges. Technological barriers, especially in sectors like aviation and heavy manufacturing, hinder scalable solutions. Economic viability is often constrained by high upfront costs, while regulatory inconsistencies across regions create uncertainty. Complex global supply chains further complicate efforts to ensure sustainability end-to-end. Nevertheless, the decarbonization of industry sectors remains essential to achieving global climate objectives. With coordinated action, continued innovation, and broad collaboration across stakeholders, industries can not only mitigate their environmental impact but also lead the charge toward a low-carbon, resilient, and sustainable future.

Corporate sustainability initiatives and commitments to reduce carbon emissions are critical to addressing the challenges of climate change. Companies across various industries are taking proactive steps to integrate ESG principles into their operations, demonstrating leadership in the global effort to reduce GHG emissions and transition to a low-carbon economy. Following are some key corporate sustainability initiatives:

A) **Carbon Neutrality and Net-Zero Commitments**
- **Carbon Neutrality**: Companies pledge to offset their emissions by supporting projects like reforestation or renewable energy generation.
- **Net-Zero Emissions**: Achieving net-zero requires balancing emitted and removed GHGs through reductions and innovative carbon removal technologies.

Examples:
- **Apple**: Committed to achieving carbon neutrality across its supply chain and products by 2030.
- **Amazon**: Pledged to reach net-zero carbon emissions by 2040 through its "Climate Pledge."

B) **Renewable Energy Adoption**
- **On-Site Renewable Energy**: Installing solar panels, wind turbines, or geothermal systems at company facilities.
- **Power Purchase Agreements (PPAs)**: Partnering with renewable energy providers to ensure a clean energy supply.

Examples:
- **Google**: Has been running on 100% renewable energy since 2017 and is working to decarbonize its energy use entirely by 2030.
- **IKEA**: Invests heavily in wind and solar energy projects, producing more renewable energy than it consumes.

C) **Energy Efficiency Programs**
- **Smart Building Technologies**: Automated systems to optimize lighting, heating, and cooling.
- **Industrial Efficiency**: Upgrading machinery and processes to use less energy and reduce waste.

Examples:
- **Walmart**: Uses advanced cooling and heating systems in its stores, aiming to reduce energy intensity by 20% by 2025.

D) **Sustainable Supply Chain Management (SCM)**
- **Supplier Engagement**: Encouraging or requiring suppliers to adopt sustainable practices.
- **Transparency Tools**: Tracking and reporting the carbon footprint of products throughout the supply chain.

Examples:
- **Unilever**: Targets a 50% reduction in the carbon footprint of its products by 2030.
- **Nike**: Works with suppliers to transition to renewable energy sources.

E) **Circular Economy Practices**
- **Product Lifecycle Management**: Designing products for durability, repairability, and recyclability.
- **Waste Reduction**: Implementing systems to reuse or recycle manufacturing waste.

Examples:
- **Patagonia**: Encourages customers to repair and recycle products through its Worn Wear program.
- **Dell Technologies**: Uses recycled plastics and other sustainable materials in its products.

F) **Carbon Pricing and Internal Carbon Budgets**
- **Carbon Pricing**: Establishing an internal cost for carbon emissions to encourage reductions and guide investments.
- **Offsets and Credits**: Participating in carbon credit markets to offset unavoidable emissions.

Examples:
- **Microsoft**: Operates an internal carbon fee to fund sustainability projects and achieve carbon negativity by 2030.

G) **Innovation in Carbon Capture and Sequestration**
- **Investment in R&D**: Developing or funding technologies to capture and store carbon emissions.
- **Pilot Projects**: Deploying small-scale carbon capture systems in industrial settings.

Examples:

- **ExxonMobil**: Invests in CCS technologies to mitigate emissions from its operations.

1) **Corporate Reporting and Transparency**
   - **ESG Reporting**: Publishing sustainability reports aligned with frameworks like the Global Reporting Initiative (GRI) or the Task Force on Climate-Related Financial Disclosures (TCFD).
   - **Science-Based Targets Initiative (SBTi)**: Aligning corporate emissions reduction targets with global temperature goals.
2) **Challenges in Corporate Sustainability**
   - **Economic Barriers**: High initial investment costs for sustainable technologies and infrastructure.
   - **Technological Gaps**: Lack of scalable solutions in some high-emission sectors.
   - **Regulatory Uncertainty**: Inconsistent policies across regions hinder global coordination.
   - **Supply Chain Complexity**: Ensuring sustainability throughout extensive and fragmented supply chains.
3) **Benefits of Corporate Sustainability**
   - **Operational Efficiency**: Reducing emissions often aligns with cost savings through energy efficiency and waste reduction.
   - **Enhanced Investor Confidence**: ESG performance is increasingly linked to financial resilience and risk management.
   - **Market Differentiation**: Sustainability commitments attract eco-conscious consumers.
   - **Risk Mitigation**: Proactively addressing climate risks reduces exposure to regulatory and market changes.

Corporate sustainability initiatives and commitments to reducing carbon emissions are essential in the fight against climate change. By adopting renewable energy, enhancing efficiency, investing in innovation, and fostering transparency, businesses can significantly contribute to global decarbonization efforts. These initiatives not only mitigate environmental impacts but also position companies as leaders in a rapidly evolving market, ensuring long-term success and societal benefit.

### 4.5.1 Business Models and Strategies for Integrating Sustainability and Decarbonization

Integrating sustainability and decarbonization into business models has become a priority for organizations across industries. Such integration not only aligns with global climate goals but also provides competitive advantages, including cost savings, customer loyalty, regulatory compliance, and long-term resilience.

1) **Circular Economy Business Models**
   - **Concept**: Shift from the traditional linear model (take, make, dispose) to a circular model where resources are reused, recycled, or repurposed.
   - **Strategies**:
     - **Product-as-a-Service**: Offering products on a rental or subscription basis (e.g., leasing equipment instead of selling it outright).
     - **Material Recycling and Upcycling**: Developing systems to reclaim and repurpose materials.
     - **Design for Longevity**: Creating products that are durable, modular, and easily repairable.
   - **Examples**:
     - Patagonia's *Worn Wear* program for repairing and reselling used clothing.
     - IKEA's furniture recycling and buy-back initiatives.
2) **Renewable Energy Integration**
   - **Concept**: Transition to renewable energy sources for operations and production processes.
   - **Strategies**:
     - Installing on-site renewable energy systems (e.g., solar panels and wind turbines).
     - Partnering with energy providers to source 100% renewable electricity.
     - Investing in renewable energy projects through green PPAs.
   - **Examples**:
     - Google achieving carbon-neutral operations through renewable energy sourcing.
     - Apple's investment in solar farms to power manufacturing facilities.
3) **Carbon Neutral and Net-Zero Business Models**
   - **Concept**: Commit to reducing carbon emissions across operations and supply chains to achieve carbon neutrality or net-zero emissions.
   - **Strategies**:
     - Implementing carbon offset programs for unavoidable emissions.
     - Using CCS technologies to mitigate emissions.
     - Transitioning to EVs and efficient logistics in transportation.
   - **Examples**:
     - Microsoft's pledge to become carbon-negative by 2030.
     - Amazon's *Climate Pledge* initiative to achieve net-zero emissions by 2040.

4) **Sustainablc Supply Chain Management**
   - **Concept**: Reduce emissions and environmental impact across the entire supply chain.
   - **Strategies**:
     - Partnering with suppliers committed to sustainability.
     - Implementing blockchain technology for supply chain transparency and tracking emissions.
     - Prioritizing local sourcing to minimize transportation-related emissions.
   - **Examples**:
     - Unilever's commitment to sourcing 100% of its agricultural raw materials sustainably.
     - Walmart's *Project Gigaton*, aimed at reducing 1 billion tons of GHG emissions from its supply chain.
5) **Green Innovation and Product Development**
   - **Concept**: Design and develop products that are sustainable, energy-efficient, and environmentally friendly.
   - **Strategies**:
     - Investing in R&D for sustainable materials and energy-efficient technologies.
     - Offering eco-labeled or certified green products.
     - Developing products that enable decarbonization for consumers (e.g., home energy storage solutions).
   - **Examples**:
     - Tesla's focus on EVs and renewable energy products.
     - Nestlé's commitment to reducing packaging waste and offering plant-based food alternatives.
6) **Digital Transformation for Sustainability**
   - **Concept**: Leverage digital technologies to optimize energy use, reduce waste, and monitor sustainability metrics.
   - **Strategies**:
     - Using AI to optimize energy use in manufacturing and operations.
     - Adopting IoT for real-time monitoring of carbon emissions and resource consumption.
     - Blockchain for tracking and verifying sustainability claims.
   - **Examples**:
     - Siemens using AI to optimize energy use in factories.
     - IBM's blockchain solutions for supply chain sustainability.

7) **Community and Stakeholder Engagement**
   - **Concept**: Foster collaboration with communities, customers, and stakeholders to drive sustainable practices.
   - **Strategies**:
     - Educating and engaging consumers on the importance of sustainability.
     - Building partnerships with nongovernmental organizations (NGOs) and governments for decarbonization projects.
     - Offering sustainability training and incentives for employees.
   - **Examples**:
     - Starbucks' *Greener Stores* framework to create sustainable retail outlets.
     - Coca-Cola's partnership with communities for water conservation projects.
8) **Green Financing and Investment Models**
   - **Concept**: Align financial strategies with sustainability goals to fund decarbonization initiatives.
   - **Strategies:**
     - Issuing green bonds to finance eco-friendly projects.
     - Attracting investments through sustainability-linked loans.
     - Allocating R&D budgets specifically for decarbonization and energy efficiency.
   - **Examples**:
     - Bank of America's issuance of green bonds to support renewable energy projects.
     - Tesla's reliance on carbon credits to support its financial model.
9) **Collaborative Ecosystems and Open Innovation**
   - **Concept**: Work with competitors, suppliers, and innovators to co-develop solutions for decarbonization.
   - **Strategies**:
     - Joining industry alliances and coalitions for sustainability.
     - Sharing best practices and data for sector-wide improvements.
     - Co-developing technologies with startups and research institutions.
   - **Examples**:
     - The Clean Energy Buyers Alliance (CEBA) enabling corporate renewable energy procurement.
     - The collaboration between major automakers for EV charging infrastructure.
10) **Regulatory and Policy Alignment**
   - **Concept**: Stay ahead of regulations by proactively integrating sustainability into business operations.
   - **Strategies**:
     - Implementing internal carbon pricing to guide decision-making.
     - Aligning with international frameworks like the Paris Agreement or the UN Sustainable Development Goals (SDGs).
     - Engaging in policy advocacy for sustainable practices.

- **Examples**:
  - BP's internal carbon pricing mechanism to incentivize emission reductions.
  - IKEA's alignment with the UN SDGs in its business strategy.

### Key Benefits of Integrating Sustainability and Decarbonization

1) **Cost Savings**: Energy efficiency and waste reduction lower operational costs.
2) **Risk Management**: Mitigates regulatory, reputational, and climate-related risks.
3) **Market Differentiation**: Builds brand value and attracts eco-conscious consumers.
4) **Employee Engagement**: Fosters a sense of purpose and attracts talent.
5) **Long-term Resilience**: Future-proofs the business against resource scarcity and climate change.

Integrating sustainability and decarbonization into business models is no longer optional but essential for long-term success. By leveraging innovative strategies, technologies, and stakeholder collaboration, businesses can not only contribute to global climate goals but also unlock significant economic and reputational benefits.

Collaboration among businesses, governments, and NGOs is critical for achieving global decarbonization objectives. These partnerships leverage the strengths of each stakeholder—business innovation, governmental policy frameworks, and NGO advocacy and expertise—to drive impactful and scalable solutions to reduce carbon emissions.

1) **Roles of Stakeholders in Decarbonization**

   **Businesses**
   - Develop and implement low-carbon technologies.
   - Innovate sustainable products and services.
   - Influence supply chains to adopt greener practices.

   **Governments**
   - Provide policy frameworks, regulations, and incentives.
   - Invest in infrastructure that supports decarbonization (e.g., renewable energy grids).
   - Promote international agreements like the Paris Accord.

   **Nongovernmental** Organizations
   - Advocate for sustainable practices and public awareness.
   - Provide research, data, and expertise for climate initiatives.
   - Act as mediators and facilitators in multi-stakeholder collaborations.

2) **Examples of Collaborative Efforts**
   **Public–Private Partnerships (PPPs)**
   - **Description**: Joint initiatives where governments and businesses pool resources for projects like renewable energy or energy efficiency.
   - **Example**: The *Mission Innovation* initiative involves 24 countries and the EU collaborating with private companies to accelerate clean energy innovation.

   **Business–NGO Alliances**
   - **Description**: Partnerships where businesses adopt sustainability strategies guided by NGOs.
   - **Example**: The partnership between Unilever and the World Wildlife Fund (WWF) to promote sustainable palm oil production.

   **Government–NGO Collaborations**
   - **Description**: Governments working with NGOs to develop and implement climate policies.
   - **Example**: Greenpeace's collaboration with the EU on promoting renewable energy policies.

   **Multi-Stakeholder Platforms**
   - **Description**: Collaborative platforms involving businesses, governments, and NGOs to drive decarbonization efforts.
   - **Example**: The *We Mean Business Coalition* brings together businesses, policymakers, and NGOs to accelerate climate action.
3) **Key Areas of Collaboration**
   A) **Renewable Energy Deployment**
      - **Partnerships**: Governments provide subsidies or incentives, businesses invest in renewable technologies, and NGOs advocate for clean energy adoption.
      - **Example**: Tesla, SolarCity, and government-backed subsidies have expanded solar and battery technologies globally.

   B) **Sustainable Supply Chains**
      - **Partnerships**: NGOs guide businesses to adopt sustainable sourcing, while governments set import/export standards.
      - **Example**: Fairtrade International's partnerships with governments and corporations to ensure sustainable agricultural practices.

   C) **Carbon Capture and Storage (CCS)**
      - **Partnerships**: Businesses develop CCS technologies, governments provide funding, and NGOs assess environmental impacts.
      - **Example**: The Gorgon Project in Australia, a collaboration between Chevron, Shell, the Australian government, and environmental groups.

D) **Policy Advocacy and Regulation**
   - **Partnerships**: NGOs push for stringent regulations, businesses support feasible policies, and governments enforce them.
   - **Example**: The C40 Cities initiative involves governments, businesses, and NGOs in developing urban climate policies.

4) **Benefits of Collaboration**
   1) **Resource Sharing**: Combining financial, technological, and human resources maximizes impact.
   2) **Innovation and Scaling**: Businesses bring innovation, while governments and NGOs help scale solutions.
   3) **Increased Credibility**: Partnerships enhance the credibility of corporate sustainability initiatives.
   4) **Comprehensive Approach**: Collaborative frameworks address economic, environmental, and social dimensions of decarbonization.

5) **Challenges in Collaboration**
   1) **Misaligned Goals**: Differing priorities between stakeholders can hinder progress.
   2) **Regulatory Barriers**: Complex regulations may slow collaborative initiatives.
   3) **Funding Gaps**: Insufficient financial resources for large-scale projects.
   4) **Accountability**: Ensuring transparency and measurable outcomes can be difficult.

6) **Strategies to Foster Effective Collaboration**
   1) **Clear Objectives and Roles**: Define each stakeholder's role and set measurable goals.
   2) **Regular Communication**: Establish transparent channels to maintain alignment.
   3) **Incentives and Funding**: Provide financial and policy incentives to encourage partnerships.
   4) **Public Engagement**: Involve communities to enhance the impact and acceptance of initiatives.

7) **Successful Collaborative Models**
   **The RE100 Initiative**

   - **Description**: A coalition of businesses committing to 100% renewable energy.
   - **Partners**: Governments and NGOs like the Climate Group.
   - **Impact**: Over 400 global companies have joined, driving renewable energy adoption.

   **The Carbon Neutral Cities Alliance (CNCA)**
   - **Description**: A partnership of global cities committed to achieving carbon neutrality.

- **Partners**: Local governments, businesses, and NGOs.
- **Impact**: Cities like Copenhagen and Melbourne have set aggressive decarbonization targets.

**The Tropical Forest Alliance (TFA)**

- **Description**: A PPP to combat deforestation.
- **Partners**: Governments, businesses, and NGOs like WWF.
- **Impact**: Reduced deforestation in supply chains for commodities like palm oil and soy.

Collaboration between businesses, governments, and NGOs is indispensable for advancing decarbonization goals. By pooling resources, expertise, and influence, these partnerships can overcome barriers, scale solutions, and accelerate the global transition to a low-carbon future. The success of these efforts depends on clear communication, shared goals, and robust accountability frameworks.

## 4.6 Conclusion

The urgency of decarbonization is undeniable in the face of escalating environmental and climate crises. The persistent rise in carbon emissions has led to unprecedented global challenges, including climate change, biodiversity loss, and ecosystem degradation. Immediate and sustained efforts to transition toward low-carbon economies and sustainable energy solutions are critical to mitigating these impacts. Governments, industries, and individuals must collectively embrace innovative policies, clean technologies, and responsible consumption patterns to drive meaningful change. By prioritizing renewable energy sources, carbon capture technologies, and circular economy models, societies can pave the way for a more resilient and climate-conscious future. The path to decarbonization is complex, yet it presents an opportunity to redefine economic growth, enhance global cooperation, and secure a sustainable planet for future generations.

## 4.7 Case Study

### 4.7.1 Case Study: The Urgency of Decarbonization—Environmental and Climate Imperatives

**Background**

Climate change has emerged as one of the most pressing global challenges, primarily driven by the excessive release of carbon dioxide ($CO_2$) and other GHGs. The increasing industrialization, reliance on fossil fuels, and deforestation have accelerated global warming, causing severe environmental disruptions. Nations worldwide are facing extreme weather events, rising sea levels, and deteriorating air quality, necessitating urgent decarbonization efforts.

**Case Context: The European Union's Green Deal**

The European Green Deal serves as a compelling example of a large-scale decarbonization initiative aimed at reducing carbon emissions while ensuring economic growth. Introduced in 2019, the EU committed to achieving net-zero emissions by 2050, with an interim goal of cutting emissions by at least 55% by 2030 compared to 1990 levels. The Green Deal encompasses various policies, including investments in renewable energy, energy efficiency, sustainable mobility, and carbon pricing mechanisms.

**Key Challenges**

Despite its ambitious targets, the EU Green Deal faces several challenges:

1) **Economic and Industrial Transition**—Shifting from fossil fuels to renewable energy sources requires substantial investments and restructuring of industries.
2) **Policy and Regulatory Hurdles**—Ensuring compliance across 27 member states with varying energy policies poses difficulties.
3) **Public Resistance and Social Impacts**—Transitioning to clean energy affects employment in traditional energy sectors, leading to socio-economic concerns.
4) **Technological Advancements**—Scaling up carbon capture, green hydrogen, and battery storage technologies remains a challenge.

**Implementation Strategies**

To address these challenges, the EU has adopted a multi-faceted approach:

- **Legislation and Carbon Pricing:** The Emissions Trading System (ETS) was strengthened to set stricter carbon limits for industries.
- **Renewable Energy Investments:** Major funding was directed toward wind, solar, and hydrogen energy projects.
- **Sustainable Transport:** The EU promoted EVs, charging infrastructure, and cleaner public transport solutions.
- **Circular Economy Practices:** Policies encouraged recycling, waste reduction, and sustainable production models.

**Impact and Future Outlook**

Early results indicate a positive shift, with EU emissions dropping by nearly 32% from 1990 levels by 2022. Investments in solar and wind energy have surged, making renewables a primary energy source. However, further global cooperation, accelerated innovation, and stringent policy implementation are crucial for achieving complete decarbonization.

## 4.8 Conclusion

The EU's experience underscores the urgency and feasibility of decarbonization, demonstrating that ambitious climate policies can drive both environmental and economic benefits. While challenges persist, strategic interventions, technological advancements, and collective action can pave the way toward a carbon-neutral and sustainable future. Other nations must adopt similar frameworks, tailoring them to their unique economic and social landscapes to effectively combat climate change.

## 4.9 Exercise

1. Define decarbonization and explain its significance in addressing climate change.
2. What are the primary objectives and goals of global decarbonization efforts?
3. Discuss the historical evolution of decarbonization initiatives. How have policies and technologies evolved over time?
4. Identify and describe the key stakeholders involved in decarbonization efforts. Why is collaboration necessary for success?
5. What are the major challenges and barriers to achieving effective decarbonization? Provide examples.
6. List the main sources of carbon emissions and their contribution to global warming.
7. Explain the climate change impacts associated with carbon emissions. How do they affect weather patterns and global temperatures?
8. How do carbon emissions contribute to environmental degradation and ecosystem disruption? Provide real-world examples.
9. Discuss the health effects of air pollution resulting from excessive carbon emissions.

10 How do regional and global disparities in carbon emissions affect environmental and economic stability?

11 Describe the economic consequences of climate change caused by carbon emissions. How do they impact industries and global economies?

12 What are the financial costs associated with adapting to climate change impacts (e.g., infrastructure resilience and disaster recovery)?

13 Discuss the economic benefits of reducing carbon emissions. How can businesses and governments gain from decarbonization?

14 Explain the concept of cost–benefit analysis in the context of climate change mitigation strategies. Provide an example.

15 Identify and explain three key technological innovations that contribute to decarbonization across different industries.

## 4.10 Objective Type Questions

1 **What is the primary goal of decarbonization?**
   a) Increase fossil fuel production
   b) Reduce greenhouse gas emissions
   c) Promote industrialization
   d) Expand urban development

2 **Which of the following best defines decarbonization?**
   a) Conversion of carbon into fuel
   b) Reduction of carbon emissions in energy and industrial systems
   c) Elimination of carbon-based life forms
   d) Increase of carbon content in the atmosphere

3 **Which of the following is NOT a key stakeholder in decarbonization efforts?**
   a) Governments
   b) Fossil fuel lobbyists
   c) Environmental NGOs
   d) International agencies

4 **What historical event is often considered the starting point for global decarbonization discourse?**
   a) The Industrial Revolution
   b) Kyoto Protocol signing
   c) World War II
   d) Invention of the steam engine

5 **One of the main challenges in decarbonization is:**
   a) Lack of carbon in the atmosphere
   b) Overdependence on renewable energy
   c) Economic and political resistance to change
   d) Excess use of hydrogen fuels

6 **Which sector contributes the most to global carbon emissions?**
   a) Agriculture
   b) Transportation
   c) Energy production
   d) Tourism

7 **Carbon emissions contribute to global warming primarily by:**
   a) Blocking solar radiation
   b) Enhancing the greenhouse effect
   c) Cooling ocean temperatures
   d) Preventing plant growth

8 **Which of the following is an example of environmental degradation linked to carbon emissions?**
   a) Desertification
   b) Volcanic eruptions
   c) Oil discovery
   d) Earthquakes

9 **Which health issue is commonly associated with air pollution caused by carbon emissions?**
   a) Malaria
   b) Asthma
   c) Diabetes
   d) Skin cancer

10 **Which region has historically contributed the most to cumulative global carbon emissions?**
a) Africa
b) Latin America
c) North America and Europe
d) Southeast Asia

11 **Which of the following is an economic impact of climate change?**
a) Increased fossil fuel reserves
b) Rising healthcare costs
c) Global population growth
d) Increased labor productivity

12 **Adaptation to climate change refers to:**
a) Denying climate impacts
b) Changing laws to promote fossil fuel use
c) Adjusting systems to minimize climate-related damages
d) Increasing emissions to speed up development

13 **Which of the following represents an economic benefit of decarbonization?**
a) Job losses in green industries
b) Reduced innovation
c) Lower health costs
d) Higher fossil fuel subsidies

14 **Economic risks of inaction on climate change include:**
a) Lower renewable energy costs
b) Stable weather patterns
c) Disruption of global supply chains
d) Reduced insurance premiums

15 **Cost–benefit analysis in climate mitigation aims to:**
a) Maximize fossil fuel extraction
b) Compare costs of mitigation to potential benefits
c) Delay policy action
d) Promote global inequality

**16 Which of the following is a breakthrough in renewable energy technology?**
- a) Gasoline-powered engines
- b) Solar photovoltaic cells
- c) Diesel generators
- d) Hydraulic fracturing

**17 What is the primary function of carbon capture and storage (CCS) technology?**
- a) Increase natural gas emissions
- b) Capture and store renewable energy
- c) Capture carbon dioxide from emission sources and store it underground
- d) Filter oxygen from the air

**18 Which digital technology is being used to optimize decarbonization efforts?**
- a) Cryptocurrency mining
- b) Artificial intelligence
- c) Gasoline sensors
- d) Print media

**19 Corporate sustainability initiatives aim to:**
- a) Maximize short-term profits only
- b) Increase fossil fuel consumption
- c) Reduce environmental footprint and carbon emissions
- d) Outsource all energy use

**20 How can industries collaborate to advance decarbonization goals?**
- a) By competing for carbon credits
- b) Through partnerships with governments and NGOs
- c) By avoiding environmental regulations
- d) By ignoring emission data

## 4.11 Glossary

1) **Decarbonization:** The process of reducing carbon dioxide ($CO_2$) and other greenhouse gas emissions from human activities, particularly energy production, industry, and transportation, to mitigate climate change.

2) **Carbon Emissions:** The release of carbon dioxide and other carbon compounds into the atmosphere, primarily from burning fossil fuels, industrial activities, and deforestation, contributing to global warming.
3) **Carbon Footprint:** The total amount of greenhouse gases, especially $CO_2$, emitted directly or indirectly by individuals, organizations, or industries, usually expressed in terms of $CO_2$-equivalent.
4) **Climate Change:** Long-term changes in global or regional climate patterns, largely driven by rising greenhouse gas concentrations, resulting in global warming, extreme weather, and environmental disruptions.
5) **Global Warming:** The increase in Earth's average surface temperature due to rising concentrations of greenhouse gases in the atmosphere, mainly from fossil fuel combustion and deforestation.
6) **Extreme Weather Events:** Severe and unusual weather conditions such as heatwaves, floods, hurricanes, and droughts, which are intensified by climate change and excessive carbon emissions.
7) **Ocean Acidification:** The reduction of ocean pH caused by the absorption of excessive carbon dioxide from the atmosphere, threatening marine ecosystems and biodiversity.
8) **Biodiversity Loss:** The decline in the variety of living species, driven by habitat destruction, climate change, pollution, and overexploitation, with carbon emissions playing a major role.
9) **Environmental Degradation:** The deterioration of the environment due to depletion of natural resources, pollution, and ecosystem destruction, often linked to unsustainable industrial and societal practices.
10) **Economic Cost of Carbon Emissions:** The financial burden resulting from climate-related impacts such as natural disasters, healthcare costs from pollution-related illnesses, and productivity losses.
11) **Low-Carbon Economy:** An economic system based on minimizing greenhouse gas emissions through the use of renewable energy, energy efficiency, sustainable industries, and green innovations.
12) **Renewable Energy:** Energy derived from naturally replenishing sources like solar, wind, hydro, and biomass, which play a vital role in reducing carbon emissions.
13) **Carbon Capture and Storage (CCS):** A technology designed to capture carbon dioxide emissions from industrial and energy sources, transport it, and store it underground to prevent its release into the atmosphere.
14) **Energy Efficiency:** Using less energy to perform the same task or produce the same outcome, thereby reducing emissions, costs, and environmental impacts.

15) **Alternative Fuels:** Non-fossil fuel energy sources such as hydrogen, biofuels, and synthetic fuels that can replace traditional carbon-intensive fuels in transport and industry.
16) **Corporate Sustainability:** Business strategies and practices that prioritize environmental stewardship, social responsibility, and economic viability to support long-term sustainability.
17) **Carbon Neutrality:** A state in which an entity balances the amount of carbon it emits with equivalent removal or offsetting measures, achieving net-zero carbon emissions.
18) **Green Investments:** Financial investments directed toward environmentally sustainable projects, such as renewable energy, green infrastructure, and climate-resilient technologies.
19) **Policymakers:** Government authorities and institutions responsible for creating and implementing policies, regulations, and frameworks to accelerate decarbonization.
20) **Integrated Approach (to Decarbonization):** A coordinated effort involving governments, industries, businesses, and civil society to achieve a sustainable, large-scale reduction in carbon emissions.
21) **Digital Technologies (in Decarbonization):** Advanced tools such as artificial intelligence, big data, and digital monitoring systems that enhance energy efficiency, optimize carbon reduction strategies, and support sustainability goals.

## References


**1** Jian, P., Xiang, S., & Sabzalian, M. H. (2024). Planning of a charging station for electric and hydrogen vehicles under hydrogen storage and fuel cell systems using a novel stochastic p-robust optimization technique. *International Journal of Hydrogen Energy*, 88, 702–712 https://doi.org/10.1016/J.IJHYDENE.2024.09.228.

**2** Khanna, D., Gemechu, E., Mahbub, N., Joy, J., & Kumar, A. (2025). Assessment of life cycle environmental impacts of materials, driving pattern, and climatic conditions on battery electric and hydrogen fuel cell vehicles in a cold region. *Environmental Impact Assessment Review*, 110, 107680 https://doi.org/10.1016/J.EIAR.2024.107680.

**3** Khare, V., & Jain, A. (2023). Predict the performance of driverless car through the cognitive data analysis and reliability analysis based approach. *e-Prime-Advances in Electrical Engineering, Electronics and Energy*, 6, 100344 https://doi.org/10.1016/J.PRIME.2023.100344.

4 Khare, V., Nema, S., & Baredar, P. (2013). Status of solar wind renewable energy in India. *Renewable and Sustainable Energy Reviews*, 27, 1–10 https://doi.org/10.1016/J.RSER.2013.06.018.

5 Khare, V., Nema, S., & Baredar, P. (2016). Optimization of hydrogen based hybrid renewable energy system using HOMER, BB-BC and GAMBIT. *International Journal of Hydrogen Energy*, 41 (38), 16743–16751 https://doi.org/10.1016/J.IJHYDENE.2016.06.228.

6 Kim, J., Kim, S., Chun, H., Sim, J., Woo, S. Y., Kang, S., & Min, K. (2025). Comprehensive analysis of polysulfone membrane humidifier in hydrogen fuel cell vehicles: experimental and theoretical approaches. *Journal of Membrane Science*, 713, 123234 https://doi.org/10.1016/J.MEMSCI.2024.123234.

7 Lei, N., Zhang, H., Chen, H., & Wang, Z. (2024). A comprehensive study of various carbon-free vehicle propulsion systems utilizing ammonia-hydrogen synergy fuel. *ETransportation*, 20, 100332 https://doi.org/10.1016/J.ETRAN.2024.100332.

8 Li, W., Toshiyuki, Y., Guan, H., & Han, Y. (2024). Assessing consumer preferences for hydrogen fuel cell vehicles: a conjoint analysis approach across multiple cities in China. *International Journal of Hydrogen Energy*, 80, 1210–1222 https://doi.org/10.1016/J.IJHYDENE.2024.07.164.

9 Liu, F., Shafique, M., & Luo, X. (2024). Quantifying delayed climate mitigation benefits in electric and hydrogen fuel cell vehicle deployment for sustainable mobility. *Sustainable Production and Consumption*, 49, 398–414 https://doi.org/10.1016/J.SPC.2024.06.001.

10 Liu, S., Hao, L., Jia, H., Xu, G., & Du, P. (2024). Study on hydrogen diffusion and combustion characteristics of hydrogen fuel cell sightseeing vehicles in open space. *International Journal of Hydrogen Energy*, 87, 50–67 https://doi.org/10.1016/J.IJHYDENE.2024.09.024.

11 Liu, W., Markert, F., Hostikka, S., & Giuliani, L. (2024). Thermal responses of a concrete slab under hydrogen fuel cell vehicle fires in a semi-open car park. *International Journal of Hydrogen Energy*, 73, 800–811 https://doi.org/10.1016/J.IJHYDENE.2024.06.096.

12 Luo, H., Xiao, J., Bénard, P., Yuan, C., Tong, L., Chahine, R., Yuan, Y., Yang, T., & Yao, C. (2024). Thermodynamic modeling and analysis of cascade hydrogen refuelling with three-stage pressure and temperature for heavy-duty fuel cell vehicles. *International Journal of Hydrogen Energy*, 63, 103–113 https://doi.org/10.1016/J.IJHYDENE.2024.03.060.

13 Nagem, N. A., Ebeed, M., Alqahtani, D., Jurado, F., Khan, N. H., & Hafez, W. A. (2024). Optimal design and three-level stochastic energy management for an interconnected microgrid with hydrogen production and storage for fuel cell electric vehicle refueling stations. *International Journal of Hydrogen Energy*, 87, 574–587 https://doi.org/10.1016/J.IJHYDENE.2024.08.415.

**14** Khare, V. (2023). Prediction of tidal energy resources at Gulf of Kutch and Gulf of Khambhat by data mining. *Cleaner Energy Systems*, 6, 100090.

**15** Khare, V., Khare, C. J., & Bhuiyan, M. A. (2023). Design, optimization, and data analysis of solar-tidal hybrid renewable energy system for Hurawalhi, Maldives. *Cleaner Energy Systems*, 6, 100088.

**16** Khare, V. (2023). Assessment of hydro energy potential from rain fall data set in India through data analysis. *Advances in Electrical Engineering, Electronics and Energy*, 6, 1–13.

**17** Khare, V., & Chaturvedi, P. (2023). Design, control, reliability, economic and energy management of microgrid: a review. *Advances in Electrical Engineering, Electronics and Energy*, 5, 100239.

**18** Khare, V., & Bhuiyan, M. A. (2023). Perspective of renewable energy in the BRICS country. *Advances in Electrical Engineering, Electronics and Energy*, 5, 100250.

**19** Khare, V., & Chaturvedi, P. (2023). Solar energy system concept change from trending technology: a comprehensive review. *Advances in Electrical Engineering, Electronics and Energy*, 4, 100183.

**20** Khare, V., & Chaturvedi, P. (2023). Design and operational management of sustainable multi-networks green energy system for study area at India. *Materials Today Proceedings*, 1–15.

**21** Oladosu, T. L., Pasupuleti, J., Kiong, T. S., Koh, S. P. J., & Yusaf, T. (2024). Energy management strategies, control systems, and artificial intelligence-based algorithms development for hydrogen fuel cell-powered vehicles: a review. *International Journal of Hydrogen Energy*, 61, 1380–1404 https://doi.org/10.1016/J.IJHYDENE.2024.02.284.

# 5

# Operations and Supply Chain Management of Green Hydrogen

## 5.1 Introduction

The operations and SCM of green hydrogen is a critical aspect of its integration into global energy markets, requiring seamless coordination across various stages, from production to delivery. The production process typically involves the electrolysis of water powered by renewable energy sources such as solar, wind, or hydroelectric power, which makes it essential to ensure that the sourcing and utilization of renewable energy are both reliable and cost-effective. Efficient operations require the optimization of electrolyzer technologies to ensure high efficiency and scalability while minimizing costs [1]. The supply chain must address several key challenges, including storage, transportation, and distribution, given hydrogen's low energy density in its gaseous form and the need for specialized infrastructure. Compression, liquefaction, and the development of pipelines or transportation systems for hydrogen are crucial components to ensure that green hydrogen can be transported safely and efficiently over long distances to various end-users. Furthermore, a robust distribution network is needed to connect hydrogen production sites to industries, fuel stations, and power plants that require it. Inventory management and demand forecasting become increasingly complex in this emerging market, as fluctuating renewable energy supply and hydrogen consumption patterns must be closely monitored. Effective SCM also requires building partnerships across the value chain, from hydrogen producers to end-users, as well as working closely with regulatory bodies to meet safety standards and ensure environmental compliance. Moreover, innovation in digital technologies, such as IoT, AI, and blockchain, is key to improving supply chain visibility, optimizing route planning, and enabling real-time monitoring of hydrogen stocks. In addition

*Hydrogen Technologies in Decarbonizing Industries and Energy Systems*, First Edition.
Vikas Khare, Sanjeet K. Dwivedi, Monica Bhatia, and K. K. Pant.

Published 2026 by John Wiley & Sons, Inc.

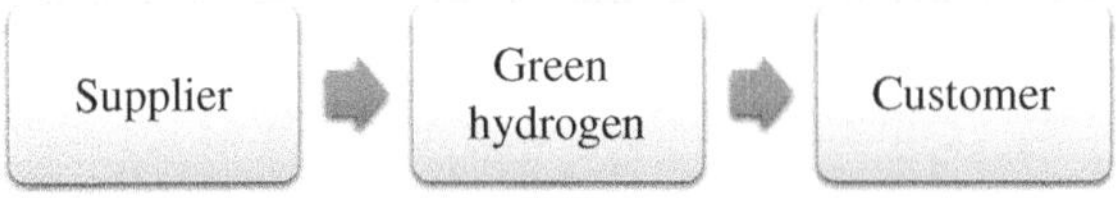


**Figure 5.1** A Conceptual Model of a Basic Supply Chain

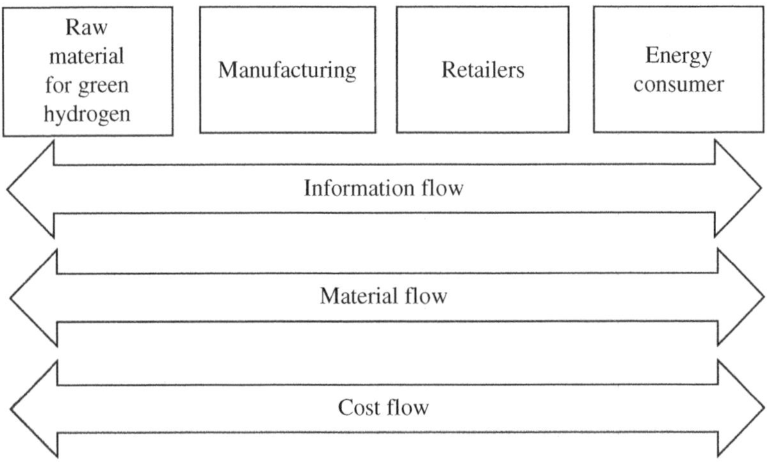


**Figure 5.2** Flows in Green Hydrogen Supply Chain

to these technical challenges, there is a need to address the economic viability of the entire supply chain, from investment in infrastructure to ensuring competitive pricing for green hydrogen, so that it can compete with other energy sources. Thus, the operations and SCM of green hydrogen is not only about logistics but also about creating a dynamic, efficient, and scalable ecosystem that can drive the large-scale adoption of green hydrogen across various sectors [2]. Figure 5.1 shows a conceptual model of a basic supply chain. Figure 5.2 shows flows in the green hydrogen supply chain.

Green hydrogen, produced using renewable energy sources like wind and solar power, is emerging as a key solution for decarbonization across industries. In the United States, the green hydrogen supply chain is gaining momentum through government incentives such as the Inflation Reduction Act (IRA) and the Hydrogen Energy Earthshot initiative, which aims to reduce hydrogen production costs to $1 per kg. The United States focuses on distributed production and regional hydrogen hubs, integrating renewable energy sources with electrolyzers to enhance supply chain efficiency. Key players like Plug Power, Air Products, and NextEra Energy are developing large-scale projects, while challenges remain in infrastructure development, storage, and distribution [3].

> ***Key: Green hydrogen:*** *Hydrogen produced using renewable energy sources (like wind, solar) through electrolysis, with no associated $CO_2$ emissions.*

In China, the world's largest hydrogen producer, green hydrogen is becoming a priority to achieve carbon neutrality by 2060. The Chinese government has integrated green hydrogen into its Five-Year Plans, and major state-owned enterprises like Sinopec and China Energy are investing in large-scale production facilities. China's supply chain emphasizes cost reduction through mass electrolyzer production, leveraging its dominance in solar panel and battery manufacturing. However, challenges such as high production costs, the dominance of coal-based hydrogen, and the need for advanced hydrogen transport infrastructure persist.

Germany, a leader in renewable energy, is driving green hydrogen adoption through its National Hydrogen Strategy and partnerships within the European Hydrogen Backbone initiative. The country focuses on developing hydrogen corridors for industrial applications and integrating green hydrogen into sectors like steel production, chemicals, and transportation. Germany also relies on international hydrogen imports from countries like Namibia and Australia, enhancing its global supply chain network. Despite its advanced infrastructure, Germany faces challenges in scaling production and reducing reliance on imports to achieve cost competitiveness [4].

Each country employs unique strategies tailored to its energy landscape, industrial needs, and policy frameworks. While the United States prioritizes regional hydrogen hubs, China emphasizes cost-effective scaling, and Germany integrates hydrogen into its existing industrial ecosystem with strong international partnerships. Overcoming infrastructure and cost challenges will be key to establishing a sustainable global green hydrogen supply chain.

## 5.2 Input Material Sourcing

Green hydrogen production requires a well-structured operations and SCM system to ensure a sustainable, cost-effective, and reliable supply of input materials. This includes renewable energy sources, water for electrolysis, electrolyzer components, and sustainable procurement strategies. Efficient sourcing practices are essential to making green hydrogen economically viable and environmentally friendly. Following is a detailed explanation of the key aspects of input material sourcing in green hydrogen production.

Since green hydrogen is produced via electrolysis, where water is split into hydrogen and oxygen using electricity, sourcing renewable energy is crucial to ensure the process remains sustainable and emission-free. The production of green hydrogen relies on renewable energy sources such as solar, wind, hydro, and geothermal power, with different countries leveraging their unique energy landscapes. In the United States, wind and solar power are the primary sources for green hydrogen production. The United States has abundant wind energy, particularly in states like Texas, Iowa, and Oklahoma, and vast solar potential in regions like California and Arizona. These renewable resources are integrated into hydrogen hubs, enabling cost-effective hydrogen production [5]. Figure 5.3 shows sources of renewable energy for green hydrogen production. The primary renewable energy sources include:

1) **Solar Energy**
   - PV solar farms generate electricity to power electrolyzers.
   - Can be deployed in areas with high solar radiation, reducing dependency on fossil fuels.
   - Challenges: Variability due to weather conditions and energy storage requirements.
2) **Wind Energy**
   - Offshore and onshore wind farms provide a stable source of renewable electricity.
   - More reliable at night compared to solar energy.

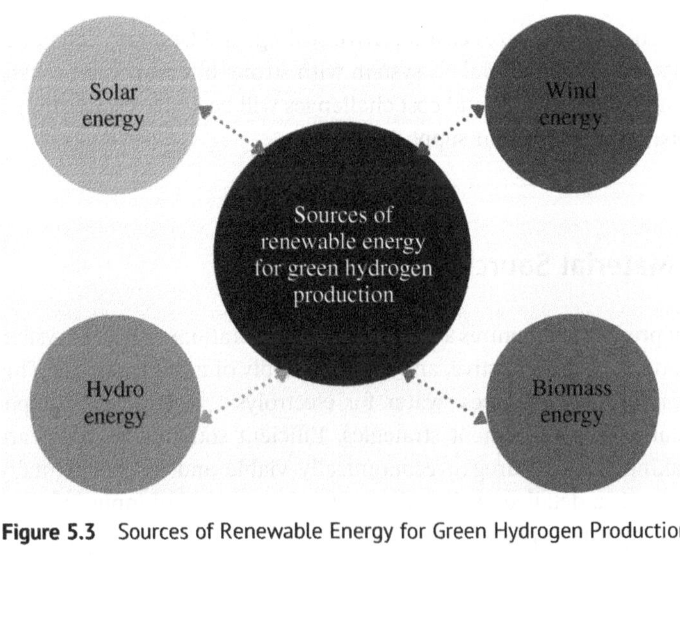


**Figure 5.3** Sources of Renewable Energy for Green Hydrogen Production

- Challenges: Requires grid infrastructure upgrades and integration with hydrogen plants.

3) **Hydropower**
   - Large-scale hydroelectric plants provide continuous electricity for electrolysis.
   - Challenges: Limited by geography and environmental concerns related to dam construction.
4) **Biomass and Geothermal Energy**
   - Biomass-derived electricity and geothermal plants can also power electrolyzers.
   - Challenges: Biomass must be sustainably sourced to remain carbon-neutral.
5) **Importance of Renewable Energy Sourcing**
   - Energy costs significantly impact the overall cost of green hydrogen.
   - Investment in smart grids and energy storage is crucial for efficient energy supply.
   - Integrating multiple renewable sources can enhance reliability and scalability.

China, as the world's largest renewable energy producer, predominantly utilizes solar and wind power for green hydrogen. The country leads in solar panel manufacturing and has extensive solar farms in regions like Inner Mongolia and Xinjiang, along with significant wind energy capacity in coastal and northern provinces. China also explores hydro-based hydrogen production, capitalizing on its massive hydroelectric infrastructure, including the Three Gorges Dam [6, 7].

Germany, with its advanced renewable energy infrastructure, focuses on wind and solar energy for hydrogen production. Offshore and onshore wind farms in the North Sea and Baltic Sea play a critical role in providing clean electricity for electrolysis. Additionally, Germany supplements its domestic production by importing green hydrogen from solar-rich countries like Morocco and Namibia. The country also integrates excess renewable energy from its power grid into hydrogen storage, ensuring a stable energy supply.

While the United States and China capitalize on domestic renewable resources for green hydrogen, Germany's approach combines domestic wind and solar power with international hydrogen imports. The successful expansion of green hydrogen production globally will depend on enhancing renewable energy capacity, storage solutions, and grid integration to ensure a stable and cost-effective hydrogen supply chain.

***Key: Supply chain management (SCM):*** *The coordination of procurement, production, logistics, and delivery activities to ensure efficient and sustainable green hydrogen flow.*

### 5.2.1 Sourcing of Water for Electrolysis in Green Hydrogen Production

Electrolysis, the process of splitting water into hydrogen and oxygen, is the most prominent method of producing green hydrogen. However, this process requires water of extremely high purity to prevent damage to the electrolyzer and to maintain efficiency. The sourcing of this water is, therefore, a fundamental challenge, as it must balance availability, sustainability, and cost-effectiveness. Various sources can be considered for supplying water to the electrolyzers, each offering unique opportunities and challenges. Figure 5.4 shows sourcing of water for electrolysis. Key considerations in water sourcing include:

**Freshwater Sources:** One of the most direct options for sourcing water is the use of freshwater from rivers, lakes, and reservoirs. These sources provide relatively easy access to water and are often located near industrial zones where hydrogen facilities could be built. However, reliance on freshwater raises sustainability concerns, particularly in regions where water scarcity already exists. Freshwater is essential for drinking, irrigation, and agricultural production, and diverting it toward industrial hydrogen production risks intensifying conflicts over resources. Thus, while freshwater remains a technically viable option, its large-scale use for hydrogen production may not be sustainable in the long run, especially in water-stressed countries [8, 9].

**Seawater Desalination:** In coastal regions, seawater desalination has emerged as a practical solution to meet the water requirements of green hydrogen projects. Through technologies like reverse osmosis and

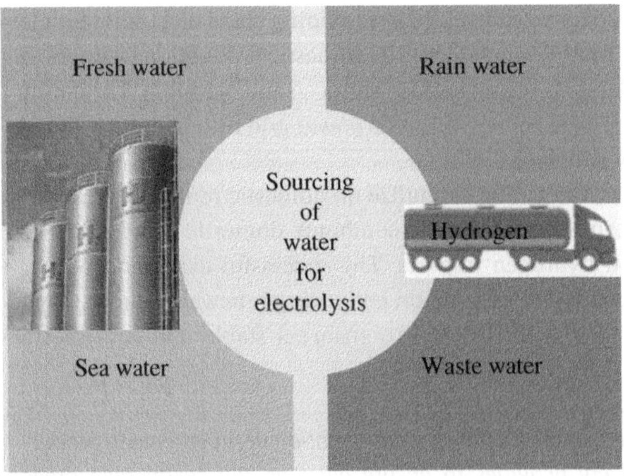


**Figure 5.4** Sourcing of Water for Electrolysis. *Source:* AA+W / Adobe Stock Photo.

membrane distillation, seawater can be purified to the required standards for electrolysis. This approach offers the advantage of tapping into the vast and virtually unlimited reserves of seawater, which eliminates the competition with drinking water or agriculture. However, desalination carries its own set of challenges. It is highly energy-intensive, thereby increasing the overall cost of hydrogen production and potentially undermining the environmental benefits of green hydrogen unless powered by renewable energy. Additionally, the disposal of concentrated brine, a by-product of desalination, poses environmental risks by harming marine ecosystems. Thus, while seawater desalination expands the geographical feasibility of hydrogen production, it must be carefully managed to minimize energy use and ecological damage.

**Wastewater Recycling:** A sustainable alternative is the recycling of treated municipal or industrial wastewater for use in electrolysis. By tapping into this resource, industries not only reduce their demand for freshwater but also contribute to the principles of a circular economy, where waste is transformed into a valuable input. Wastewater recycling is particularly appealing in urban and industrialized areas, where large volumes of wastewater are consistently generated and treated. However, this option requires advanced purification processes to ensure the water meets the purity standards necessary for electrolyzers, which adds to operational costs. Furthermore, strict regulatory frameworks often govern wastewater reuse, and compliance with these rules can be complex and costly. Despite these challenges, wastewater recycling remains an attractive and environmentally responsible approach to water sourcing for hydrogen production.

**Rainwater Harvesting:** Rainwater harvesting presents another potential source of water for electrolysis, particularly in regions that experience heavy seasonal rainfall. Capturing and storing rainwater for use in hydrogen production can reduce dependence on other freshwater supplies and provide a low-cost alternative in areas where rainfall is abundant. However, the feasibility of this method is highly dependent on geographical and climatic conditions. Regions with low or irregular rainfall patterns cannot rely on rainwater harvesting as a consistent source of water. Even in areas with sufficient rainfall, large-scale storage infrastructure is required to make this method practical for industrial hydrogen production. Thus, rainwater harvesting is best suited as a supplementary solution rather than a primary water source [10].

**Key Considerations in Water Sourcing:** Beyond identifying potential water sources, several considerations determine their suitability for electrolysis. Water purity is essential, as high-purity water minimizes the risk of electrolyzer degradation and ensures long-term efficiency. Cost-effectiveness is another important factor, as processes like desalination and wastewater treatment demand significant energy and infrastructure investments, which directly impact the cost of hydrogen production. Finally, sustainability must be prioritized; hydrogen production is inherently water-intensive, and scaling it up without careful planning could worsen local water stress. Therefore, sourcing strategies must carefully balance technical, economic, and environmental factors to ensure that green hydrogen truly remains sustainable. The electrolyzer is the core equipment in green hydrogen production, and sourcing its components involves:

**Key Electrolyzer Components and Materials**

1) **Electrodes** (platinum, iridium, nickel-based materials)
2) **Membranes** (PEMs)
3) **Bipolar Plates** (titanium, graphite, or stainless steel)
4) **Power Electronics** (rectifiers, inverters, transformers)
5) **Cooling and Gas Separation Systems**

**Procurement Strategies:** Different procurement strategies are employed to secure these electrolyzer components, each with its advantages and limitations. Local sourcing can significantly reduce transportation costs and carbon emissions, while also strengthening domestic industries; however, many regions lack the supply base for specialized materials such as platinum or iridium. In contrast, global supply chains provide access to advanced technologies and specialized components but are exposed to geopolitical risks, trade restrictions, and transportation-related emissions. Strategic partnerships with Original Equipment Manufacturers (OEMs) offer another pathway by ensuring long-term stability of supply through collaboration and joint ventures (JVs). Some companies even pursue vertical integration, investing directly in manufacturing their own components to reduce dependence on external suppliers. Each of these strategies must be tailored to the company's resources, regional conditions, and long-term vision for scaling hydrogen production [11, 12].

**Challenges in Procurement:** Despite advances in procurement strategies, the industry faces several persistent challenges. The high cost of critical materials like platinum and iridium remains a significant barrier to scaling up green hydrogen technologies, as these metals are scarce and expensive. Supply chain disruptions, often caused by geopolitical instability, trade disputes, or concentrated production in certain countries, further increase the risks associated with global sourcing. Moreover, the rapid pace of technological advancement in electrolyzer design demands flexibility in procurement practices, as companies must quickly adapt to new materials, designs, and standards. Addressing these challenges will require innovation in material science, diversification of supply sources, increased recycling of scarce elements, and stronger international cooperation to secure a stable supply chain for the green hydrogen industry [13].

### 5.2.2 Ethical and Sustainable Sourcing Practices for Input Materials with Challenges and Opportunity

Sustainability and ethics play a crucial role in green hydrogen SCM, particularly in the procurement of critical materials and energy sources.

**Sustainable Sourcing Approaches**

- Responsible Mining Practices: Ensuring raw materials (e.g., nickel, platinum, and iridium) come from ethically managed mines.
- Recycling and Circular Economy: Using recycled metals in electrolyzer components to reduce environmental impact.
- Supply Chain Transparency: Implementing blockchain technology to track material origin.
- Labor Rights Compliance: Ensuring fair wages and ethical labor practices in supplier networks.

**Challenges**

1) Raw Material Scarcity: Limited availability of platinum and iridium for electrolyzers.
2) High Production Costs: Expensive manufacturing and logistical hurdles in sourcing components.
3) Geopolitical Risks: Trade restrictions and regional instability affecting material supply.
4) Infrastructure Limitations: Need for large-scale renewable energy projects to power electrolysis.
5) Water Scarcity: Competing water demands in drought-prone regions.

**Opportunities**

1) Material Innovation: Developing low-cost, abundant catalysts as alternatives to platinum/iridium.
2) Circular Economy Practices: Recycling and repurposing materials to reduce dependency on virgin resources.
3) Advancements in Water Recycling: Utilizing wastewater for electrolysis, reducing freshwater consumption.
4) Green Supply Chains: Strengthening local production to minimize carbon footprint and enhance supply chain resilience.
5) Public–Private Partnerships (PPPs): Collaborating with governments and industries to invest in sustainable sourcing practices.

## 5.3 Infrastructure Development

Designing and constructing green hydrogen production facilities is a complex but rewarding process, critical to advancing renewable energy solutions. Following is a step-by-step framework for approaching this endeavor:

i) **Feasibility Study and Site Selection**

A feasibility study for green hydrogen production involves a comprehensive analysis of technical, economic, and environmental factors to ensure the project's viability and sustainability. Key considerations include the availability of renewable energy sources such as solar, wind, or hydropower, which are critical for powering the electrolysis process to produce green hydrogen. The study evaluates the proximity to water resources, as electrolysis requires a substantial water supply. Infrastructure readiness, such as grid connectivity, transportation networks, and access to pipelines or storage facilities, also plays a pivotal role. Furthermore, the regulatory framework, incentives for renewable energy projects, and potential market demand for hydrogen influence the site selection process. Ideal sites are strategically located near renewable energy hubs, industrial clusters, or export facilities to minimize costs and optimize logistics. Environmental impact assessments ensure the project aligns with sustainability goals, considering land use, biodiversity, and carbon footprint. An integrated feasibility study and careful site selection lay the groundwork for efficient and economically viable green hydrogen production facilities.

- **Energy Source Assessment**: Identify renewable energy sources (e.g., solar, wind, and hydropower) near the site for powering the hydrogen production process.
- **Location**: Select a site with access to renewable energy, water supply, and proximity to infrastructure for storage and distribution.
- **Regulatory Compliance**: Ensure compliance with local environmental and zoning laws.
- **Economic Analysis**: Evaluate costs, potential subsidies, and long-term economic viability.

ii) **Technology Selection**

The selection of technology for green hydrogen production facilities is a critical decision that directly impacts efficiency, cost, and environmental sustainability. The core technology, electrolysis, splits water into hydrogen and oxygen using electricity from renewable energy sources. Among the primary electrolyzer technologies—AELs, PEM electrolyzers, and solid oxide electrolyzers (SOEs—each has unique advantages. AELs are mature, cost-effective, and suitable for large-scale applications, while PEM electrolyzers offer higher efficiency, rapid response times, and compatibility with variable renewable energy sources like wind and solar. SOEs, though still in the developmental phase, promise high efficiencies at elevated temperatures, leveraging industrial waste heat. In addition to the electrolyzer choice, technologies for hydrogen storage, purification, and transportation must be considered. Integration with renewable energy systems, automation, and digitalization for real-time monitoring and optimization is vital for operational success. A thorough assessment of technology readiness, capital and operational costs, scalability, and compatibility with existing infrastructure ensures the selected technology aligns with the project's goals and long-term sustainability objectives. Figure 5.5 shows the infrastructure development of green hydrogen production.

- **Electrolysis Technology**:
  - **Alkaline Electrolyzers**: Mature technology, suitable for large-scale operations.
  - **PEM Electrolyzers**: More efficient for intermittent renewable sources.
  - **Solid Oxide Electrolyzers**: Emerging technology with high efficiency but requires high temperatures.
- **Renewable Energy Integration**: Design systems to integrate renewable sources with the electrolyzers for optimal efficiency.

iii) **Design and Engineering**

The design and engineering of green hydrogen production facilities encompass meticulous planning to ensure efficiency, scalability, and environmental sustainability. The process begins with conceptualizing a layout that integrates key components such as renewable energy sources, electrolyzers, storage tanks, purification units, and distribution systems. Advanced simulation tools and computational models are employed to optimize the facility's energy flows, minimize losses, and ensure seamless integration with renewable energy inputs like solar or wind. Materials selection plays a crucial role, with a focus on durability and corrosion resistance, especially for equipment exposed to water and hydrogen. Safety is paramount, requiring adherence to stringent engineering standards for handling hydrogen's flammability and high diffusivity. Modular and flexible designs are increasingly adopted to accommodate future expansions or technological upgrades. The inclusion of smart systems for automation, real-time monitoring, and data analytics enhances operational efficiency and reliability. Moreover, the design emphasizes minimizing the environmental footprint through energy-efficient practices, effective water recycling, and sustainable construction materials. A well-engineered facility not only optimizes production but also positions the project as a benchmark in the transition to clean energy.

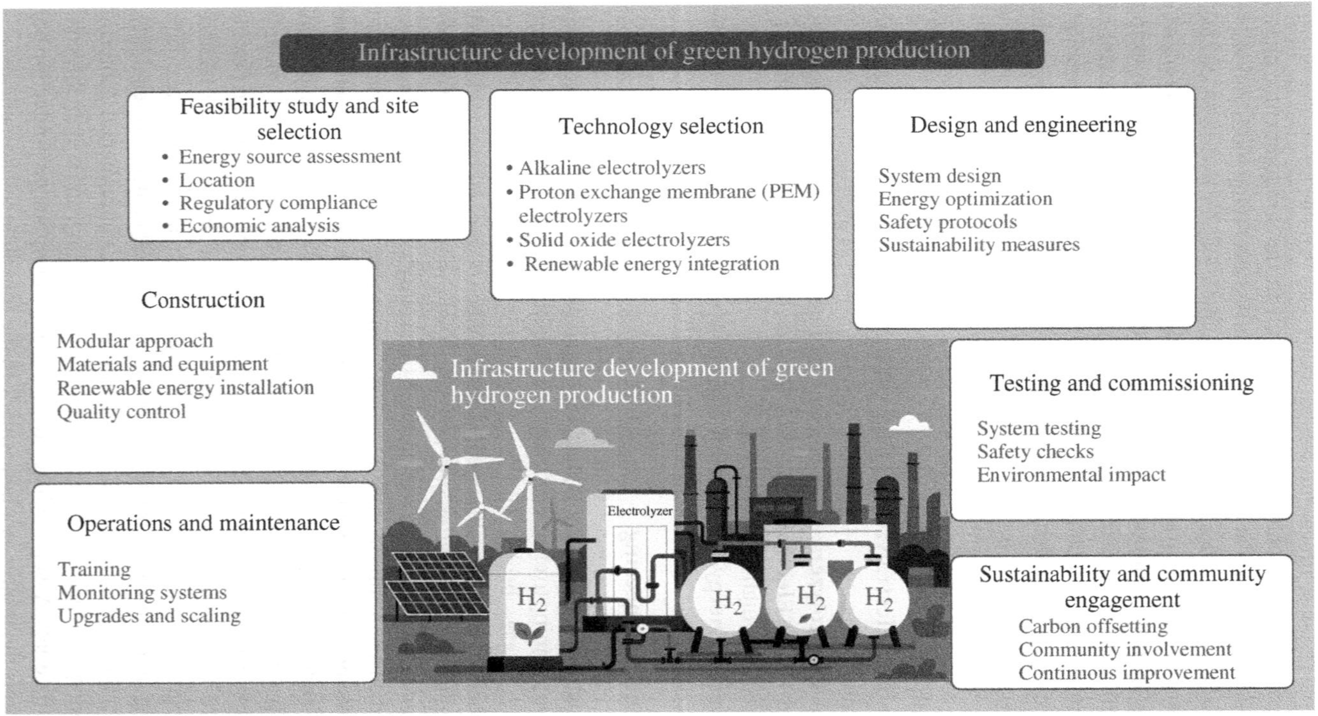


**Figure 5.5** Infrastructure Development of Green Hydrogen Production

- **System Design**:
  - Determine the scale of production based on demand.
  - Include components for water purification, electrolysis, compression, and storage.
- **Energy Optimization**:
  - Implement EMS for efficient use of renewable energy.
  - Use battery storage for renewable energy surplus.
- **Safety Protocols**:
  - Adhere to hydrogen safety standards (e.g., National Fire Protection Association (NFPA) 2).
  - Design fail-safe systems and emergency response mechanisms.
- **Sustainability Measures**:
  - Incorporate circular water usage systems.
  - Minimize carbon footprint during construction and operation.

iv) **Construction**

The construction of green hydrogen production facilities requires precise execution to transform detailed designs into operational infrastructure. It involves the assembly of core components, including electrolyzers, renewable energy systems, hydrogen storage tanks, and auxiliary systems like cooling, water purification, and safety equipment. Site preparation is a critical initial step, ensuring a stable foundation and proper alignment with environmental regulations. Specialized construction materials and techniques are employed to withstand the demands of hydrogen production, such as corrosion-resistant coatings and robust piping systems. Collaboration with skilled engineers and contractors is essential for the installation of advanced technologies, electrical connections, and control systems. Safety measures are rigorously implemented during construction, addressing risks associated with hydrogen's flammable nature. Adherence to project timelines, budget constraints, and quality standards is managed through effective project management strategies. Integration testing is conducted during construction to ensure the seamless operation of all systems before commissioning. By focusing on precision, sustainability, and safety, the construction phase lays the groundwork for a facility that meets the highest operational and environmental standards.

- **Modular Approach**: Opt for modular construction for scalability and reduced initial investment.
- **Materials and Equipment**: Use corrosion-resistant materials for durability and long-term reliability.
- **Renewable Energy Installation**: Install solar panels, wind turbines, or hydropower connections as per the design.
- **Quality Control**: Conduct rigorous quality checks during construction to ensure safety and performance.

v) **Testing and Commissioning**

Testing and commissioning are critical phases in the development of green hydrogen production facilities, ensuring that all systems operate safely, efficiently, and in alignment with design specifications. The process begins with pre-commissioning activities, including inspections, calibration of instruments, and verifying the integrity of components such as electrolyzers, storage tanks, and pipelines. Functional testing follows, focusing on individual systems to confirm their performance under various operational scenarios. Renewable energy inputs are integrated, and the electrolysis process is tested to assess hydrogen production efficiency and purity. Safety systems, including leak detection and emergency shutdown mechanisms, are rigorously evaluated to address the inherent risks of hydrogen handling. Once individual systems are validated, integrated testing is conducted to ensure seamless interaction between subsystems, including energy supply, hydrogen storage, and distribution. Operational trials under simulated full-load conditions verify the facility's reliability and readiness for continuous operation. Detailed documentation of test results and compliance with regulatory standards are essential for final approval. Successful commissioning transitions the facility into commercial operation, marking the culmination of meticulous planning, design, and construction efforts.

- **System Testing**:
  - Test electrolyzers, storage tanks, and pipelines under simulated operational conditions.
  - Verify the integration of renewable energy sources.
- **Safety Checks**: Inspect all safety measures and protocols.
- **Environmental Impact**: Ensure compliance with environmental standards.

vi) **Operations and Maintenance**

The operations and maintenance (O&M) of green hydrogen production facilities are crucial for ensuring sustained efficiency, safety, and profitability. Operations involve monitoring the performance of electrolyzers, renewable energy inputs, hydrogen storage systems, and distribution networks. Advanced digital tools like IoT-enabled sensors and AI-powered analytics are employed for real-time monitoring, predictive diagnostics, and optimization of energy flows. Maintenance activities are divided into preventive, predictive, and corrective categories, focusing on minimizing downtime and prolonging equipment lifespan. Preventive measures include regular inspections, cleaning of electrolyzers, and calibration of control systems, while predictive maintenance leverages data-driven insights to anticipate failures before they occur. Proper handling of hydrogen, adherence to safety protocols, and periodic safety drills are essential to mitigate risks associated with its flammable nature. Training of personnel in the latest O&M practices and fostering a culture of safety and innovation are vital. By prioritizing efficient operations and proactive maintenance, green hydrogen facilities can achieve reliability, reduce costs, and contribute effectively to the clean energy ecosystem.

- **Training**: Provide training for operators on hydrogen safety, system operation, and emergency response.
- **Monitoring Systems**:
  - Implement IoT and AI-based monitoring for predictive maintenance.
  - Ensure real-time tracking of production efficiency and energy usage.
- **Upgrades and Scaling**: Plan for future scaling and integration of advanced technologies.

vii) **Sustainability and Community Engagement**

Sustainability and community engagement are integral to the success and acceptance of green hydrogen production facilities. These facilities prioritize environmental stewardship by leveraging renewable energy, minimizing carbon emissions, and adopting water-efficient technologies to produce hydrogen in an eco-friendly manner. LCAs are conducted to evaluate and mitigate environmental impacts, ensuring alignment with global sustainability goals. Community engagement begins with transparent communication about the project's benefits, such as job creation, clean energy contributions, and local economic development. Involving community stakeholders in the planning and decision-making processes fosters trust and collaboration. Educational initiatives, such as workshops and awareness programs, highlight the role of green hydrogen in combating climate change and promote community support. Additionally, facilities often invest in local infrastructure, environmental restoration, and social initiatives to enhance their positive impact. By embedding sustainability into operations and building strong community partnerships, green hydrogen projects can drive inclusive growth and reinforce their role as catalysts for a greener future. Table 5.1 shows the key attributes of design and construction for green hydrogen production facilities.

- **Carbon Offsetting**: Document the carbon reduction achieved and consider partnerships for carbon credit trading.
- **Community Involvement**: Educate local communities on the benefits of green hydrogen and create job opportunities.
- **Continuous Improvement**: Regularly update the facility with innovations to maintain sustainability and cost-effectiveness.

### 5.3.1 Expansion and Integration of Existing Infrastructure for Hydrogen Distribution

Expanding and integrating existing infrastructure for hydrogen distribution is vital for ensuring the efficient and widespread adoption of hydrogen as a clean and sustainable energy source. A comprehensive approach begins with the careful

**Table 5.1** The Key Attributes of Design and Construction for Green Hydrogen Production Facilities

| Attribute | Description |
|---|---|
| Location selection | Choosing locations with abundant renewable energy sources (solar, wind, hydro) and proximity to end-users |
| Electrolyzer Technology | Selection of suitable electrolyzers (PEM, alkaline, or solid oxide) for optimal efficiency and scalability |
| Renewable Energy Integration | Integration of renewable energy sources to power the electrolysis process, ensuring sustainability and cost-effectiveness |
| Energy storage systems | Implementation of energy storage solutions to balance intermittent renewable energy availability, such as batteries or hydrogen storage |
| Grid connection and infrastructure | Establishing reliable connections to the electrical grid or renewable energy plants, and ensuring flexibility in energy sourcing |
| Safety standards | Adherence to strict safety protocols and regulations for hydrogen production, storage, and handling, ensuring safety during all operations |
| Hydrogen storage solutions | Design and installation of safe and efficient hydrogen storage methods, including high-pressure tanks or cryogenic storage systems |
| Piping and distribution systems | Design of robust piping systems for transporting hydrogen safely within the facility and to external distribution networks |
| Facility size and scalability | Designing facilities with scalability in mind to accommodate future growth as demand for green hydrogen increases |
| Environmental impact assessment | Conducting thorough environmental impact assessments to minimize ecological disruption during construction and operation |
| Construction materials | Selection of durable, sustainable construction materials that support the longevity and efficiency of the facility |
| Water supply management | Ensuring an adequate and sustainable water supply for the electrolysis process, considering water availability and quality |
| Cost efficiency | Designing facilities to maximize cost efficiency in both construction and operation, while ensuring competitive production costs for green hydrogen |
| Modular design | Incorporating modular designs to allow for easier expansion and upgrades as technology advances or demand grows |
| Digitalization and automation | Utilizing digital technologies (IoT, AI, sensors) for monitoring, control, and optimization of production processes |
| Maintenance and upkeep | Planning for regular maintenance schedules, ensuring operational continuity and minimizing downtime |

**Table 5.1** (Continued)

| Attribute | Description |
|---|---|
| Regulatory compliance | Ensuring the facility complies with local, regional, and international regulations and standards for green hydrogen production |
| Community engagement | Engaging with local communities and stakeholders to ensure their support and address any concerns related to the facility's impact |

assessment of current infrastructure. This involves conducting inventory analyses to identify natural gas pipelines, storage facilities, and refueling stations that could potentially be retrofitted for hydrogen use. Suitability assessments are critical, as pipelines and equipment must be inspected for risks such as hydrogen embrittlement, while upgrades or replacements may be required to ensure long-term safety. At the same time, understanding energy demand and distribution needs across regions is essential, with analyses focusing on transportation, industry, and power sectors to identify gaps in infrastructure coverage and prioritize areas for development.

Pipeline expansion and retrofitting play a central role in scaling hydrogen distribution. One important strategy is retrofitting existing natural gas pipelines to allow hydrogen blending, typically up to 20% without major modifications. For 100% hydrogen transport, sections of pipelines may need to be replaced or coated internally with hydrogen-compatible materials, along with the installation of hydrogen compressors and advanced leak detection systems. In parallel, dedicated hydrogen pipelines can be constructed in regions with high demand or along major energy corridors. These pipelines should use materials like polyethylene or fiber-reinforced polymers to prevent embrittlement and minimize leakage, while incorporating smart monitoring systems to provide real-time operational data and ensure reliability.

HRSs are another critical component of distribution infrastructure. Existing compressed natural gas (CNG) or LNG stations can be retrofitted to accommodate hydrogen by adding storage tanks, compressors, and cryogenic systems for $LH_2$. Additionally, new stations should be strategically built in high-demand urban areas and transit corridors, following modular designs that enable scalability as demand increases. Adhering to international safety standards, such as ISO 19880, is essential to ensure safe operations, supported by automated safety systems for leak detection and emergency shutdowns.

Integration with renewable energy sources enhances both efficiency and sustainability. Co-locating hydrogen production facilities near renewable energy plants reduces transmission losses and enables direct use of surplus renewable power for hydrogen production through Power-to-Gas technologies. This setup can be further strengthened by incorporating battery storage systems to balance intermittent renewable supply and provide a stable input for hydrogen generation. Figure 5.6 shows the integration of existing infrastructure for hydrogen distribution.

Storage infrastructure must also be expanded to support growing distribution networks. This includes the installation of compressed gas and cryogenic liquid storage tanks at distribution hubs, as well as the development of underground storage in geological formations such as salt caverns for large-scale, long-term capacity. Modular storage systems further improve flexibility, allowing hydrogen to be easily transported and integrated into local and regional networks.

In terms of transportation infrastructure, road, rail, and maritime systems all play vital roles. For road transport, expanding fleets of hydrogen transport vehicles, including tube trailers for compressed hydrogen and cryogenic tankers for $LH_2$, is necessary, alongside standardization of container sizes for intermodal efficiency. Rail networks can be adapted for bulk hydrogen transport over long

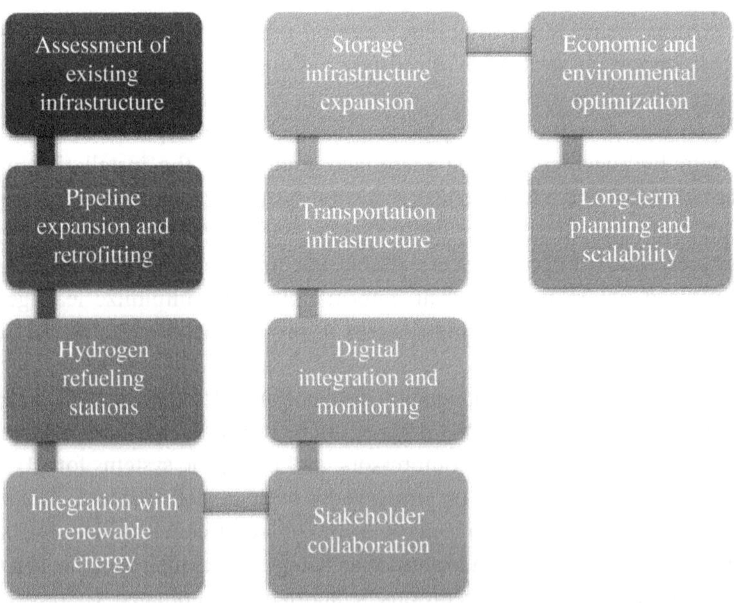


**Figure 5.6** Integration of Existing Infrastructure for Hydrogen Distribution

distances, with the potential introduction of hydrogen-powered locomotives for sustainable logistics. Maritime transport, meanwhile, can benefit from developing hydrogen carrier ships equipped with cryogenic storage for global trade, or utilizing alternative transport methods such as ammonia or LOHCs.

Digital integration and monitoring enhance system efficiency and safety. Deploying IoT and AI technologies allows for real-time monitoring of pipelines, storage facilities, and refueling stations, while predictive analytics can be applied to optimize distribution routes and reduce energy consumption. Centralized control systems can coordinate hydrogen flows across interconnected networks, ensuring reliability and adaptability.

Collaboration among stakeholders is essential for success. Governments play a key role by offering funding, subsidies, and regulatory support while developing hydrogen-specific codes and zoning regulations. Partnerships with private sector energy companies, gas utilities, and infrastructure developers can accelerate deployment through JVs. Equally important is community engagement, where educating local populations about hydrogen's benefits and safety measures fosters public trust and acceptance.

Economic and environmental optimization strategies ensure that hydrogen distribution remains sustainable and cost-effective. Logistics must be optimized to reduce costs, while economies of scale can be achieved through regional hubs and integrated supply chains. Carbon footprint reduction is equally vital, with emphasis on using green hydrogen produced from renewable energy sources and offsetting emissions through carbon credit programs.

Finally, long-term planning and scalability are essential for building a resilient hydrogen economy. Hydrogen distribution corridors should be developed to connect major production centers with consumption hubs, while strategies should be designed to integrate with emerging global hydrogen trade networks. Incorporating future technologies such as solid-state hydrogen storage systems and advanced pipeline materials will further strengthen distribution networks and ensure adaptability to evolving energy needs.

> ***Key: Electrolyzer:*** *A device that splits water into hydrogen and oxygen using electricity; central to green hydrogen production.*

### 5.3.2 Investment and Financing Options for Infrastructure Development

Developing infrastructure for green hydrogen production and distribution requires substantial capital investment. Investment in green hydrogen infrastructure is crucial for scaling up production, distribution, and utilization, with different countries adopting varied financing mechanisms. In the United States, funding comes from both government initiatives and private sector investments. The IRA provides tax credits and subsidies, while the DOE's Hydrogen Hubs Program allocates billions for regional hydrogen development. Private companies like Plug Power, Air Products, and BP are also investing in large-scale hydrogen projects, supported by venture capital (VC) and corporate partnerships. Additionally, public–private collaborations play a significant role in financing hydrogen infrastructure, including electrolyzer manufacturing and pipeline networks. Following are various investment and financing options tailored for such projects:

**Government Support and Public Funding:** Government support plays a foundational role in financing infrastructure development for renewable energy and green hydrogen projects. Public funding can take the form of grants and subsidies, which directly reduce the upfront costs of projects while improving their financial viability. Governments often introduce schemes such as Investment Tax Credits (ITCs) and Production Tax Credits (PTCs), enabling investors and developers to benefit from reduced tax burdens while promoting cleaner technologies. In addition, direct public sector investments are common in critical infrastructure, where governments partner with project developers to share the financial risks associated with large-scale deployment. Governments also issue green public bonds that provide long-term, low-interest financing for projects dedicated to renewable energy and hydrogen, thereby attracting institutional investors who are focused on sustainable growth.

**Private Sector Financing:** The private sector has emerged as a significant source of funding for infrastructure development, particularly through VC, private equity, and corporate financing. VC firms that specialize in renewable energy and green technology are particularly interested in innovative projects with high growth potential, scalability, and long-term profitability. Private equity funds, on the other hand, typically focus on larger capital infusions, acquiring equity stakes in projects to accelerate growth and operational efficiency. Corporate financing is also an important mechanism, as large multinational corporations with net-zero and decarbonization goals often collaborate with hydrogen developers to co-finance infrastructure. In such cases, co-development agreements and shared ownership models allow corporations to invest directly in the hydrogen value chain, ensuring alignment with their sustainability targets while reducing capital burdens on project developers.

**Loans and Debt Financing:** Debt financing offers another pathway for funding green hydrogen infrastructure, with multiple instruments tailored to sustainable projects. Green loans, specifically designed for environmentally responsible investments, are becoming increasingly popular, provided they adhere to Green Loan Principles (GLP) that ensure transparency and accountability. Project financing is another widely used model, allowing developers to leverage the future cash flows of the project itself to repay loans, thereby reducing the risks for sponsors and lenders. For very large projects, syndicated loans from multiple financial institutions provide the necessary capital infusion, spreading risk across several lenders. Export Credit Agencies (ECAs) also play a pivotal role by offering loans and guarantees to support the import of hydrogen-related technology and expertise, particularly benefiting countries aiming to develop domestic hydrogen capabilities but lacking local manufacturing strength.

**Public–Private Partnerships (PPPs):** PPPs are increasingly used to accelerate the rollout of hydrogen infrastructure while sharing both risks and benefits between stakeholders. One approach is through JVs, where government agencies and private investors collaborate to co-own and co-manage projects. Another widely used model is Build-Operate-Transfer (BOT), which allows private companies to finance, construct, and operate infrastructure projects for a defined period before transferring ownership to the government. Concession agreements also facilitate the use of public land and infrastructure, allowing private companies to build hydrogen facilities under negotiated terms. These frameworks not only mobilize private capital but also ensure that governments maintain long-term control and oversight over strategically important energy infrastructure.

**Green Bonds and Climate Financing:** Green bonds and international climate funds are vital instruments for financing environmentally friendly infrastructure. By issuing green bonds, project developers can attract investors who are guided by ESG principles, ensuring a steady flow of capital into sustainable projects. Beyond private markets, access to international climate funds such as the Green Climate Fund (GCF) and the Global Environment Facility (GEF) provides financial resources specifically targeted at renewable energy transitions and low-carbon infrastructure. These funds offer concessional loans, grants, or guarantees that make hydrogen projects more financially viable, particularly in developing countries where the upfront costs of such infrastructure might otherwise be prohibitive.

**International Development and Multilateral Organizations:** Multilateral development banks (MDBs) and international organizations are increasingly supporting the hydrogen economy through technical assistance, concessional financing, and policy advisory. Institutions such as the World Bank, the Asian Development Bank (ADB), and the European Investment Bank (EIB) provide long-term loans and guarantees for hydrogen-related projects. Additionally, organizations like the IRENA and the IEA offer both financial support and technical expertise to accelerate project deployment. Bilateral agreements between governments also create opportunities for shared

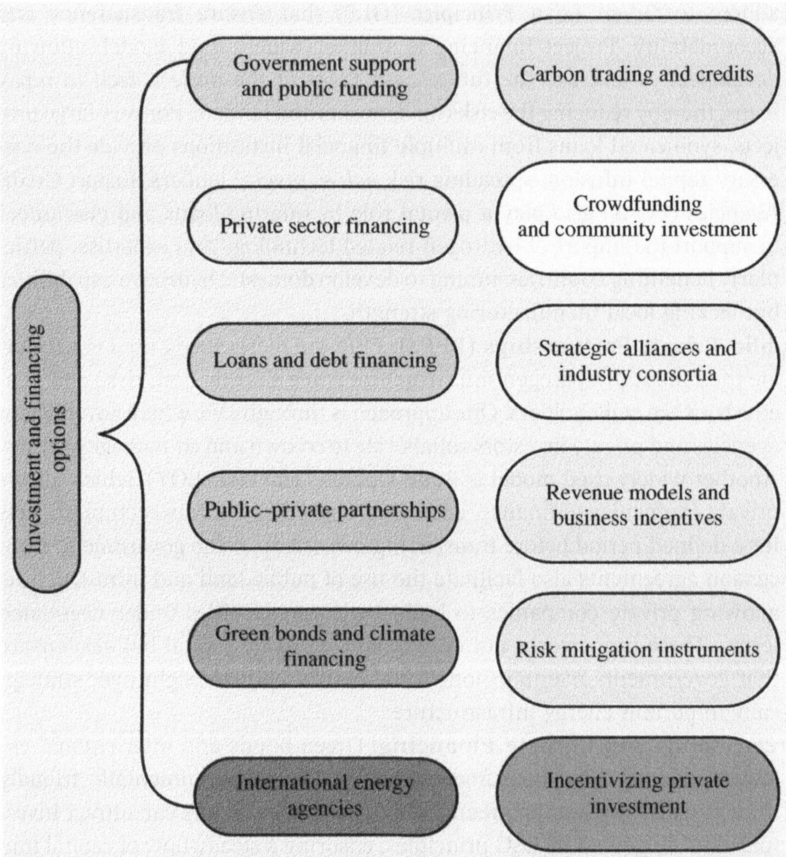


**Figure 5.7** Investment and Financing Options

development of hydrogen infrastructure, often combining financial assistance with technology transfer. These collaborations help countries overcome barriers related to high costs, limited expertise, and early-stage market risks. Figure 5.7 shows investment and financing options.

**Carbon Trading and Credits:** Carbon trading and credit mechanisms provide innovative financing streams by monetizing emissions reductions. Projects that reduce GHG emissions can generate carbon credits, which are sold through carbon offset programs to organizations aiming to meet their sustainability targets. Participation in cap-and-trade systems further allows developers to sell surplus credits in regulated markets, creating an additional revenue stream that improves project economics. This mechanism not only enhances financial feasibility but also strengthens the role of hydrogen as a key enabler of global decarbonization efforts.

**Crowdfunding and Community Investment:** Crowdfunding platforms and community-based financing models offer alternative approaches to raising capital for hydrogen projects, particularly those with strong local impact. Through crowdfunding platforms dedicated to renewable energy, individuals can contribute directly to project development, thereby democratizing access to investment opportunities. Community bonds are another mechanism where bonds are issued to local residents, allowing them to invest in region-specific projects while earning returns. Cooperative ownership models further involve community members as stakeholders, aligning project success with local benefits and ensuring stronger public support for hydrogen infrastructure deployment.

**Strategic Alliances and Industry Consortia:** Forming alliances with industry partners is an effective way to share infrastructure development costs and reduce risks. Partnerships with energy companies, utilities, and transport firms allow for pooled resources and shared ownership of large-scale projects such as hydrogen pipelines or refueling stations. R&D collaborations with universities and technology providers also play an important role, as they allow companies to co-fund innovation, share intellectual property (IP), and accelerate technological breakthroughs. Such consortia not only spread financial risks but also build strong ecosystems that advance hydrogen deployment more rapidly than isolated efforts.

**Revenue Models and Business Incentives:** A sustainable financial strategy for hydrogen projects requires robust revenue models. PPAs are particularly effective, as they secure long-term contracts for selling hydrogen or renewable energy at fixed prices, thereby ensuring stable cash flows. Another innovative model is Hydrogen-as-a-Service (HaaS), where industries or transportation sectors pay subscription fees for a reliable hydrogen supply, reducing their upfront investment requirements. Additionally, developers can generate revenue by charging for infrastructure usage, such as pipeline transportation, hydrogen storage, and refueling services. These business models improve profitability and attract investors by providing clear revenue streams.

**Risk Mitigation Instruments:** To attract large-scale investment, it is essential to deploy risk mitigation instruments that safeguard projects from uncertainties. Credit enhancement mechanisms, such as guarantees from multilateral agencies, increase investor confidence by lowering perceived risks. Insurance products that cover risks like construction delays, operational failures, or market volatility also enhance project bankability. Furthermore, hedging strategies, including derivatives, help project developers protect against fluctuations in energy and commodity prices, thereby ensuring predictable financial outcomes over the long term.

**Incentivizing Private Investment:** Finally, governments and policymakers can incentivize private sector participation through targeted measures that improve the returns on investment. Offering accelerated depreciation benefits for hydrogen infrastructure assets allows investors to reduce their tax liabilities, making projects more attractive. Performance-based incentives, which reward developers for achieving specific production targets or emission reductions, further enhance the commercial viability of hydrogen infrastructure. Such measures align private investment with public policy objectives, ensuring that the growth of hydrogen infrastructure contributes directly to environmental and economic goals.

China follows a state-led investment model, where major government-backed enterprises such as Sinopec, China Energy, and State Grid Corporation finance green hydrogen projects. The government provides direct subsidies, low-interest loans, and preferential policies to accelerate hydrogen adoption. Additionally, China's Five-Year Plans integrate green hydrogen development with strategic industries, ensuring sustained investment. Private companies and international investors are also entering the market, particularly in the production of electrolyzers and FCVs. However, financial risks remain due to high initial costs and competition with coal-based hydrogen.

In Germany, investment in green hydrogen is driven by the National Hydrogen Strategy, which includes €9 billion in government funding for hydrogen projects. The country also benefits from EU-level funding under programs like the European Green Deal and IPCEI (Important Projects of Common European Interest). Germany encourages private sector involvement through incentives, grants, and PPPs, with companies like Siemens Energy, Thyssenkrupp, and Linde leading hydrogen investments. Additionally, Germany secures financing through international collaborations, signing agreements with countries like Australia and Namibia for hydrogen imports. While the United States relies on tax incentives and private sector funding, China's investment model is state-driven, and Germany combines national funding with EU support and international partnerships. Overcoming financial barriers and ensuring long-term profitability remain critical for scaling up green hydrogen infrastructure globally.

## 5.4 Logistics, Transportation, Distribution, and Delivery of Green Hydrogen

Efficient logistics and transportation are critical for the widespread adoption and commercialization of green hydrogen. Due to hydrogen's low density and unique properties, it requires specialized infrastructure and technologies for safe and cost-effective movement. Compressed hydrogen gas, $LH_2$, and chemical carriers such as ammonia or LOHCs are the primary transportation methods. Each approach demands tailored solutions, such as high-pressure storage tanks, cryogenic equipment, or chemical conversion and reconversion systems. Pipelines offer an economical option for large-scale, continuous transportation, especially for industrial hubs, while trucks and ships equipped with hydrogen-compatible tanks are essential for distribution to remote locations. Safety measures, including rigorous leak detection, pressure regulation, and adherence to international standards, are vital to mitigate risks. Advancements in hydrogen transportation technologies, coupled with strategic planning of logistics networks, are necessary to lower costs and support the green hydrogen supply chain's reliability and scalability. By optimizing logistics and embracing innovation, green hydrogen can seamlessly integrate into global energy systems. Efficient logistics are vital for the large-scale distribution of green hydrogen, ensuring safe and cost-effective delivery.

A well-designed network ensures efficient delivery from production to end-users, balancing cost, demand, and reliability. The design and optimization of distribution networks for green hydrogen delivery are pivotal to ensuring efficient, reliable, and cost-effective supply chains. These networks connect production facilities to end-users across various sectors, including transportation, industry, and energy storage. Designing an effective distribution network involves strategic placement of hydrogen hubs, pipelines, and storage facilities to minimize transportation distances and costs while maximizing coverage. Optimization techniques leverage advanced modeling and simulation tools to evaluate factors such as demand patterns, geographical constraints, and infrastructure costs. Key considerations include selecting the most suitable transportation mode—compressed gas, $LH_2$, or chemical carriers—based on distance and volume requirements. Integration with renewable energy sources, safety standards, and environmental regulations further shapes the network's design. Digital technologies, such as IoT and AI, enable real-time monitoring and dynamic optimization, enhancing operational efficiency. By focusing on scalability, flexibility, and sustainability, optimized distribution networks play a crucial role in accelerating the adoption of green hydrogen as a cornerstone of the clean energy transition. Table 5.2 shows the design and optimization of distribution networks for green hydrogen delivery.

Ensuring reliable and safe hydrogen delivery at the final stage is critical for customer satisfaction and system viability. Last-mile delivery of green hydrogen to end-users is a critical component in ensuring the successful deployment of hydrogen as a clean energy source across various sectors. This stage involves the final leg of transportation, where hydrogen must be efficiently delivered from distribution hubs to individual users, such as industrial plants, fueling stations, or power generation facilities. The challenge lies in overcoming logistical hurdles such as hydrogen's low density and the need for specialized infrastructure, including storage tanks, pipelines, and refueling stations. For small-scale applications, high-pressure cylinders or portable tanks are often used, while larger-scale users may rely on dedicated pipelines or hydrogen transport trucks. Ensuring cost-effective, safe, and reliable delivery requires careful optimization of routes, vehicle fleets, and storage solutions, especially in urban or remote areas. Additionally, the integration of advanced monitoring systems, including IoT sensors and AI-driven analytics, can help track hydrogen levels, maintain safety protocols, and improve efficiency.

**Table 5.2** Design and Optimization of Distribution Networks for Green Hydrogen Delivery

| Attribute | Description |
|---|---|
| Objective | Designing efficient distribution networks for delivering green hydrogen, ensuring reliability, cost-effectiveness, and sustainability |
| Key components | Hydrogen production facilities |
| | Storage systems |
| | Transport infrastructure (pipelines, trucks) |
| | Delivery points (e.g., refueling stations) |
| Optimization criteria | Minimization of operational costs |
| | Maximization of hydrogen delivery efficiency |
| | Minimization of carbon footprint |
| | Optimal infrastructure layout |
| Technology integration | Green hydrogen production methods (e.g., electrolysis) |
| | Renewable energy integration |
| | Smart grid technologies for network control and monitoring |
| Challenges | Intermittency of renewable energy sources |
| | Long-distance transportation of hydrogen |
| | Storage and handling of hydrogen under extreme conditions |
| Regulatory and policy considerations | Compliance with safety and environmental regulations |
| | Government incentives for green hydrogen projects |
| | International standards for hydrogen delivery |
| Environmental impact | Reduction in greenhouse gas emissions |
| | Minimization of ecological footprint in production, storage, and distribution of hydrogen |
| Economic impact | Cost-effectiveness of hydrogen delivery |
| | Impact on hydrogen market prices |
| | Potential for creating jobs in green hydrogen sectors |
| Energy efficiency | Optimization of energy use in hydrogen production, storage, and distribution |
| | Energy losses during transportation and storage |
| Supply chain logistics | Efficient scheduling and routing of hydrogen delivery vehicles |
| | Integration with existing energy infrastructure |
| | Risk management and contingency plans |
| Infrastructure design | Pipeline design for hydrogen transport |
| | Compression and storage technologies |
| | Distributed network model versus centralized hub-based delivery system |

**Table 5.2** (Continued)

| Attribute | Description |
|---|---|
| Modeling and simulation | Use of optimization algorithms (e.g., mixed-integer linear programming) |
| | Simulation of network performance under varying demand and supply scenarios |
| Stakeholder collaboration | Partnerships with energy producers, infrastructure providers, and government bodies |
| | Collaboration with research institutions for technological advancements |
| Scalability and flexibility | Design of scalable and flexible networks to accommodate future increases in hydrogen demand |
| | Ability to integrate with future renewable energy sources |
| Hydrogen storage systems | Selection of storage technologies (e.g., high-pressure tanks and liquefied hydrogen) |
| | Optimal placement of storage facilities within the distribution network |
| Risk assessment | Identification of risks related to supply disruptions, safety hazards, and environmental impacts |
| | Development of contingency plans and mitigation strategies |

By overcoming the complexities of last-mile delivery, green hydrogen can reach end-users seamlessly, driving its adoption across industries and contributing to a sustainable energy future.

> ***Key: Hydrogen distribution network:*** *The infrastructure, including pipelines, transport vehicles, and refueling stations, used to deliver green hydrogen from production to end-users.*

## 5.5 Lifecycle Assessment of Green Hydrogen Production, Distribution, and Utilization

LCA evaluates the environmental impacts of green hydrogen throughout its lifecycle stages, from production to utilization. The goal is to identify areas for improvement, minimize environmental burdens, and ensure sustainability.

**Key Lifecycle Stages**

1) Production: Green hydrogen generation via water electrolysis using renewable energy.
2) Distribution: Transportation, storage, and delivery to end-users.
3) Utilization: Use of hydrogen in industrial, transport, or energy applications.

1) **Green Hydrogen Production**
   **Inputs and Outputs:**
   - **Inputs**:
     - Electricity (from renewable sources like wind, solar, or hydropower).
     - Water (for electrolysis).
   - **Outputs**:
     - Green hydrogen.
     - Oxygen as a by-product.

   **Environmental Impacts:**
   - **Carbon Footprint**:
     - Near-zero emissions if powered entirely by renewable energy.
     - Indirect emissions from renewable energy infrastructure (e.g., solar panel or wind turbine production).
   - **Resource Use**:
     - Water Consumption: Requires approximately 9 L of water per kilogram of hydrogen produced.
     - Land use for renewable energy installations.
   - **Energy Efficiency**:
     - Electrolyzer efficiency (~60–80%) affects overall energy consumption.

   **Mitigation Strategies:**
   - Increase electrolyzer efficiency.
   - Use desalinated water in water-scarce regions.
   - Co-locate hydrogen production with renewable energy facilities to reduce transmission losses. Figure 5.8 shows LCA of green hydrogen.
2) **Distribution**
   **Logistics and Storage:**
   - **Modes**:
     - Pipelines (compressed hydrogen gas)
     - Cryogenic tankers ($LH_2$)
     - Carriers (ammonia, LOHCs)
   - **Storage**:
     - High-pressure tanks for gaseous hydrogen.
     - Cryogenic tanks for $LH_2$.

   **Environmental Impacts:**
   - **Energy Use**:
     - Compression and liquefaction are energy-intensive.
     - Losses during transportation (e.g., boil-off in cryogenic storage).
   - **Infrastructure**:
     - Materials used for pipelines and tanks (e.g., steel and composite materials).

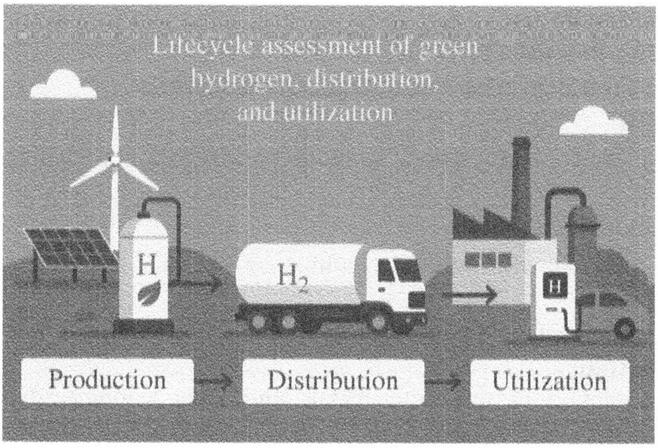


**Figure 5.8** Lifecycle Assessment of Green Hydrogen

- **Leakage**:
  - Hydrogen leakage contributes to indirect warming due to its role in atmospheric chemistry.

**Mitigation Strategies:**

- Develop high-efficiency compressors and liquefaction systems.
- Use leak-proof, durable materials in pipelines and storage.
- Optimize distribution networks to minimize transport distances.

3) **Utilization**

**Applications:**

- **Transport**: Hydrogen FCVs.
- **Power Generation**: Hydrogen turbines or fuel cells for electricity.
- **Industrial Use**: Steelmaking, ammonia synthesis, and refineries.

**Environmental Impacts:**

- **Carbon Footprint**:
  - Zero emissions at the point of use for fuel cells.
  - Reduced emissions in industrial processes compared to fossil-based methods.
- **Resource Use**:
  - Dependence on rare-earth metals (e.g., platinum) for fuel cells.

**Mitigation Strategies:**

- Advance recycling and recovery of rare-earth metals.
- Improve fuel cell efficiency and durability.

**Lifecycle Impact Metrics**

1) **Global Warming Potential (GWP)**:
   - Measure $CO_2$-equivalent emissions throughout the lifecycle.
2) **Water Footprint**:
   - Quantify total water consumption, including electrolysis and cooling processes.
3) **Energy Payback Ratio (EPR)**:
   - Evaluate the ratio of energy output to input.
4) **Land Use**:
   - Assess land requirements for renewable energy and hydrogen infrastructure.
5) **Resource Depletion**:
   - Analyze consumption of rare-earth metals and other nonrenewable resources.

**Strategies for Sustainability**

1) **Energy Efficiency**:
   - Use advanced electrolyzers, such as solid oxide or PEM systems.
   - Optimize renewable energy integration for continuous hydrogen production.
2) **Circular Economy**:
   - Recycle materials from decommissioned systems.
   - Develop scalable solutions for the recovery of rare-earth metals.
3) **Decarbonized Logistics**:
   - Electrify hydrogen transport fleets.
   - Invest in localized production to reduce distribution distances.
4) **Lifecycle Innovations**:
   - Implement digital twins and AI for monitoring lifecycle performance.
   - Explore new materials for electrolyzers and storage.

Evaluating the environmental impacts of hydrogen energy involves assessing emissions, resource use, and ecological effects throughout its lifecycle. This spans production, distribution, storage, and utilization.

1) **Production Stage**

   The environmental impact of hydrogen production depends significantly on the production method:

**Green Hydrogen (Electrolysis Using Renewables):**

- **Positive Impacts**:
  - Near-zero GHG emissions.
  - Co-production of oxygen, which can be utilized in industrial or medical applications.

- **Negative Impacts**:
  - High water consumption (~9 L per kg of hydrogen).
  - Indirect emissions from the construction and disposal of renewable energy infrastructure.
  - Land use for solar panels, wind turbines, or hydroelectric plants.

**Blue Hydrogen (Natural Gas Reforming with Carbon Capture):**

- **Positive Impacts**:
  - Reduced GHG emissions due to CCS.
- **Negative Impacts**:
  - Methane leakage during natural gas extraction (a potent GHG).
  - Limited CCS efficiency (~85–90%), leading to residual emissions.

**Gray Hydrogen (Conventional Natural Gas Reforming):**

- **Negative Impacts**:
  - Significant $CO_2$ emissions (~10 kg $CO_2$ per kg of hydrogen).
  - Methane leakage and fossil fuel dependence.

2) **Distribution Stage**

Hydrogen's low volumetric energy density poses challenges during distribution.

**Transport Impacts:**

- **Pipelines**:
  - Positive: Efficient for large-scale transport over short to medium distances.
  - Negative: High upfront costs and potential for hydrogen embrittlement in materials.
- **Tankers**:
  - Compressed Hydrogen: Energy-intensive compression (up to 700 bar) with ~15% energy loss.
  - Liquid Hydrogen: Requires cryogenic cooling (−253 °C), leading to ~30% energy loss during liquefaction.

**Environmental Concerns:**

- Hydrogen leakage: Contributes to indirect warming due to its interaction with atmospheric methane and ozone.
- Energy consumption: Significant losses during compression, liquefaction, and transport.

3) **Storage Stage**
   - Storing hydrogen efficiently and safely is critical to minimizing environmental impacts.

**Compressed Gas Storage:**

- High-pressure tanks require energy-intensive compression processes.
- Material Concerns: Use of steel or carbon fiber composites increases resource demand.

**Cryogenic Liquid Storage:**

- High energy demand for maintaining extremely low temperatures.
- Risk of boil-off losses, especially during long-term storage.

**Innovative Storage Solutions:**

- **Ammonia**: Easier to store and transport but requires reconversion to hydrogen, which is energy-intensive.
- **Liquid Organic Hydrogen Carriers (LOHCs)**: Reusable but involves complex chemical processes.

4) **Utilization Stage**

- Hydrogen's end use has diverse environmental implications based on the application.

**Transportation:**

- **Fuel Cell Vehicles (FCVs)**:
  - Positive: Zero tailpipe emissions, water vapor as the only by-product.
  - Negative: Dependence on rare-earth elements (e.g., platinum in fuel cells).

**Power Generation:**

- Positive: Hydrogen turbines and fuel cells produce electricity with no direct emissions.
- Negative: Lower efficiency compared to some conventional renewable technologies (e.g., wind or solar).

**Industrial Applications:**

- Positive: Potential to decarbonize hard-to-abate sectors like steelmaking and ammonia production.
- Negative: High energy input required for hydrogen-based processes compared to conventional methods.

**Cross-Cutting Environmental Impacts**

**Global Warming Potential (GWP):**

- **Green Hydrogen**: ~1 kg $CO_2$ equivalent per kg of hydrogen.
- **Blue Hydrogen**: ~2–3 kg $CO_2$ equivalent per kg (with CCS).
- **Gray Hydrogen**: ~10 kg $CO_2$ equivalent per kg.

**Water Use:**

- Electrolysis demands high-quality water (~9 L per kg hydrogen), posing challenges in water-scarce regions.
- Solutions include desalination or wastewater treatment, though these add energy costs.

**Energy Use:**

- Efficiency losses occur at each stage:
  - Electrolysis: 20–40% energy loss.
  - Compression/Liquefaction: 10–30% energy loss.
  - Fuel Cells: 40–60% efficiency, depending on the application.

**Land Use:**

- Renewable energy infrastructure for green hydrogen production can compete with agriculture or natural habitats.

**Material Use:**

- High demand for critical raw materials like rare-earth elements and advanced alloys for fuel cells, storage tanks, and pipelines. Table 5.3 shows a comparison of environmental impacts between green hydrogen and conventional hydrogen production.

**Table 5.3** Comparison of Environmental Impacts Between Green Hydrogen and Conventional Hydrogen Production

| Aspect | Green Hydrogen (via Electrolysis using Renewable Energy) | Conventional Hydrogen (e.g., Steam Methane Reforming–SMR) |
|---|---|---|
| Primary energy source | Renewable energy (solar, wind, hydro) | Fossil fuels (natural gas, coal) |
| Greenhouse gas emissions | Near zero during production | High $CO_2$ emissions due to fossil fuel combustion |
| Carbon footprint | Very low; dependent on renewable source lifecycle | Very high; approx. 9–12 kg $CO_2$ per kg $H_2$ produced |
| Water usage | High (for electrolysis), but with clean return | Lower water input, but risk of thermal pollution |
| Air pollution | No direct emissions | Emission of $CO_2$, NO , CO, and VOCs |
| Land use | Moderate (for renewable installations) | Moderate (for refineries, pipelines) |
| Resource dependency | Dependent on availability of renewable energy and pure water | Dependent on nonrenewable fossil fuel reserves |
| Waste generation | Minimal solid or hazardous waste | Potential chemical and thermal waste by-products |
| Sustainability | Highly sustainable and circular with proper planning | Unsustainable due to finite fossil fuels and emissions |
| Lifecycle impact | Low when powered by green energy | High lifecycle emissions and degradation impacts |

**Key Metrics for Environmental Impact Evaluation**

1) **Carbon Footprint**:
   - Total $CO_2$-equivalent emissions across all lifecycle stages.
2) **Water Footprint**:
   - Water consumption, especially in electrolysis and cooling systems.
3) **Energy Efficiency**:
   - Energy required per unit of hydrogen delivered to end-users.
4) **Resource Sustainability**:
   - Availability and recyclability of materials used in hydrogen technologies.

**Strategies to Minimize Environmental Impacts**

1) **Optimize Renewable Energy Use**:
   - Maximize integration of solar, wind, or hydro energy in green hydrogen production.
2) **Improve Energy Efficiency**:
   - Develop advanced electrolyzers and fuel cells to minimize energy losses.
3) **Leakage Prevention**:
   - Use advanced sealing technologies and hydrogen sensors in pipelines and storage.
4) **Circular Economy Approaches**:
   - Recycle rare-earth materials and repurpose retired hydrogen infrastructure components.
5) **Localized Production**:
   - Produce hydrogen close to demand centers to reduce transportation energy costs.
6) **Public–Private Collaboration**:
   - Foster policies and investments to develop sustainable hydrogen value chains.

***Key: Energy landscape:*** *The comprehensive view of energy resources, technologies, and systems within a specific region or sector, reflecting the distribution, utilization, and impact of energy sources and infrastructure.*

## 5.6 Market Dynamics and Demand Forecasting

The green hydrogen market is influenced by multiple factors, from technological advancements to regulatory frameworks. The current market dynamics for green hydrogen are shaped by a combination of technological advancements, policy support, and evolving market demand. As the world moves toward decarbonization, green hydrogen is gaining traction as a key solution for industries that are difficult to electrify, such as heavy transportation, steel production, and chemical manufacturing. Market growth is driven by a significant decrease in renewable energy costs, making electrolysis for hydrogen production more competitive. Governments across the globe are offering incentives, subsidies, and regulatory frameworks to accelerate the adoption of green hydrogen, recognizing its potential to decarbonize hard-to-abate sectors. However, the market is still in its nascent stages, with challenges such as high production costs, scalability of infrastructure, and the need for widespread adoption of hydrogen-based technologies. The integration of green hydrogen into existing energy systems and its ability to compete with other forms of energy, such as natural gas or blue hydrogen, will determine its long-term success. Additionally, the development of international hydrogen trade routes and cross-border collaborations is expected to unlock new market opportunities. As these factors evolve, the green hydrogen market is poised for significant expansion, but it requires continued investment in technology, infrastructure, and regulatory frameworks to realize its full potential.

**Key Drivers of Market Dynamics:**

- **Government Policies and Regulations**:
  - Incentives and subsidies for green hydrogen production (e.g., tax credits, grants, and low-carbon fuel standards).
  - National hydrogen strategies, such as the EU's Green Deal and hydrogen roadmaps from various countries.
- **Technological Advancements**:
  - Development in electrolyzer efficiency and cost reduction.
  - Progress in renewable energy generation (solar, wind) lowering the cost of green hydrogen production.
- **Private Sector Investment**:
  - Corporations in energy, transportation, and industrial sectors are increasing investments in green hydrogen, attracted by decarbonization goals.
- **Global Energy Transition**:
  - The move toward decarbonizing heavy industries (e.g., steel and cement), transport (e.g., trucks, ships, and trains), and power generation. Figure 5.9 shows market dynamics of green hydrogen.

**Challenges:**

- **High Production Costs**:
  - Although costs are falling, green hydrogen is still more expensive than gray hydrogen, limiting its competitiveness.
- **Infrastructure Gaps**:
  - Lack of sufficient storage, distribution, and refueling infrastructure.
- **Market Uncertainty**:
  - Volatility in renewable energy prices and hydrogen adoption rates.

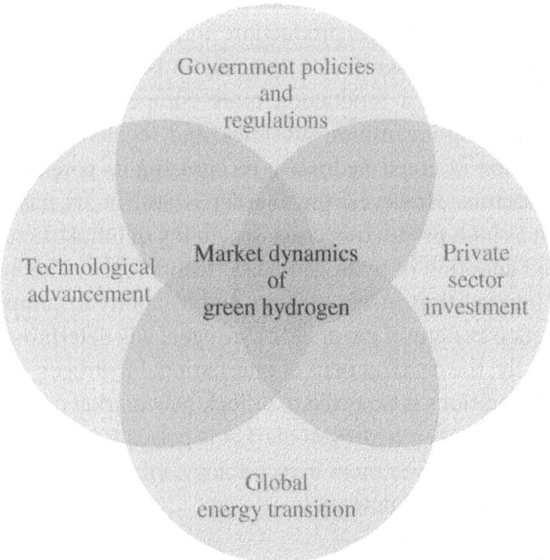


**Figure 5.9** Market Dynamics of Green Hydrogen

As of 2024, the global green hydrogen market is valued between USD 6.26 billion and USD 8.78 billion, according to various industry reports. This market is projected to experience significant growth, with forecasts indicating a CAGR ranging from 38.5% to 61.5% over the next decade. By 2033, estimates suggest that the market could reach between USD 62.4 billion and USD 165.84 billion. This rapid expansion is driven by increasing global efforts to reduce carbon emissions and the growing adoption of renewable energy sources. Regionally, Europe currently dominates the green hydrogen market, holding a significant share in 2024. The Asia-Pacific region, particularly China, is also emerging as a key player, with China accounting for approximately one-third of global green hydrogen production. In the United States, the market was valued at USD 931.8 million in 2023 and is expected to grow at a CAGR of 38.8%. Despite the optimistic projections, challenges such as high production costs, infrastructure development, and regulatory uncertainties persist. However, with continued technological advancements and supportive government policies, the green hydrogen market is poised for substantial growth in the coming years.

➢ **Factors Influencing Market Demand for Green Hydrogen**

Several factors are driving the market demand for green hydrogen, with sustainability, technological advancements, and policy support playing pivotal roles. The global push for decarbonization, particularly in hard-to-abate sectors such as heavy industry, transportation, and power generation, is one of the primary catalysts for green hydrogen demand. As governments set more ambitious carbon reduction goals and impose stricter emissions regulations, industries are increasingly seeking alternative, clean energy sources like green hydrogen to meet these targets. The decreasing cost of renewable energy technologies, such as solar and wind, which are essential for hydrogen production through electrolysis, further enhances the economic viability of green hydrogen. Additionally, advancements in electrolysis technology and hydrogen storage solutions are improving efficiency and reducing production costs, making green hydrogen more competitive with traditional fuels. Market demand is also influenced by the rise of hydrogen-powered vehicles, with investments in fuel cell technology and infrastructure driving its adoption in transportation. Moreover, the establishment of government policies, subsidies, and international agreements to promote hydrogen adoption, such as the EU's Green Deal and hydrogen roadmaps in various countries, play a critical role in boosting market growth. Collectively, these factors are creating a favorable environment for the expansion of green hydrogen markets, positioning it as a key player in the global transition to clean energy. The demand for green hydrogen is primarily shaped by several factors:

1) **Decarbonization Goals:**
   - Countries with strong climate goals (e.g., carbon neutrality by 2050) are driving the demand for green hydrogen as a key tool for decarbonizing hard-to-abate sectors such as heavy industry and transportation.
2) **Energy Transition in Key Sectors:**
   - **Heavy Industry**: Demand from industries like steel production, cement, and chemicals is growing as they seek to reduce reliance on fossil fuels.
   - **Transportation**: Hydrogen is seen as a viable alternative to fossil fuels for heavy-duty vehicles, long-distance transport, and public transport.
   - **Power Generation**: Hydrogen's potential as a storage medium for renewable energy is pushing demand, particularly in regions with high renewable energy generation variability.
3) **Technological Maturity:**
   - As electrolyzer technology becomes more efficient and cost-effective, the production of green hydrogen becomes more economically viable, thereby driving demand.
4) **Infrastructure Development:**
   - The establishment of HRSs, pipelines, and storage systems is key to enabling demand growth, especially in the transportation and energy sectors.

5) **Policy Support:**
   - Government mandates, low-carbon fuel standards, and emission reduction targets are creating favorable conditions for hydrogen adoption, stimulating demand in both energy and industrial applications.
6) **Corporate Sustainability:**
   - Growing pressure on corporations to meet sustainability targets is increasing interest in green hydrogen as part of corporate decarbonization strategies.

➢ **Forecasting Future Demand for Green Hydrogen Across Various Sectors:**

1) **Industrial Demand:**
   - **Steel Production**: Green hydrogen is a potential alternative for replacing coke in steel production. The demand for hydrogen in this sector is expected to rise significantly as steelmakers commit to reducing $CO_2$ emissions.
     - **Forecast**: Global demand for hydrogen in the steel industry could reach 10–15 million tons annually by 2030.
   - **Chemicals and Fertilizers**: Hydrogen is critical for ammonia production, and green hydrogen will increasingly replace gray hydrogen.
     - **Forecast**: Demand from the chemical sector is expected to grow at a rate of 10–12% annually.
2) **Transportation:**
   - **Heavy-duty transport (trucks, buses, ships, trains)**: Green hydrogen is gaining traction in the heavy-duty transport segment, especially in countries like Japan, Germany, and the United States.
     - **Forecast**: The hydrogen-powered transport market could account for 10–20% of total vehicle sales in the next 10 years, with growth accelerating post-2025 as infrastructure expands.
   - **Aviation**: Hydrogen-powered aviation is still in the early stages, but potential demand exists for long-haul flights.
     - **Forecast**: Commercial adoption of hydrogen-powered aircraft could begin by 2035, with demand gradually ramping up.
3) **Power Generation:**
   - **Grid Balancing and Energy Storage**: Hydrogen offers a solution for storing excess renewable energy, particularly in regions with large wind or solar capacity.
     - **Forecast**: The use of hydrogen for energy storage and grid balancing could account for 10–15% of global hydrogen demand by 2030.
   - **Cogeneration and Distributed Power**: Decentralized hydrogen-based power generation systems are expected to see growth, especially in off-grid applications.

4) **Residential and Commercial Use**:
   - **Fuel Cells for Heating**: As fuel cells become more affordable, hydrogen is expected to find a market for residential heating and commercial energy applications.
     - **Forecast**: This market will likely grow at a slower pace but could represent a significant share of demand in regions with strong decarbonization policies.

### 5.6.1 Market Segmentation, Collaboration, and Partnerships of Target Markets for Green Hydrogen

Market segmentation for green hydrogen involves identifying specific industries and sectors that can benefit most from its adoption, based on their energy demands, emissions reduction potential, and technological readiness. The primary target markets for green hydrogen include heavy industries such as steel, cement, and chemical manufacturing, which are challenging to decarbonize through electrification alone. These sectors require high-temperature heat and chemical processes, making hydrogen a viable clean alternative to fossil fuels. Another key market is transportation, particularly in heavy-duty vehicles like trucks, buses, and trains, where hydrogen fuel cells offer a longer range and faster refueling times compared to BEVs. Additionally, the power generation sector, especially for backup or off-grid applications, is a promising market for green hydrogen, where it can serve as a storage solution for renewable energy, converting excess electricity into hydrogen for later use. Emerging markets in residential and commercial heating, as well as green hydrogen's role in hydrogen hubs or refueling stations, present additional growth opportunities. Identifying and targeting these market segments allows for tailored strategies to address specific needs, with early adoption driving the scaling of green hydrogen production and distribution infrastructure across industries.

**Market Segments:**

1) **By Application**:
   - **Transport**: Hydrogen-powered vehicles, including buses, trucks, trains, and ships.
   - **Industry**: Steel production, refining, chemicals, and ammonia production.
   - **Energy**: Grid balancing, energy storage, and distributed power generation.
   - **Residential/Commercial**: Fuel cell systems for heating and off-grid power.

2) **By Region**:
   - **Europe**: Leading in the adoption of green hydrogen, driven by EU hydrogen strategies.
   - **Asia-Pacific**: Strong demand from Japan, South Korea, and China for transportation and industrial applications.

- **North America**: Rapid growth in transportation and industrial sectors, with increasing government support.
- **Middle East and Africa**: Emerging markets for green hydrogen production, particularly in the Gulf Cooperation Council (GCC) countries aiming for energy diversification.

3) **By Technology**:
   - **Electrolysis**: Dominant production method for green hydrogen.
   - **SMR with Carbon Capture**: A transitional technology for blue hydrogen. Figure 5.10 shows market segmentation of green hydrogen.

**Target Markets:**

- **Heavy Industries**: Especially steel, cement, and chemicals, where hydrogen can replace coal and natural gas.
- **Transport and Mobility**: Long-haul freight, aviation, and marine transport markets are key targets.
- **Energy Storage Providers**: Power producers seeking solutions for grid balancing and renewable energy integration.
- **Regions with High Renewable Energy Potential**: Countries with significant renewable resources (e.g., solar and wind) will be key markets for green hydrogen production.

The successful development and scaling of green hydrogen as a clean energy source requires a concerted effort from all sectors of society, particularly through collaboration between government, industry, and academia. This synergistic

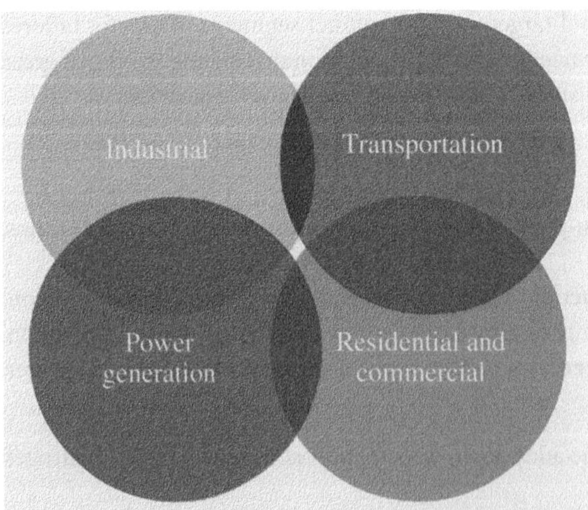


**Figure 5.10** Market Segmentation of Green Hydrogen

approach can accelerate technological innovation, create policies to foster market growth, and bridge the gap between research and commercial application. Following are the detailed analysis of how stakeholders can collaborate to advance green hydrogen initiatives:

1) **Government Role in Green Hydrogen Initiatives**

Governments play a crucial role in creating the enabling environment for green hydrogen to thrive. Their support, both in terms of policy and funding, is critical for reducing barriers to adoption and scaling up production.

**Key Areas of Government Contribution:**

- **Policy and Regulatory Frameworks**:
  - Developing clear and consistent policies that set long-term targets for hydrogen adoption, such as national hydrogen strategies, carbon neutrality goals, and clean energy transition roadmaps.
  - Introducing regulations and incentives that reduce the cost gap between green hydrogen and conventional hydrogen, such as subsidies, tax incentives, and grants.
- **Funding and Investment**:
  - Providing grants, low-interest loans, and other financial mechanisms to support hydrogen research and the construction of production, storage, and distribution infrastructure.
  - Funding PPPs and pilot projects to demonstrate the feasibility of green hydrogen in various sectors (transportation, industry, etc.).
- **Creating Market Demand**:
  - Government purchasing programs for hydrogen-powered vehicles, clean energy projects, and industrial decarbonization.
  - Setting binding renewable energy targets that incentivize hydrogen use for storage, grid balancing, and energy production.

**Examples:**

- The EU has committed to making green hydrogen a key part of its energy transition under the EU Hydrogen Strategy.
- Japan's Hydrogen Roadmap targets becoming a "hydrogen society" by integrating hydrogen in various sectors, including transport, energy, and industry.

2) **Industry's Contribution to Green Hydrogen Initiatives**

The industrial sector is essential for scaling up the production and application of green hydrogen. Industry can drive technological innovations, establish large-scale production plants, and integrate hydrogen into various industrial processes.

**Key Areas of Industry Contribution:**

- **Technology Development and Commercialization**:
  - o Private companies are leading the charge in developing more efficient and cost-effective electrolyzers, fuel cells, and other hydrogen technologies.
  - o Developing integrated hydrogen systems for transportation (hydrogen vehicles), energy storage, and industrial processes (e.g., green steel).

- **Infrastructure Development**:
  - o Companies can help build the infrastructure required for hydrogen production, storage, and distribution, including electrolyzers, refueling stations, pipelines, and storage facilities.
  - o Key industry players are also investing in hydrogen hubs to create a network of production, storage, and distribution.

- **Market Demand Creation**:
  - o Industry's adoption of green hydrogen in its operations (e.g., using hydrogen in steel production or as a fuel for heavy-duty transport) can provide significant demand and drive market growth.
  - o Collaborations between hydrogen producers and large-scale industrial users (e.g., automotive, aerospace, and cement) can set the foundation for a sustainable market.

**Examples:**

- **Shell**, **Siemens**, and **Air Products** are investing in large-scale green hydrogen production projects globally, including hydrogen plants in the Middle East and Europe.
- **Toyota** and **Hyundai** have been leading the way in hydrogen FCVs, while **Nikola** focuses on hydrogen trucks for logistics. Figure 5.11 shows stakeholder collaboration to green hydrogen initiatives.

3) **Academia's Contribution to Green Hydrogen Initiatives**

Academia plays a vital role in advancing the scientific understanding and technological capabilities required for the successful development of green hydrogen. Research and innovation driven by universities and research institutions can address technical barriers, optimize production processes, and improve storage and distribution systems.

**Key Areas of Academia's Contribution:**

**Research and Development**: Universities conduct cutting-edge research on electrolysis technology, fuel cells, and hydrogen storage systems, aiming to improve efficiency, reduce costs, and expand applications of green hydrogen. Research in advanced materials (e.g., catalysts and membranes) is critical for making green hydrogen more cost-competitive.

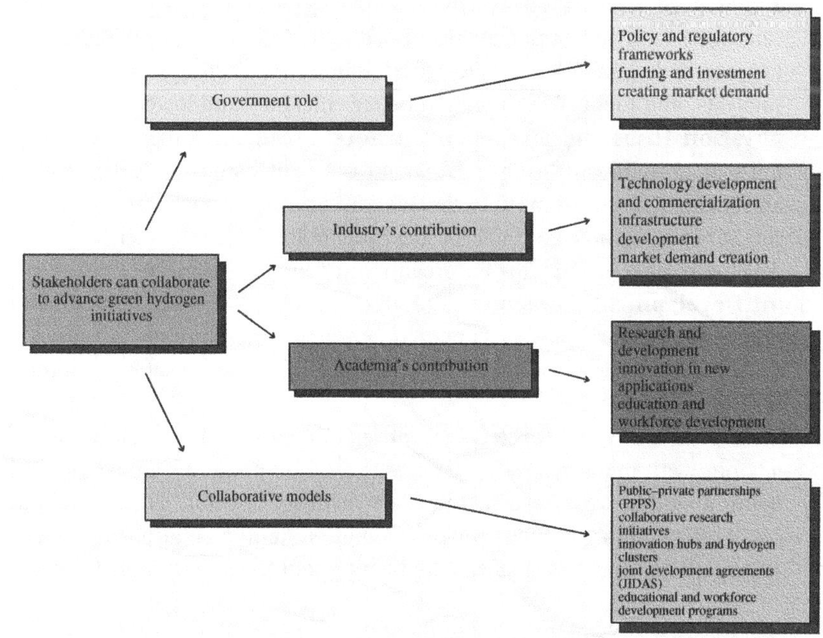


**Figure 5.11** Stakeholder Collaboration to Green Hydrogen Initiatives

**Innovation in New Applications**: Academia is exploring novel applications of hydrogen in various sectors, such as hydrogen-powered aircraft, maritime transport, and off-grid power systems. Research on integrating hydrogen with other renewable energy sources (e.g., wind and solar) to create hybrid systems can improve the reliability and sustainability of the energy supply.

**Education and Workforce Development**: Universities offer specialized programs in hydrogen technologies, renewable energy, and energy systems, helping develop a skilled workforce capable of supporting the green hydrogen industry. Academics can foster a collaborative ecosystem by engaging in joint projects, conferences, and research networks with industry partners.

4) **Collaborative Models for Advancing Green Hydrogen**

To maximize the potential of green hydrogen, collaboration between government, industry, and academia is essential. Following are some collaborative models that can drive the growth of green hydrogen:

**Public–Private Partnerships (PPPs):** Governments can collaborate with private companies to co-fund large-scale hydrogen projects. This includes joint funding for R&D, as well as shared infrastructure investments such as hydrogen production plants, storage facilities, and refueling stations.

**Collaborative Research Initiatives:** Universities and research institutions can work with industry players to focus on technological breakthroughs in green hydrogen. For example, research into new electrolysis technologies, catalysts, and fuel cells can lead to better efficiency and lower costs.

**Innovation Hubs and Hydrogen Clusters:** Creating hydrogen innovation hubs or clusters can bring together government bodies, industry leaders, and academic institutions to foster collaboration and innovation. These hubs can serve as testing grounds for new technologies, while also helping to develop market ecosystems for green hydrogen.

**Joint Development Agreements (JDAs):** JDAs between academic institutions and industry players can help accelerate the commercialization of research innovations, particularly in hydrogen production, storage, and utilization technologies.

**Educational and Workforce Development Programs:** Universities can collaborate with industry and government to create programs aimed at developing a skilled workforce in the green hydrogen sector. These programs could offer training in hydrogen technology, system integration, and energy management, ensuring a steady supply of talent to support industry needs.

Collaboration between government, industry, and academia is vital for driving the adoption of green hydrogen and achieving the ambitious decarbonization goals set by nations globally. Governments provide the regulatory frameworks and funding mechanisms necessary to support green hydrogen development. The industry is responsible for scaling up production, developing infrastructure, and creating market demand. Academia plays an essential role in advancing the technology and ensuring a skilled workforce for the green hydrogen economy. Through collaborative efforts, green hydrogen can become a cornerstone of the global energy transition, helping to mitigate climate change and create a sustainable energy future.

## 5.7 Partnerships Between Renewable Energy Developers and Hydrogen Producers

The integration of renewable energy sources with hydrogen production offers significant potential for decarbonizing various sectors, such as transportation, industry, and power generation. Renewable energy developers and hydrogen producers are increasingly forming strategic partnerships to leverage the complementary nature of their technologies. These partnerships are essential for scaling up green hydrogen production, optimizing the use of renewable energy, and enhancing the overall economic viability of hydrogen as a clean energy carrier.

1) **Importance of the Partnership:** Renewable energy developers provide the necessary infrastructure to generate clean electricity (e.g., solar, wind, and hydropower) that is essential for producing green hydrogen through electrolysis. Hydrogen producers convert the clean electricity into hydrogen by splitting water molecules in an electrolyzer (using renewable energy). This process ensures the hydrogen produced is green and carbon-free. Renewable energy, particularly solar and wind, is intermittent, meaning it can produce excess energy when conditions are favorable and insufficient energy during low-generation periods. Green hydrogen production offers an opportunity to store excess renewable energy and convert it into a storable and transportable form. Hydrogen production can act as a buffer, ensuring that renewable energy is not wasted and can be used later, either for electricity generation, storage, or for industrial and transport applications.
2) **Key Benefits of Partnerships:** By jointly developing large-scale green hydrogen production plants, both renewable energy developers and hydrogen producers can share the costs of infrastructure, reducing the initial capital expenditures required. This also leads to economies of scale, lowering production costs. Renewable energy developers can generate additional revenue by integrating hydrogen production into their business model. Hydrogen producers benefit from access to cheaper renewable electricity, which reduces the cost of hydrogen production.
3) **Maximizing Renewable Energy Utilization:** During periods of high renewable energy generation (e.g., windy days or sunny afternoons), green hydrogen production can absorb surplus energy and avoid curtailment, thereby improving the overall efficiency of renewable energy systems. Hydrogen production offers grid services by stabilizing fluctuating renewable energy generation. It helps balance supply and demand by acting as a flexible, dispatchable source of energy.
4) **Regulatory and Policy Support:** Partnerships may align with national energy transition goals, such as decarbonization targets, which could qualify them for government support in the form of subsidies, grants, or tax incentives. This regulatory support can make green hydrogen projects more financially attractive for both developers and producers. Joint projects can create synergies that help meet the renewable energy targets while promoting a cleaner energy system.

I) **Types of Partnerships Between Renewable Energy Developers and Hydrogen Producers**

  1) **Co-Location of Renewable Energy and Hydrogen Production Facilities:**
     - **What It Entails**: Renewable energy developers and hydrogen producers can co-locate their facilities, such as building hydrogen production plants next to wind or solar farms. This ensures that renewable energy can be

directly used to produce hydrogen without the need for long-distance transmission, thus improving efficiency and reducing losses.

- **Example**: **Ørsted**, a leading renewable energy company, has entered partnerships with hydrogen producers to co-locate hydrogen production units with its offshore wind farms. This allows for the use of surplus wind power to produce green hydrogen.

2) **PPAs for Green Hydrogen:**
   - **What It Entails**: Renewable energy developers can enter into long-term PPAs with hydrogen producers to guarantee a steady supply of renewable energy at competitive rates. These agreements can help hydrogen producers lock in predictable energy prices, which is crucial for the economic viability of green hydrogen production.
   - **Example**: **$H_2$GO Power**, a U.K.-based hydrogen producer, has secured PPAs with renewable energy developers to supply green electricity for its hydrogen production facilities. This arrangement provides stability in energy pricing and enhances the financial feasibility of hydrogen projects.
3) **JVs and Strategic Alliances:**
   - **What It Entails**: Renewable energy developers and hydrogen producers can form JVs to develop large-scale green hydrogen production projects. These JVs can combine expertise from both sectors and allow the sharing of risk, resources, and technologies.
   - **Example**: **Siemens Energy** and **Green Hydrogen Systems** have formed a strategic alliance to develop green hydrogen projects. Siemens Energy provides renewable energy systems and hydrogen infrastructure, while Green Hydrogen Systems focuses on producing the electrolyzers that convert renewable electricity into hydrogen.
4) **Integrated Hydrogen and Renewable Energy Hubs:**
   - **What It Entails**: Creating hydrogen hubs where renewable energy developers, hydrogen producers, and industrial users collaborate within a single integrated system. These hubs can facilitate the production, storage, transportation, and utilization of hydrogen at scale.
   - **Example**: **The HyNet North West Project** in the United Kingdom is a collaboration involving renewable energy developers, hydrogen producers, and industrial users. The project focuses on creating a hydrogen hub to decarbonize industries such as chemicals, cement, and steel by integrating renewable energy with hydrogen production and storage.

II) **Challenges and Considerations in Partnerships**

1) **Infrastructure Development and Integration:**
   - **Challenge**: Building the necessary infrastructure for green hydrogen production, storage, and transportation can be capital-intensive and

technically challenging. Ensuring smooth integration between renewable energy facilities and hydrogen production systems is critical.

- **Solution**: Close collaboration between developers and producers early in the project lifecycle can lead to better planning and execution of infrastructure development.

2) **Energy Price Volatility:**
   - **Challenge**: The price of renewable energy can be volatile due to fluctuations in weather patterns, grid congestion, and other factors. This volatility could affect the cost of green hydrogen production.
   - **Solution**: Long-term PPAs and hedging mechanisms can help manage price risks and ensure stable pricing for hydrogen production.
3) **Regulatory and Policy Alignment:**
   - **Challenge**: Policies and regulations related to renewable energy and hydrogen can vary by region and may evolve over time. This can create uncertainty and risk for long-term partnerships.
   - **Solution**: Collaborating with policymakers and engaging in advocacy efforts can help shape regulatory frameworks that support green hydrogen development and reduce policy-related risks.

As green hydrogen production matures and economies of scale are realized, partnerships between renewable energy developers and hydrogen producers will become more common and increasingly essential. Such collaborations will play a vital role in driving down the cost of green hydrogen, improving market access, and fostering the creation of hydrogen hubs that integrate various sectors, including transportation, power, and industry.

The ongoing global transition to cleaner energy and the increasing demand for decarbonization across industries will fuel the expansion of these partnerships. As hydrogen becomes a key component of energy systems worldwide, the combined efforts of renewable energy developers and hydrogen producers will be critical in shaping a sustainable, low-carbon future.

JVs and consortia play a critical role in the development and execution of large-scale green hydrogen projects. These collaborative partnerships allow stakeholders to pool resources, share risks, and leverage complementary expertise, thereby enhancing the feasibility and scalability of green hydrogen initiatives. Green hydrogen production, distribution, and utilization involve high capital costs, complex technological innovations, and a broad range of stakeholders, making collaboration essential for achieving both technical and financial success.

Following are a detailed look at how JVs and consortia can be structured and leveraged for large-scale green hydrogen projects:

1) **Structure and Formation of JVs and Consortia**

**Joint Ventures (JVs)**

A JV is a partnership where two or more entities come together to form a new business entity to undertake a specific project or business activity. In the case of green hydrogen, JVs can involve a combination of:

- **Energy Companies**: Large energy companies may partner with hydrogen producers to share infrastructure, power plants, and renewable energy resources.
- **Technology Providers**: Electrolyzer manufacturers, storage companies, and fuel cell developers can be part of the JV to provide the necessary equipment for production, storage, and transport of hydrogen.
- **Utilities and Grid Operators**: Utility companies may collaborate with green hydrogen producers to integrate hydrogen into the energy grid or use it for energy storage purposes.
- **Industrial End-Users**: Large-scale industrial users, such as steel manufacturers or chemical companies, can join JVs to ensure a steady supply of green hydrogen as a raw material or energy source.

**Consortia**

A consortium is a group of organizations, typically from diverse sectors, that come together for a shared purpose. Unlike JVs, consortia do not form new entities but operate as a loose collective. Consortia for green hydrogen projects often consist of:

- **Multiple Industry Stakeholders**: A mix of energy companies, tech developers, infrastructure builders, and end-users.
- **Research Institutions**: Academic and research organizations contribute to technology innovation, efficiency improvements, and environmental assessments.
- **Government Agencies**: Governments provide funding, regulatory support, and policy incentives to enable large-scale green hydrogen projects.

2) **Benefits of JVs and Consortia for Green Hydrogen Projects**

**Risk Sharing and Cost Reduction**

- Green hydrogen projects involve high capital investments, technological uncertainties, and regulatory challenges. By forming JVs and consortia, partners can share the financial burden and spread risks, making large-scale projects more financially viable.

- JVs and consortia allow for economies of scale in procurement, manufacturing, and operational costs, which reduces the per-unit cost of hydrogen production and accelerates the transition to competitive green hydrogen pricing.

**Technology and Knowledge Synergy**

- **Technology Transfer**: Partners in a JV or consortium can share their expertise and technological innovations. For example, a tech company specializing in electrolyzers can collaborate with energy firms experienced in renewable power generation to optimize hydrogen production efficiency.
- **Cross-Sector Innovation**: JVs and consortia allow for the exchange of knowledge across sectors, enabling innovative solutions to emerge from diverse perspectives. For instance, integrating hydrogen into the maritime or aviation sectors may require specialized expertise in those industries.

**Access to Funding and Resources**

- Large-scale green hydrogen projects often require significant investment. JVs and consortia can secure funding from a broader base of investors and government programs. For example, governments may offer grants or subsidies to multi-party initiatives that align with national energy transition goals.
- Partners may also bring their own financial resources and operational capacity, further enabling the project to secure capital and meet milestones on time.

**Enhanced Market Penetration and Customer Reach**

- Collaboration with industrial end-users in a JV ensures that green hydrogen has an immediate market and can be integrated into supply chains. For instance, a steel manufacturer entering a JV can guarantee demand for hydrogen as a reducing agent in their production process.
- JVs and consortia allow for geographical expansion and the ability to reach global markets by pooling networks of partners and distributors across different regions.

**Regulatory and Policy Advocacy**

- Governments and regulators often have an important role in fostering the green hydrogen ecosystem through favorable policies, incentives, and subsidies. By forming consortia or JVs, project participants can engage in collective lobbying and advocacy for policies that promote hydrogen production and infrastructure development.
- Regulatory compliance can be better managed as consortia often bring together legal and compliance experts from various sectors to ensure adherence to safety, environmental, and operational standards.

3) **Key Areas for Collaboration in Green Hydrogen JVs and Consortia**

**Hydrogen Production Facilities**

- **Renewable Power Providers + Electrolyzer Manufacturers**: A JV between renewable energy developers (solar or wind farms) and electrolyzer manufacturers can lead to integrated green hydrogen production hubs, where renewable electricity is directly converted to hydrogen. This can drastically reduce transmission losses and increase efficiency.

**Hydrogen Storage and Transportation**

- **Energy Companies + Logistics Providers**: A JV or consortium involving energy companies, transportation and logistics providers, and storage solution developers can work together to design and optimize hydrogen storage and distribution infrastructure, such as pipelines, liquefaction plants, and storage terminals.

**Infrastructure Development**

- **Governments + Infrastructure Developers**: Partnerships with public bodies can unlock incentives for the development of HRSs, pipeline networks, and storage facilities. Governments can help reduce infrastructure costs through subsidies, tax breaks, and grants.

**Industrial Application**

- **Green Hydrogen Producers + Heavy Industries**: Collaborations between hydrogen producers and heavy industries, such as cement, steel, and chemical industries, can ensure that green hydrogen is effectively used as a feedstock or energy source in production processes, driving demand for hydrogen.

4) **Examples of Successful JVs and Consortia in Green Hydrogen**

**HyDeal Ambition**

This European consortium aims to produce and distribute green hydrogen at scale. The consortium includes energy companies, utilities, industrial partners, and investors, working together to reduce the cost of green hydrogen production by integrating renewable energy with hydrogen production and storage technologies. It focuses on utilizing solar energy in the south of Spain for the production of green hydrogen.

**$H_2$Global**

A JV involving several German companies and institutions, including hydrogen producers and industrial partners, $H_2$Global aims to establish a global green hydrogen supply chain. The JV focuses on large-scale production and import of green hydrogen into Europe to support sectors like chemicals and steel, integrating hydrogen production, transportation, and consumption.

5) **Challenges and Considerations in JVs and Consortia**

**Aligning Objectives**
Different partners may have differing financial, operational, and strategic goals. Clear governance structures and decision-making frameworks must be established to ensure smooth collaboration and alignment of objectives.

**IP Management**
In technology-driven sectors like green hydrogen, managing IP and proprietary knowledge is crucial. JVs and consortia must outline clear agreements on IP ownership and usage to avoid disputes.

**Regulatory Compliance**
Green hydrogen projects operate in highly regulated environments, especially in terms of safety, emissions, and energy production standards. Ensuring that all JV or consortium partners adhere to relevant regulations and standards is essential.

**Capital Allocation**
Funding large-scale hydrogen projects requires substantial investment. Ensuring that each partner contributes adequately and aligns with long-term capital requirements is vital to the financial success of the JV or consortium.

JVs and consortia represent powerful models for advancing large-scale green hydrogen projects. By pooling resources, sharing risks, and leveraging complementary expertise, stakeholders can accelerate the development of the green hydrogen ecosystem, reduce costs, and promote sustainability. Strategic partnerships across production, infrastructure, and end-use sectors will drive the successful commercialization of green hydrogen, contributing to the global energy transition and the achievement of carbon neutrality goals.

### 5.7.1 International Collaboration and Partnerships to Promote Global Adoption of Green Hydrogen

International collaboration and partnerships are crucial to accelerating the global transition to green hydrogen as a key energy source. The adoption of green hydrogen faces various challenges, such as high production costs, limited infrastructure, and regulatory barriers. Through coordinated global efforts, these challenges can be addressed, paving the way for large-scale deployment of green hydrogen. Following is an overview of how international collaboration and partnerships can promote the widespread adoption of green hydrogen.

1) **Role of International Organizations in Green Hydrogen Development**

**International Energy Agency (IEA)**

- The IEA has been instrumental in promoting the transition to clean energy and plays a pivotal role in coordinating global efforts for hydrogen adoption.

It provides policy recommendations, publishes annual hydrogen reports, and coordinates collaborative efforts across nations to align on hydrogen strategies, technologies, and standards.

**Hydrogen Council**

- The Hydrogen Council, a global initiative of leading companies in the energy and industrial sectors, advocates for hydrogen as a key enabler of the global energy transition. It brings together stakeholders from the public and private sectors to accelerate the development and scaling up of hydrogen technologies, including green hydrogen. Its global reach helps unify efforts and amplify the voices of businesses, governments, and innovators supporting hydrogen.

**United Nations (UN)**

- The UN supports international partnerships for sustainable development through frameworks such as the SDGs. Green hydrogen plays a role in SDGs, especially in energy access, climate action, and affordable clean energy. The UN can facilitate international dialogues, partnerships, and projects, particularly in developing countries where green hydrogen could contribute to economic development and clean energy access.

2) **Bilateral and Multilateral Partnerships for Green Hydrogen**

**Europe and the Middle East**

- **Germany and the UAE**: Germany has partnered with the UAE, which has abundant renewable energy resources, to produce and export green hydrogen to Europe. This collaboration includes the development of hydrogen infrastructure, including production facilities, transportation, and storage.
- **EU Green Hydrogen Partnerships**: The EU has been working with various countries in the Mediterranean and North Africa to establish green hydrogen production hubs. These collaborations are part of the EU's strategy to diversify its hydrogen supply and reduce dependence on fossil fuels. The EU collaborates with countries like Morocco, Egypt, and Algeria, utilizing solar and wind energy to produce green hydrogen for export to Europe.

**Australia and Japan**

- Australia has been a leader in green hydrogen research and production, particularly leveraging its vast renewable energy resources (e.g., solar and wind). In collaboration with Japan, which is a leading hydrogen consumer, Australia is working on large-scale green hydrogen production and export projects. These projects focus on using hydrogen in Japan's industrial sectors, including steel manufacturing and transportation.

**Australia and South Korea**

- South Korea has also entered into partnerships with Australia to secure green hydrogen supplies. South Korea is investing heavily in green hydrogen technologies for applications in transport and heavy industries. Australia's role as a producer of green hydrogen complements South Korea's demand for low-carbon energy solutions.

3) **Cross-Regional Partnerships for Green Hydrogen Export and Trade**

**Africa and Europe**

- Africa holds substantial potential for green hydrogen production due to its vast renewable energy resources. European nations, such as Germany and Netherlands, are developing partnerships with African countries like Namibia, South Africa, and Morocco to create large-scale green hydrogen production and export facilities. These partnerships focus on harnessing solar and wind energy in Africa to produce hydrogen for export to European markets.
- Africa's role in green hydrogen production could significantly contribute to global energy security while enabling economic growth in regions that are abundant in renewable energy resources but have limited access to global energy markets.

**South America and North America**

- Chile is another country that is emerging as a leader in green hydrogen production. Chile, with its exceptional renewable energy resources, has entered partnerships with U.S. companies to develop green hydrogen technologies and infrastructure. This includes the development of large solar-powered hydrogen production plants and creating supply chains for hydrogen export to North America and Asia.

4) **Joint Research and Technology Development**

**Public–Private Partnerships (PPP)**

- Countries and companies can jointly fund **R&D** projects to drive innovations in green hydrogen production, storage, and transportation technologies. PPPs can pool resources from governments, industry leaders, and research institutions. This reduces the financial burden and accelerates the commercialization of new hydrogen technologies.
- **Examples**: The European Clean Hydrogen Alliance promotes joint R&D projects across Europe. Additionally, Japan's Hydrogen Roadmap involves

both public and private sector actors collaborating on hydrogen-related technology developments and infrastructure creation.

**International Research Collaborations**

- Collaborative R&D efforts between universities and research institutions from different countries can accelerate the development of cutting-edge hydrogen technologies. For example, the HESC project, a collaboration between Australian, Japanese, and U.S. research institutions, focuses on developing a scalable supply chain for transporting liquefied hydrogen by sea.

5) **Standardization and Regulatory Cooperation**

**International Standards for Hydrogen**

- A unified set of international standards for green hydrogen production, storage, transportation, and utilization is necessary to ensure safe and efficient operations across borders. Standardization ensures that hydrogen infrastructure and technologies can be interoperable, making it easier to establish a global hydrogen economy.
- International collaborations can push for ISO standards specific to green hydrogen, ensuring that protocols are followed in production facilities, transportation networks, and end-use industries.

**Harmonizing Regulatory Frameworks**

- Harmonizing regulatory frameworks between countries will reduce barriers to the cross-border movement of green hydrogen. Regulatory collaboration can cover aspects such as safety, emissions, infrastructure requirements, and pricing mechanisms.
- The International Maritime Organization (IMO) is already working with various stakeholders to develop safety and regulatory standards for the maritime transport of hydrogen, which will be key to facilitating international trade in green hydrogen.

6) **Financing and Investment for International Green Hydrogen Projects**

**Global Financial Institutions**

- MDBs such as the World Bank, Asian Infrastructure Investment Bank (AIIB), and EIB play a key role in financing cross-border hydrogen initiatives, especially in developing countries. They provide concessional loans and grants to support green hydrogen projects and infrastructure development, ensuring that funding is available to make these projects financially feasible.
- Private Sector Investments: Private sector companies, including major oil and gas firms, have started to invest in hydrogen production and

infrastructure. As a result, cross-border investment flows are becoming more common. For example, Shell, BP, and TotalEnergies are investing in international hydrogen projects and forming partnerships with local governments and businesses.

7) **Global Policy Frameworks and Initiatives**

**International Hydrogen Collaboration Agreements**
- Countries can enter into bilateral agreements to promote mutual goals for green hydrogen development, such as reducing carbon emissions and increasing energy security. These agreements can include commitments to produce, transport, and consume green hydrogen in line with national climate targets.

**The Clean Hydrogen Partnership**
- A part of the Horizon Europe program, the Clean Hydrogen Partnership brings together the European Commission, industry stakeholders, and research organizations to fund R&D in hydrogen technologies. The partnership aims to advance the adoption of green hydrogen within the EU and with international partners.

International collaboration and partnerships are essential for scaling up green hydrogen production, infrastructure, and global trade. By sharing knowledge, pooling resources, standardizing technologies, and harmonizing regulations, countries and industry players can create a robust green hydrogen economy. These collaborations will not only help meet climate and energy goals but also drive economic growth, energy security, and job creation in both developed and developing nations. Global partnerships will unlock the potential of green hydrogen as a sustainable, low-carbon energy solution and pave the way for its widespread adoption worldwide.

## 5.8 Conclusion

The evolution of green hydrogen as a cornerstone of the global energy transition necessitates a strategic rethinking of operations and SCM. This chapter has outlined the multifaceted nature of managing green hydrogen across its lifecycle from ethical and sustainable sourcing of input materials to its safe and efficient delivery. Establishing robust sourcing practices ensures that the production of green hydrogen aligns with environmental and social governance standards, forming the foundation for truly sustainable energy systems.

The design and construction of green hydrogen production facilities play a pivotal role in determining operational efficiency and scalability. These facilities must be integrated thoughtfully into the existing infrastructure to enable seamless distribution and utilization. Equally, optimizing logistics and transportation networks is crucial, given hydrogen's unique properties and the technical complexities of its movement in various states.

Ensuring safety and compliance with regulatory frameworks is not merely a technical necessity but a public trust imperative. As outlined, the environmental superiority of green hydrogen over conventional methods like SMR is evident in its significantly reduced emissions and ecological footprint. However, this environmental promise must be backed by rigorous LCAs and long-term commitment to clean energy inputs.

The current market dynamics reveal both opportunities and challenges. While global interest and investments are growing rapidly, the sector requires deliberate alignment of policies, market incentives, and technological innovation. This can only be achieved through proactive collaboration between government, industry, and academia, fostering an ecosystem of knowledge sharing, co-creation, and capacity building.

In essence, green hydrogen's success hinges not just on technological advancement but also on a resilient, ethical, and environmentally responsible supply chain. The path forward must be collaborative, inclusive, and strategic ensuring green hydrogen can fulfil its potential as a pillar of a clean and sustainable future.

## 5.9 Case Study

### ➤ Green Hydrogen Implementation in Europe

The EU has initiated the "Hydrogen Strategy for a Climate-Neutral Europe" to scale up green hydrogen production. A key project under this initiative is the North Sea Wind Hydrogen Hub, which integrates offshore wind farms with hydrogen electrolyzers.

**Operational Insights**

- Offshore wind energy is directly supplied to electrolysis units, reducing dependency on grid electricity.
- A network of hydrogen pipelines is being developed to transport hydrogen to industrial clusters.
- AI-based predictive analytics optimize production efficiency and minimize energy wastage.

**Supply Chain Innovations**

- Establishing on-site hydrogen liquefaction plants for easier transportation.

- Implementing blockchain technology for real-time tracking and transparency in hydrogen transactions.
- Collaborative efforts with automotive and industrial sectors for demand-driven supply chain planning.

**Detailed Implementation Plan**

1) **Infrastructure Development**:
   - Investment in large-scale electrolyzers integrated with offshore wind farms.
   - Construction of dedicated hydrogen transport pipelines to connect production sites with key industrial hubs.
   - Expansion of refueling stations across major transport routes.
2) **Cost Reduction Strategies**:
   - Leveraging economies of scale to reduce electrolyzer costs.
   - Utilizing government incentives and subsidies to make green hydrogen price-competitive with fossil fuels.
   - Optimizing logistics through hydrogen storage hubs strategically located near demand centers.
3) **Industrial and Commercial Applications**:
   - Supply of hydrogen to steel and chemical industries to replace fossil fuel-based processes.
   - Development of hydrogen-powered heavy transport, including trucks, buses, and trains.
   - Integration with power grids to serve as an energy storage solution.
4) **Policy and Regulatory Framework**:
   - The EU's Fit for 55 package provides financial support and regulatory guidelines to boost green hydrogen adoption.
   - Standardization of safety protocols for hydrogen production, storage, and transportation.
   - Carbon pricing mechanisms to encourage industries to shift toward green hydrogen solutions.

**Expected Impact and Future Prospects**

- Reduction in carbon emissions by replacing fossil fuel-based industrial processes.
- Strengthening of energy security by reducing reliance on imported fossil fuels.
- Job creation in the renewable energy and hydrogen production sectors.
- Development of a global green hydrogen export market, positioning Europe as a leader in hydrogen innovation.

Green hydrogen holds immense potential in achieving global sustainability goals. However, its widespread adoption depends on advancements in operations and SCM. Strategic investments in infrastructure, technological innovation, and policy support are crucial to making green hydrogen a commercially viable energy source. Companies and governments must collaborate to streamline production, enhance logistics, and optimize the supply chain to ensure a sustainable hydrogen economy.

➢ **Case Study: Implementation of Green Hydrogen in India**

India has committed to achieving net-zero carbon emissions by 2070, and green hydrogen is expected to play a crucial role in this transition. Green hydrogen, produced using renewable energy sources such as wind and solar power, offers a sustainable alternative to fossil fuels. This case study explores India's progress in implementing green hydrogen, key initiatives, challenges, and future prospects.

**Government Initiatives and Policies**

**National Green Hydrogen Mission**

Launched in 2023 with an initial budget of INR 19,744 crore, the National Green Hydrogen Mission aims to:

- Produce 5 million metric tons of green hydrogen annually by 2030.
- Reduce fossil fuel dependence and GHG emissions.
- Develop domestic manufacturing capabilities for electrolyzers and fuel cells.
- Facilitate R&D in hydrogen technologies.
- Promote PPPs to scale up green hydrogen production.

**Renewable Energy Integration**

- India is expanding its renewable energy capacity to power electrolysis units.
- Solar and wind energy projects are being integrated into green hydrogen production plants.
- The government offers incentives for renewable energy developers to set up hydrogen infrastructure.

**Key Industry Players and Projects**

**Reliance Industries Limited (RIL)**

- Plans to invest $75 billion in renewable energy and hydrogen production.
- Developing giga-scale electrolyzers and HRSs.

**Indian Oil Corporation (IOC)**

- Setting up pilot projects for green hydrogen in refineries.

- Partnering with National Thermal Power Corporation (NTPC) to establish large-scale hydrogen production facilities.

**Adani Group**

- Collaborating with global companies to develop hydrogen production and export hubs.

**National Thermal Power Corporation (NTPC)**

- Running hydrogen-powered bus trials in Ladakh and Delhi.
- Developing hydrogen-based energy storage solutions.

**Challenges in Implementation**

**High Production Costs**

- Electrolysis technology is expensive, making green hydrogen costlier than gray hydrogen.
- Capital investment required for infrastructure development is substantial.

**Storage and Transportation Issues**

- Hydrogen has low energy density and requires advanced storage solutions.
- Pipeline and distribution network for hydrogen are underdeveloped in India.

**Policy and Regulatory Gaps**

- Lack of clear hydrogen pricing mechanisms and subsidies.
- Need for safety and regulatory frameworks for hydrogen production, transport, and usage.

**Future Prospects**

- India aims to become a global green hydrogen hub by leveraging its renewable energy potential.
- Collaboration with international stakeholders for technology transfer and investment.
- Advancements in electrolysis technology and cost reductions are expected to drive large-scale adoption.
- Green hydrogen export opportunities, particularly to Europe and Japan, could strengthen India's position in the global clean energy market.

India's green hydrogen initiative is a critical step toward a sustainable energy future. While challenges remain, strategic investments, government support, and industry collaborations can help overcome barriers. With continued progress, India can emerge as a leader in the global green hydrogen economy, contributing to both domestic energy security and climate change mitigation.

## 5.10 Exercise

1. What are the major sources of renewable energy used for green hydrogen production, and how do they impact production efficiency?

2. Discuss the key considerations in sourcing water for electrolysis in green hydrogen production. What challenges can arise in water procurement?

3. Explain the different procurement strategies for electrolyzer components and materials. How do they affect the overall cost and efficiency of green hydrogen production?

4. Why is ethical and sustainable sourcing of input materials important in the green hydrogen industry? Provide examples of best practices in sustainable sourcing.

5. What are the major challenges in securing reliable and cost-effective input materials for green hydrogen production? Suggest potential solutions to overcome these challenges.

6. Describe the key design and construction considerations for a green hydrogen production facility. How does site selection impact efficiency and cost?

7. What are the major components of hydrogen storage infrastructure? Explain the advantages and disadvantages of different hydrogen storage methods.

8. Discuss the role of existing infrastructure in the expansion and integration of hydrogen distribution networks. What modifications might be necessary?

9. What factors should be considered when planning the scale-up of green hydrogen production and distribution infrastructure? Provide examples of successful scale-up strategies.

10. Identify different investment and financing options available for green hydrogen infrastructure development. What are the risks and benefits associated with each option?

11. What are the key logistical challenges associated with transporting green hydrogen? How can transportation efficiency be improved?

12. Explain how the design and optimization of distribution networks impact the efficiency and cost of green hydrogen delivery.

13 What are the major safety and regulatory considerations in the distribution and delivery of hydrogen? How do they impact supply chain operations?

14 Compare the lifecycle environmental impacts of green hydrogen production with those of conventional hydrogen production methods. Which areas show the most significant differences?

15 How can collaboration between government, industry, and academia accelerate the adoption of green hydrogen? Provide examples of successful partnerships.

## 5.11 Objective Type Questions

1 **Which of the following is a primary source of energy for green hydrogen production?**
a) Natural gas
b) Coal
c) Solar or wind power
d) Diesel

2 **What is the main water requirement in the electrolysis process for green hydrogen?**
a) Saltwater
b) Distilled or deionized water
c) Rainwater
d) Wastewater

3 **What is a critical factor in sourcing electrolyzer components?**
a) Brand popularity
b) Aesthetic design
c) Availability and reliability of materials
d) Packaging style

4 **Ethical sourcing of green hydrogen input materials includes:**
a) Using the cheapest suppliers regardless of impact
b) Ignoring supply chain transparency
c) Ensuring fair labor practices and sustainability
d) Prioritizing cost over compliance

**5 Which is a major challenge in securing input materials for green hydrogen?**

a) Lack of demand
b) Over-supply of renewable energy
c) High costs and global supply chain disruptions
d) Simple logistics

**6 What is essential in the design of a green hydrogen production facility?**

a) Proximity to shopping malls
b) Aesthetic appearance
c) Access to renewable energy sources and water supply
d) High traffic visibility

**7 Hydrogen is most commonly stored in:**

a) Liquid nitrogen tanks
b) High-pressure cylinders or underground caverns
c) Paper bags
d) Wooden barrels

**8 What is a key goal in integrating existing infrastructure for hydrogen distribution?**

a) Increasing fossil fuel dependency
b) Replacing all pipelines
c) Enhancing efficiency and cost-effectiveness
d) Discontinuing renewable sources

**9 Which of the following is a factor in infrastructure planning for scaling hydrogen production?**

a) Distance to urban centers
b) Social media presence
c) Future energy demand and grid integration
d) Color coding of buildings

**10 What is crucial for financing green hydrogen infrastructure?**

a) Investment from fossil fuel companies only
b) Ignoring cost–benefit analysis
c) Public–private partnerships and government subsidies
d) Excluding financial institutions

**11 A challenge in hydrogen transportation is:**

a) Lack of interest
b) Its low density and need for compression or liquefaction

c) Simplicity of hydrogen logistics
d) Its flow like oil

12 **Last-mile delivery refers to:**
a) Transporting hydrogen within the production plant
b) Delivery of hydrogen to the final consumer
c) Exporting hydrogen internationally
d) Long-distance freight delivery

13 **Why are safety regulations critical in hydrogen distribution?**
a) Hydrogen is nonflammable.
b) Regulations increase production time.
c) Hydrogen is highly flammable and explosive.
d) It lowers market demand.

14 **Which innovation supports safer and efficient hydrogen distribution?**
a) AI-powered sensors and IoT-enabled valves
b) Manual leak checks
c) Open tanks
d) Paper records

15 **Lifecycle assessment of green hydrogen focuses on:**
a) Only transportation impacts
b) The visual appearance of plants
c) Environmental impacts from production to use
d) Only capital investment

16 **Which method has the least environmental impact for hydrogen production?**
a) Steam methane reforming
b) Green hydrogen via electrolysis
c) Coal gasification
d) Diesel cracking

17 **Hotspot identification in lifecycle analysis aims to:**
a) Identify low-cost suppliers
b) Reduce marketing expenses
c) Pinpoint areas with high environmental impact
d) Optimize office layout

18 **What influences market demand for green hydrogen the most?**
a) TV advertisements
b) Policy support, energy needs, and sustainability goals

c) Number of vendors
d) Social media followers

**19 Demand forecasting in green hydrogen helps in:**
a) Setting up distribution only in rural areas
b) Estimating future requirements for production planning
c) Ignoring new technologies
d) Reducing renewable investments

**20 International collaboration in green hydrogen helps in:**
a) Increasing national restrictions
b) Sharing best practices, technologies, and standardization
c) Creating isolated energy hubs
d) Avoiding global trade

## 5.12 Glossary

1) **Green Hydrogen**: Hydrogen produced using renewable energy sources (such as solar, wind, or hydropower) through water electrolysis, without emitting carbon dioxide, thus considered a clean energy vector.
2) **Decarbonization Strategies**: Global initiatives aimed at reducing or eliminating carbon emissions from energy production, industrial activities, and transportation to combat climate change.
3) **Ethical Sourcing**: The practice of procuring raw materials responsibly, ensuring fair labor practices, human rights, and minimizing negative social and environmental impacts.
4) **Sustainable Sourcing**: Acquiring resources in a manner that preserves environmental integrity, reduces ecological footprint, and ensures long-term availability of essential materials such as water and renewable electricity.
5) **Transparency**: Open and accountable communication in supply chains regarding sourcing, production methods, and sustainability impacts.
6) **Environmental Stewardship**: The responsibility of managing natural resources carefully to maintain ecosystem health, reduce pollution, and support sustainability goals.
7) **Production Facilities**: Specialized plants and infrastructure where green hydrogen is produced, often incorporating electrolysis and powered by renewable electricity.
8) **Innovative Technologies**: Cutting-edge advancements in engineering and process design that enhance the efficiency, scalability, and sustainability of hydrogen production.

9) **Sustainable Engineering Practices**: Design and construction approaches that minimize environmental impact, conserve resources, and promote long-term viability in hydrogen operations.
10) **Infrastructure Development**: The creation and expansion of systems such as pipelines, storage units, refueling stations, and integration with existing networks to support green hydrogen distribution.
11) **Legacy Systems**: Existing infrastructure and technologies originally designed for conventional fuels or natural gas, which may be adapted or integrated with green hydrogen solutions.
12) **Logistics**: The planning, implementation, and management of hydrogen transportation and storage, ensuring timely and efficient delivery.
13) **Transportation Challenges**: Difficulties related to moving hydrogen, including compression (gas form), liquefaction (liquid form), and pipeline transport, while ensuring safety and efficiency.
14) **Compression**: The process of pressurizing hydrogen gas to enable efficient storage and transportation.
15) **Liquefaction**: Cooling hydrogen to extremely low temperatures (−253 °C) to convert it into liquid form for storage and long-distance transport.
16) **Pipeline Transport**: Using pipeline networks to move hydrogen from production sites to end-users, often requiring material adaptations due to hydrogen's properties.
17) **Safety Frameworks**: Policies, standards, and procedures to ensure secure handling, transport, and storage of hydrogen, minimizing the risks of leaks, explosions, or accidents.
18) **Regulatory Frameworks**: Laws, guidelines, and policies that govern the production, distribution, and use of hydrogen, ensuring safety, quality, and sustainability compliance.
19) **Environmental Implications**: The overall impact of green hydrogen production on ecosystems, carbon emissions, and resource use, compared to fossil fuel-based hydrogen.
20) **Lifecycle Benefits**: The long-term environmental and economic advantages of green hydrogen over its entire production, distribution, and consumption cycle.
21) **Market Dynamics**: The forces of supply, demand, competition, and investment trends that influence the growth and development of the green hydrogen sector.
22) **Demand Trends**: Patterns of consumer, industrial, and national demand for green hydrogen across sectors such as energy, mobility, and industry.
23) **Cost Trajectories**: The projected changes in production, storage, and distribution costs of green hydrogen over time due to technological innovation and economies of scale.

24) **Investment Patterns**: The flow of financial resources into green hydrogen projects, infrastructure, and technologies by governments, industries, and private investors.
25) **Collaboration**: Joint efforts by governments, industries, academia, and research institutions to accelerate green hydrogen development and strengthen supply chains.
26) **Resilient Supply Chain**: A flexible and adaptive supply chain that can withstand disruptions, respond to demand fluctuations, and support long-term hydrogen deployment.
27) **Responsive Supply Chain**: A supply chain designed to quickly adapt to market and technological changes, ensuring efficient production and distribution of green hydrogen.

## References


1 Sahin, H. (2024). Hydrogen refueling of a fuel cell electric vehicle. *International Journal of Hydrogen Energy*, 75, 604–612 https://doi.org/10.1016/J.IJHYDENE.2024.04.021.

2 Shi, Y., Qiu, Z., Snyder, S. W., & Wang, Y. (2024). A perspective on broad deployment of hydrogen-fueled vehicles for ground freight transportation with a comparison to electric vehicles. *Energy Strategy Reviews*, 56, 101565 https://doi.org/10.1016/J.ESR.2024.101565.

3 Shin, D. H., Hwang, H. K., & Kim, S. J. (2024). Effect of plasma ion nitriding on hydrogen embrittlement and wear resistance of martensitic stainless steel for hydrogen valve core of fuel cell electric vehicles. *Engineering Fracture Mechanics*, 306, 110247 https://doi.org/10.1016/J.ENGFRACMECH.2024.110247.

4 Shin, D.-H., & Kim, S.-J. (2024). Effect of hydrogen charging on the wear resistance of CrN-coated Al alloy for hydrogen valve of fuel cell electric vehicles. *International Journal of Hydrogen Energy*, 89, 279–292 https://doi.org/10.1016/J.IJHYDENE.2024.09.251.

5 Singla, M. K., Gupta, J., Safaraliev, M., Nijhawan, P., & Oberoi, A. S. (2024). Hydrogen storage in activated carbon for fuel cell-powered vehicles: a cost-effective and sustainable approach. *International Journal of Hydrogen Energy*, 58, 446–458 https://doi.org/10.1016/J.IJHYDENE.2024.01.117.

6 Song, B., Wang, X., Kang, Y., & Li, H. (2024). Research on hydrogen leakage diffusion and safety analysis in hydrogen fuel cell vehicles with regard to leakage location and ventilation ports. *International Journal of Hydrogen Energy*, 83, 173–187 https://doi.org/10.1016/J.IJHYDENE.2024.08.105.

7 Sun, L., Jia, J., Wang, Q. L., & Zhang, Y. (2024). A novel multiphase DC/DC boost converter for interaction of solar energy and hydrogen fuel cell in hybrid electric vehicles. *Renewable Energy*, 229, 120779 https://doi.org/10.1016/J.RENENE.2024.120779.

8 Tuncer, D., & Yilmaz Ulu, E. (2024). Contribution of regenerative suspension module to charge efficiency and range in hydrogen fuel cell electric vehicles. *International Journal of Hydrogen Energy*, 75, 547–556 https://doi.org/10.1016/J.IJHYDENE.2024.03.219.

9 Vilberger, M. E., Popov, N. S., Domakhin, E. A., Anibroev, V. I., & Mosin, M. E. (2024). Increasing the energy efficiency of an electric vehicle powered by hydrogen fuel cells. *International Journal of Hydrogen Energy*, 85, 406–415 https://doi.org/10.1016/J.IJHYDENE.2024.08.099.

10 Wang, G., Wang, J., Hu, C., Lv, W., & Mao, H. (2024). Simulation of hydrogen leakage and diffusion in an enclosed garage of mixed parking of fuel-cell vehicles. *Next Research*, 1, 100073 https://doi.org/10.1016/J.NEXRES.2024.100073.

11 Wang, Q., Wang, Y., & Chen, Z. (2025). Stackelberg-equilibrium-based collaborative charging management strategy for renewable fuel vehicles in regional integrated electricity-hydrogen system. *Applied Energy*, 377, 124617 https://doi.org/10.1016/J.APENERGY.2024.124617.

12 Wu, Y., Gu, W., Huang, S., Wei, X., & Safaraliev, M. (2025). A four-layer business model for integration of electric vehicle charging stations and hydrogen fuelling stations into modern power systems. *Applied Energy*, 377, 124630 https://doi.org/10.1016/J.APENERGY.2024.124630.

13 Xia, W., Ren, Z., Li, H., & Pan, Z. (2024). A data-driven probabilistic evaluation method of hydrogen fuel cell vehicles hosting capacity for integrated hydrogen-electricity network. *Applied Energy*, 376, 123895 https://doi.org/10.1016/J.APENERGY.2024.123895.

# 6

# Inventory Control and Total Quality Management

## 6.1 Introduction

In the evolving landscape of sustainable energy, green hydrogen has emerged as a transformative solution for reducing carbon emissions and achieving energy security. However, the success of green hydrogen production and distribution depends significantly on two critical factors: inventory control and total quality management (TQM). Effective inventory control ensures optimal resource allocation, minimizes waste, and enhances supply chain efficiency, while TQM focuses on maintaining high production standards, ensuring safety, and meeting regulatory compliance [1].

The production and distribution of green hydrogen involve complex logistics, requiring precise forecasting and real-time monitoring to balance demand and supply efficiently. Traditional inventory management models may not be sufficient due to the unique challenges associated with hydrogen storage, transportation, and perishability concerns. Therefore, modern data-driven inventory techniques, such as predictive analytics, AI, and blockchain-based tracking, are being explored to enhance operational efficiency.

Similarly, TQM principles play a pivotal role in maintaining the quality and reliability of green hydrogen. From the selection of raw materials to the final delivery of hydrogen to end-users, continuous monitoring, process standardization, and compliance with international safety protocols are essential to ensure the long-term viability of green hydrogen as an alternative energy source [2].

This chapter explores the integration of inventory control strategies with TQM principles to develop a sustainable and efficient green hydrogen supply chain. It examines best practices, technological advancements, and case studies that

*Hydrogen Technologies in Decarbonizing Industries and Energy Systems*, First Edition.
Vikas Khare, Sanjeet K. Dwivedi, Monica Bhatia, and K. K. Pant.

Published 2026 by John Wiley & Sons, Inc.

illustrate how effective management techniques can enhance the reliability, affordability, and accessibility of green hydrogen. By addressing key challenges and proposing innovative solutions, this chapter aims to provide valuable insights into the strategic management of green hydrogen production and distribution.

## 6.2 Inventory Management for Green Hydrogen

With the global push toward clean energy transition, green hydrogen has gained prominence as a sustainable fuel alternative. Produced through electrolysis using renewable energy sources, green hydrogen is poised to revolutionize industries such as transportation, power generation, and industrial manufacturing. However, its widespread adoption faces significant challenges, particularly in inventory management, which plays a critical role in ensuring a stable, cost-effective, and efficient supply chain.

Unlike conventional fuels, green hydrogen presents unique inventory challenges, including high storage costs, volatility, perishability, and complex logistics. Traditional inventory management models are often inadequate due to the high-pressure storage requirements, cryogenic conditions, and the need for specialized infrastructure. This makes it imperative to develop advanced inventory control strategies that integrate real-time monitoring, predictive analytics, and blockchain-based tracking systems to optimize supply chain operations. Effective inventory management for green hydrogen involves demand forecasting, storage optimization, transportation logistics, and risk mitigation strategies. It requires a balance between production scalability, storage limitations, and distribution efficiency to minimize losses and meet fluctuating market demands. Additionally, regulatory compliance, safety standards, and sustainability goals must be incorporated into inventory strategies to support the long-term viability of green hydrogen.

> ***Key: Inventory management:*** *The process of planning, tracking, and controlling green hydrogen stock (e.g., storage tanks and cylinders) to ensure supply availability and minimize waste.*

### 6.2.1 Inventory Planning and Forecasting for Green Hydrogen Production

Inventory planning and forecasting for green hydrogen production involves several key components, particularly in the context of optimizing resources for efficient and sustainable production. Following are the points how anyone can approach this:

**Demand Forecasting for Green Hydrogen:** Forecasting the demand for green hydrogen is critical. This can be influenced by factors such as energy policies, government incentives, industrial adoption (e.g., for fuel cells and steel production), and competition from other energy sources. Demand for green hydrogen can be forecast based on sectors such as transport, manufacturing, and power generation. Predicting demand in these sectors requires understanding their specific adoption rates and technological developments. Understanding how price changes affect demand in various markets can help refine the forecasting models.

**Production Forecasting:** Forecasting the output from electrolyzers, which use renewable electricity to split water into hydrogen and oxygen, is key. The forecast should consider the availability of renewable energy sources like solar or wind power. As technology improves, forecasting production should account for increasing efficiency in the electrolyzer process and decreasing costs of production. Renewable energy production has seasonal variability, so this needs to be factored into the production forecast. For instance, wind energy might be higher in certain seasons, affecting hydrogen production capacity [3].

**Raw Material Inventory Management:** A reliable source of water is necessary for hydrogen production via electrolysis. Managing the availability of water, especially in areas with fluctuating availability, is important for inventory planning. Forecasting and managing the inventory of renewable energy resources required for the production of solar or wind energy will ensure that hydrogen production aligns with available energy supply.

**Storage and Distribution:** Since hydrogen is difficult to store, particularly in large quantities, a forecasting model should plan for the storage capabilities available. This could include pressurized tanks, underground storage, or $LH_2$ storage. Managing the logistics of distributing hydrogen to various sectors (e.g., fuel stations, factories and power plants) requires forecasting demand at distribution points and managing inventory in transit [4].

**Supply Chain Management:** The green hydrogen production supply chain involves sourcing electrolyzers, renewable energy equipment, and other components. Forecasting demand for these materials, alongside lead times, is crucial to avoid delays in production. Ensuring that the supply chain remains green is important, so forecasting the carbon footprint of raw materials and transportation is increasingly necessary. Figure 6.1 illustrates inventory planning and forecasting for green hydrogen production.

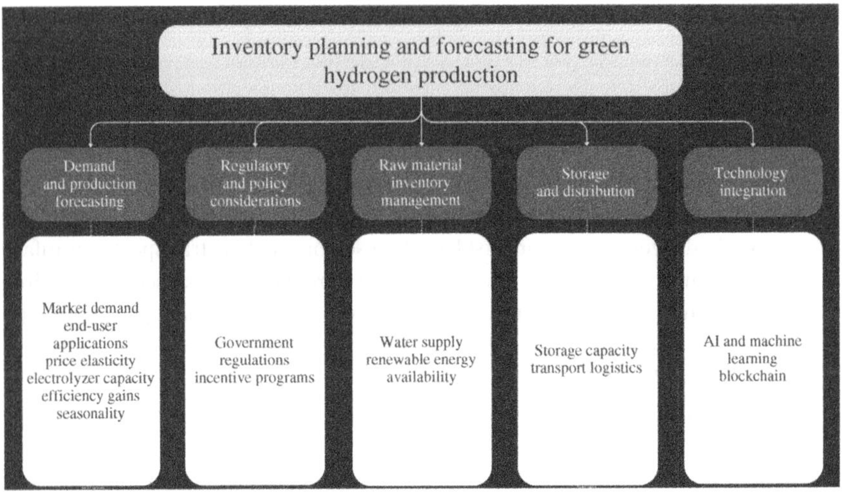


**Figure 6.1** Inventory Planning and Forecasting for Green Hydrogen Production

**Technology Integration:** Using AI and machine learning algorithms for forecasting can significantly improve inventory planning by predicting demand patterns based on historical data, energy trends, and other factors. Integrating blockchain technology can help track and ensure transparency in the sourcing and usage of green hydrogen, which is essential for sustainability certification.

**Regulatory and Policy Considerations:** Anticipating regulatory changes or new policies around green hydrogen production and storage can affect both demand and production forecasts. Changes in government incentives for renewable energy and green hydrogen production can influence both demand forecasts and the economic feasibility of various production methods.

By integrating these factors into a cohesive inventory planning and forecasting model, you can ensure that green hydrogen production is aligned with demand, cost-effective, and sustainable [5].

### 6.2.2 Just-in-Time (JIT) Inventory Management

JIT inventory management can be an effective approach for green hydrogen distribution, especially when balancing production, storage, and demand. Given the unique challenges of hydrogen production and its storage, applying JIT in this context requires careful planning to optimize efficiency, minimize costs, and ensure availability while reducing waste. Following are several JIT approaches that can be applied to green hydrogen distribution:

1) **Real-Time Demand Monitoring:** Green hydrogen distributors should closely monitor real-time demand across various industries (e.g., FCVs, industrial applications, and power generation). This helps in predicting the short-term demand and minimizing overproduction or underproduction. Use data from consumption patterns in different sectors, coupled with weather forecasts (for renewable energy), to predict when demand spikes might occur (e.g., colder months or peak industrial usage). Adjust hydrogen distribution and production schedules accordingly to match forecasted demand.
2) **Dynamic Production Scheduling:** With JIT, production is closely aligned with demand. For green hydrogen, this means adjusting the electrolyzer production capacity to match the required hydrogen volumes, thus avoiding excess inventory or understock. Implement smart systems that dynamically adjust production schedules based on real-time demand signals. Renewable energy sources (solar, wind) are variable, so scheduling should allow for flexibility in hydrogen production to align with available energy, avoiding waste in times of energy surplus.
3) **Collaborative Forecasting with Key Customers:** Close collaboration between green hydrogen producers, suppliers, and key customers (such as industrial users or fuel cell companies) is essential to predict their consumption needs accurately. Establish long-term agreements or regular meetings with major customers to forecast their hydrogen needs with high accuracy. For instance, collaborating with hydrogen-powered transportation fleets to know exactly when they will need hydrogen refueling allows the producer to prepare just enough supply. Figure 6.2 illustrates JIT Inventory Management System (IMS).
4) **Supplier-Managed Inventory:** In a JIT system, VMI allows the supplier to monitor stock levels and automatically replenish inventory when needed, reducing excess stock and storage costs. For green hydrogen, a VMI strategy could involve hydrogen suppliers working closely with distribution points (e.g., hydrogen refueling stations and industrial users) to monitor their hydrogen stock levels and order replenishments only when supply reaches a predetermined low threshold, ensuring efficiency and reducing waste [6].
5) **Flexible Transport and Storage Systems:** JIT inventory relies on timely transportation and storage systems that can quickly react to demand shifts. Hydrogen, being a bulky and sometimes difficult-to-store gas, requires adaptable systems. Develop flexible transport options (e.g., high-pressure storage tanks and cryogenic transportation) that allow for rapid redistribution of

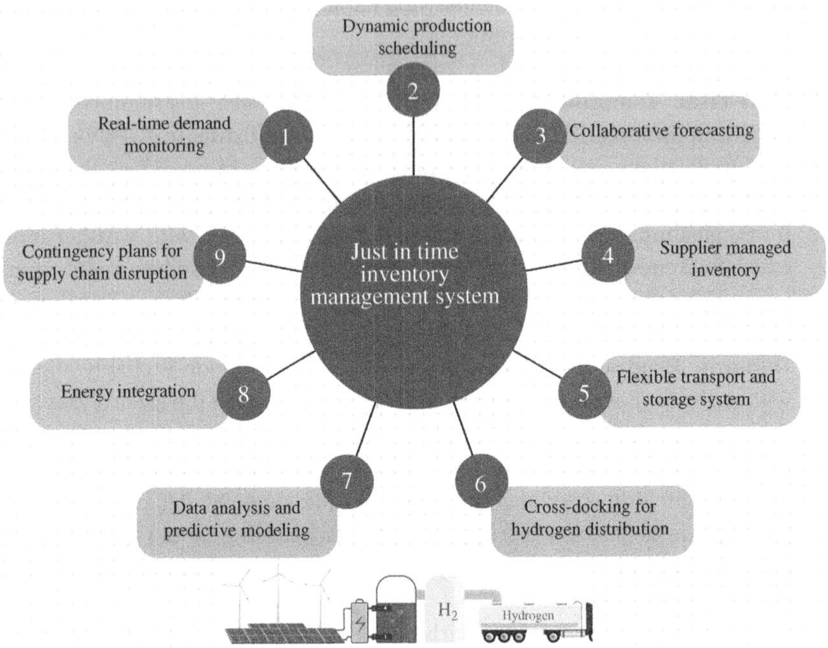


**Figure 6.2** Just-in-Time Inventory Management System. *Source:* scharfsinn86/Adobe Stock Photos.

hydrogen based on shifting regional demands. The storage systems must be strategically located near high-demand areas and should be able to store hydrogen temporarily if production and demand are not perfectly synchronized.

6) **Cross-Docking for Hydrogen Distribution:** Cross-docking involves moving hydrogen directly from the production or storage site to the end-user without long-term storage. It reduces storage costs and ensures that only necessary inventory is held. For green hydrogen, a network of distribution hubs could be set up where hydrogen is moved from production sites or centralized storage directly to customers without being stored for extended periods. This helps maintain minimal inventory and reduces waste. Advanced logistics and transportation planning would be crucial to ensure that this strategy works efficiently.
7) **Advanced Data Analytics and Predictive Modeling:** Data analytics and machine learning can help predict when demand for green hydrogen will spike or drop, allowing for better synchronization of supply with real-time

demand. Use predictive models that take into account renewable energy production patterns, weather data, industry usage forecasts, and consumer trends to predict the exact amount of hydrogen required at any given time. This will optimize production and storage capacity to minimize both overproduction and underproduction [7].

8) **Energy Integration for Flexible Supply:** Hydrogen production is heavily reliant on renewable energy sources, which can be intermittent. A JIT approach must account for these fluctuations in energy availability to ensure that hydrogen production can respond to real-time demand. Create hybrid systems where surplus renewable energy can be stored (e.g., in batteries or other energy storage systems) and used for hydrogen production when renewable energy is available in abundance. This way, hydrogen production can scale according to demand, even if renewable energy availability fluctuates.
9) **Localized Production and Distribution Networks:** Instead of centralized hydrogen production, a JIT system might involve decentralized, localized production units closer to high-demand regions. This reduces the need for long-distance transportation and optimizes storage space at central locations. Establish smaller-scale hydrogen production units located near industrial hubs, transport networks, or urban centers with high hydrogen demand. This ensures a faster response to local demand fluctuations, reduces transportation costs, and minimizes storage needs.
10) **Minimal Storage with Backup Buffer:** While JIT minimizes storage, having a small buffer or safety stock is often necessary to handle sudden demand spikes or supply chain disruptions. Establish buffer storage with a minimal amount of hydrogen at distribution points to handle unexpected demand surges. This buffer should be carefully calculated based on historical demand variability and potential delays in production or transportation [8].
11) **Contingency Plans for Supply Chain Disruptions:** JIT requires robustness against supply chain disruptions. Developing contingency plans for unexpected events (such as natural disasters, technical failures, or supply shortages) is critical for minimizing the risk of inventory shortages. Work with suppliers and distributors to create plans that can quickly adjust to disruptions. For example, have backup suppliers for electrolyzers or transport options, or set up agreements for emergency hydrogen storage to meet critical demand during a shortfall.

By applying these JIT inventory management strategies in green hydrogen distribution, companies can reduce storage costs, minimize waste, and improve supply chain responsiveness while ensuring hydrogen is available where and when it is needed [9].

> ***Key: Total quality management (TQM):*** *A holistic approach to long-term success through customer satisfaction, involving all members of an organization in improving processes and products, including hydrogen purity and safety.*

### 6.2.3 Inventory Tracking and Monitoring Systems for Green Hydrogen Storage Facilities

Effective inventory tracking and monitoring systems for green hydrogen storage facilities are crucial to ensure efficient storage, safety, and availability of hydrogen. Given hydrogen's unique characteristics and the complexities of its storage (e.g., high-pressure gas and cryogenic conditions), robust systems are needed to ensure proper management and avoid overstocking, understocking, and safety risks. Following are some of the key inventory tracking and monitoring systems that can be used in green hydrogen storage facilities:

1) **Real-Time IMSs**
   - **Concept**: A real-time IMS allows for the continuous tracking of hydrogen levels, production, and consumption. This system can automate updates on inventory levels, enabling better decision-making and proactive management of the storage facility.
   - **Features**:
     - Live data on hydrogen storage tank levels (both pressure and volume).
     - Automated notifications when inventory reaches predefined thresholds (e.g., minimum and maximum storage levels).
     - Integration with production systems to align hydrogen production and consumption.
   - **Implementation**: Use IoT-enabled sensors to monitor the hydrogen levels in storage tanks, whether they are high-pressure or cryogenic. These sensors can feed data into the IMS, allowing facility operators to track inventory in real-time [10].
2) **IoT Sensors and Devices**
   - **Concept**: IoT devices, including pressure sensors, temperature sensors, flow meters, and leak detectors, are essential for monitoring hydrogen in storage facilities, given its volatile nature.
   - **Features**:
     - Sensors that monitor pressure, temperature, and flow rates in hydrogen storage tanks.
     - Leak detection to ensure safety and prevent loss of hydrogen.
     - Remote monitoring capabilities for continuous data access.

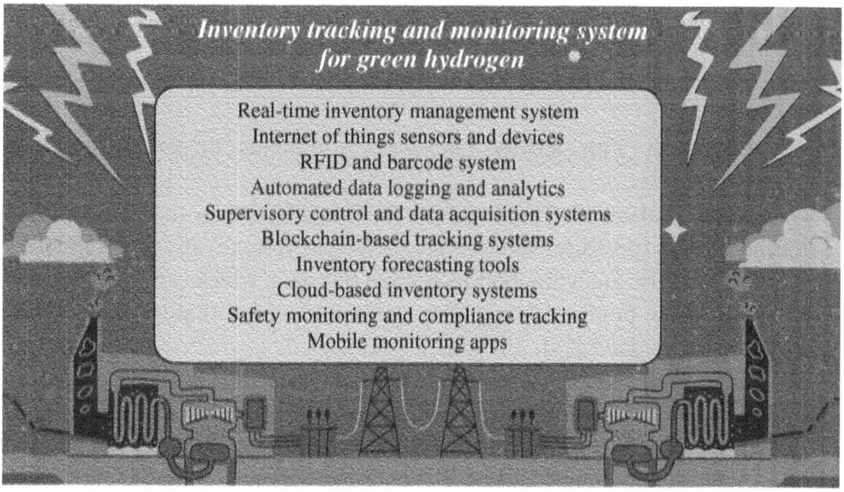


**Figure 6.3** Inventory Tracking and Monitoring System

- **Implementation**: Deploy IoT-based sensors to track critical parameters like tank pressure, ambient temperature, and hydrogen purity. These sensors can be integrated with a central system for real-time data collection and alerting if any anomalies occur, improving both safety and inventory tracking. Figure 6.3 illustrates Inventory Tracking and Monitoring System [11].

3) **Radio Frequency Identification (RFID) and Barcode Systems**
   - **Concept**: RFID and barcode systems can be used to track the movement and status of hydrogen storage containers, valves, or other associated equipment.
   - **Features**:
     - RFID tags or barcodes attached to storage containers and key components.
     - Scanning capabilities at key entry and exit points within the storage facility.
     - Automated data logging and integration with inventory management software.
   - **Implementation**: Use RFID tags on storage cylinders, tanks, or delivery trucks to automatically log their movements and storage status within the facility. Barcodes can also be used on maintenance equipment or spare parts, ensuring they are accounted for during inventory audits and reducing the chances of loss or misplacement [12].
4) **Automated Data Logging and Analytics**
   - **Concept**: Automated data logging systems collect real-time data and use analytics to provide insights into hydrogen usage, storage efficiency, and potential issues. This system can also forecast future hydrogen needs.

- **Features**:
  - o Automated data collection from various sensors and systems.
  - o Predictive analytics to forecast future inventory needs and identify trends.
  - o Integration with facility management systems to ensure seamless operation.
- **Implementation**: Set up an automated logging system that collects data from sensors (pressure, temperature, hydrogen volume) and environmental conditions. Use analytics tools to analyze this data, which can help optimize storage levels, predict maintenance needs, and ensure that inventory levels align with demand [13].

5) **Advanced Supervisory Control and Data Acquisition (SCADA) Systems**
   - **Concept**: SCADA systems provide centralized control and monitoring of processes in real-time. For hydrogen storage facilities, SCADA can be used to monitor equipment, tank levels, and production in a highly automated manner.
   - **Features**:
     - o Integration with IoT sensors for real-time monitoring of hydrogen storage.
     - o Visualization tools that show hydrogen levels, pressure, and temperature data.
     - o Alarm systems that notify operators of critical conditions (e.g., leaks, overpressure, and low stock).
   - **Implementation**: Deploy a SCADA system that integrates data from hydrogen storage tanks and production units. The SCADA system can provide real-time updates on inventory levels, track usage patterns, and alert operators to any irregularities.

6) **Blockchain-Based Tracking Systems**
   - **Concept**: Blockchain can be used for transparent and immutable tracking of hydrogen from production to storage and distribution. This ensures traceability, accountability, and enhanced security.
   - **Features**:
     - o Immutable records of hydrogen transactions, from production to distribution.
     - o Secure data logging, making the tracking process transparent and tamper-proof.
     - o Automated updates for stakeholders across the supply chain.
   - **Implementation**: Implement blockchain to track each unit of hydrogen from production through storage and into its final distribution. The blockchain system can be accessed by all stakeholders (producers, storage facility managers, and end-users) to verify hydrogen's authenticity and ensure it has been properly handled [14].

7) **Inventory Forecasting Tools**
   - **Concept**: Inventory forecasting tools use historical data and demand predictions to estimate future hydrogen requirements, enabling better planning and stock management.
   - **Features**:
     - Demand forecasting algorithms based on production rates, market trends, and seasonal fluctuations.
     - Integration with IMS to adjust production and storage strategies.
     - Visualization tools to forecast inventory levels and make informed decisions.
   - **Implementation**: Use advanced forecasting models that incorporate historical demand data, production schedules, and other factors influencing hydrogen consumption. These tools help storage facilities plan for changes in demand, minimizing overstocking or shortages [15].
8) **Cloud-Based Inventory Systems**
   - **Concept**: Cloud-based systems allow for remote and centralized management of hydrogen inventory across multiple storage facilities. This improves scalability and accessibility, especially for large or multi-location operations.
   - **Features**:
     - Cloud storage for real-time data from multiple storage locations.
     - Remote access and monitoring for facility managers and decision-makers.
     - Integration with other systems (e.g., production and logistics) for seamless operation.
   - **Implementation**: Implement a cloud-based IMS that connects all storage sites, providing a central hub for tracking inventory levels, safety conditions, and real-time production and consumption data [16].
9) **Safety Monitoring and Compliance Tracking Systems**
   - **Concept**: Hydrogen storage facilities are subject to strict safety and regulatory requirements. Monitoring and compliance systems ensure that safety standards are met while tracking hydrogen inventory.
   - **Features**:
     - Real-time monitoring of safety parameters (e.g., pressure limits, leak detection, and temperature).
     - Compliance tracking to ensure adherence to local regulations and standards.
     - Automated reporting tools to generate compliance documentation.
   - **Implementation**: Integrate safety and compliance sensors with the inventory monitoring systems. Automated alerts can notify operators of any deviations from safety standards, and reports can be generated to demonstrate compliance with regulations (e.g., local safety standards and environmental regulations) [17].

10) **Mobile Monitoring Apps**
    - **Concept**: Mobile apps enable on-the-go monitoring of hydrogen inventory and storage conditions, providing flexibility and quick response times for storage facility managers.
    - **Features**:
        - Mobile access to real-time data from sensors and control systems.
        - Alert notifications for potential issues (e.g., tank overfill and pressure fluctuations).
        - Remote access for on-site or off-site managers.
    - **Implementation**: Deploy mobile applications that allow operators and managers to access real-time data on hydrogen inventory, tank conditions, and safety alerts. This provides flexibility in managing the facility and allows for quicker decision-making.

By implementing these tracking and monitoring systems, green hydrogen storage facilities can ensure efficient inventory management, maintain safety standards, and optimize storage utilization. These systems also provide real-time data, enabling proactive management and informed decision-making throughout the hydrogen supply chain [18, 19].

Integrating IMS with supply chain operations is essential for the efficient production, storage, and distribution of green hydrogen. Green hydrogen poses unique challenges due to its dependence on renewable energy sources, fluctuating demand, and complexities in transportation and storage. Seamless integration of IMS across the supply chain can significantly enhance operational efficiency, reduce costs, and minimize risks. Real-time data exchange is a foundational element, allowing all stakeholders, producers, storage facilities, and customers to share information instantly through centralized, cloud-based platforms. IoT devices play a critical role in collecting data on production output, storage conditions, and demand indicators, enabling automated updates and alerts that help synchronize production schedules with actual demand.

Accurate demand forecasting is crucial and can be achieved by integrating IMS with predictive analytics and machine learning tools that consider historical trends, weather conditions, and customer requirements. This synchronization allows dynamic adjustment of production and inventory plans, minimizing overstock or understock scenarios. Optimized production and distribution scheduling further supports this efficiency by leveraging IMS integration with SCM software to align hydrogen production with renewable energy availability and optimize transport routes based on real-time conditions. Collaboration and data sharing across stakeholders—ranging from renewable energy suppliers to end-users—are vital for improving visibility and coordination, ensuring that inventory levels, lead times, and production capacities are known and balanced throughout the chain.

Enhanced inventory visibility and control can be achieved through barcode/RFID systems and IoT sensors that monitor hydrogen units across the supply chain, including their condition and location. Centralized dashboards offer real-time insights that support better production and logistics planning. A lean inventory approach, supported by JIT principles, reduces unnecessary stockpiling, cuts waste, and ensures that hydrogen is produced and delivered based on immediate consumption needs. IMS integration also strengthens supply chain risk management by enabling real-time alerts, contingency planning, and simulation of potential disruptions to develop backup strategies and buffers.

Blockchain technology adds a layer of transparency and security by creating immutable records of inventory movements and transactions. Smart contracts can automate actions such as releasing inventory based on predefined triggers, thereby improving accountability and trust among stakeholders. Automated replenishment systems integrated with IMS can maintain optimal inventory levels by initiating orders automatically when stocks fall below certain thresholds, while analytics fine-tune reorder points to ensure continuous supply without overstocking. Finally, performance and key performance indicator (KPI) tracking across the integrated system allows real-time monitoring of order fulfilment, lead times, production efficiency, and inventory turnover, identifying areas for continuous improvement.

The integration of IMSs into the green hydrogen supply chain is instrumental in building a sustainable, efficient, and responsive ecosystem. It enables real-time responsiveness, demand alignment, cost-effectiveness, and risk mitigation, supporting the broader goal of developing a resilient and environmentally sound hydrogen economy.

> ***Key: Quality assurance (QA):*** *A systematic process of ensuring that green hydrogen meets defined quality standards across its production and supply chain.*

## 6.3 Quality Control Measures and Testing and Inspection of Green Hydrogen

As the world transitions toward sustainable energy solutions, green hydrogen has emerged as a promising alternative to fossil fuels. Produced through electrolysis using renewable energy sources, green hydrogen offers a zero-emission energy solution for industries such as transportation, power generation, and manufacturing. However, ensuring its purity, safety, and efficiency requires the implementation of rigorous quality control measures throughout the production, storage, and distribution processes.

Unlike conventional fuels, hydrogen quality is highly sensitive to impurities such as moisture, oxygen, nitrogen, and carbon-based compounds, which can compromise fuel cell performance and industrial applications. Additionally, storage conditions, transportation methods, and regulatory compliance play crucial roles in maintaining the integrity of green hydrogen. Therefore, advanced quality monitoring techniques such as gas chromatography, spectroscopy, and real-time sensor-based analysis are essential for maintaining high-purity standards.

Effective quality control measures involve a combination of standardized testing protocols, real-time monitoring systems, and adherence to global certification frameworks such as ISO 14687 and SAE J2719, which specify hydrogen purity levels for fuel cell applications. Furthermore, TQM principles, process automation, and AI-driven analytics are increasingly being integrated into quality assurance (QA) processes to enhance reliability and minimize contamination risks.

Quality standards and specifications for green hydrogen production are essential to ensure the purity, safety, and efficiency of hydrogen produced via renewable energy sources. These standards ensure that green hydrogen meets the required parameters for various industrial, commercial, and transportation applications. Table 6.1 illustrates key quality standards and specifications for green hydrogen production

**Table 6.1** Key Quality Standards and Specifications for Green Hydrogen Production

| Parameter | Specification/ Standard | Description | Relevance/Application |
|---|---|---|---|
| Purity ($H_2$ concentration) | ≥99.97% (ISO 14687:2019) | Minimum hydrogen content required for fuel cell applications (Grade D) | Critical for mobility (e.g., fuel cells in vehicles) |
| Oxygen ($O_2$) | ≤5 ppm (ISO 14687:2019) | High $O_2$ content can lead to poor combustion or corrosion. | Affects fuel cell performance and safety |
| Moisture ($H_2O$) | ≤5 ppm (ISO 14687:2019) | Water vapor can degrade fuel cell membranes. | Ensures durability of fuel cell systems |
| Total hydrocarbons | ≤2 ppm (ISO 14687:2019) | Includes methane and other hydrocarbons that can poison catalysts | Impacts fuel cell efficiency |
| Carbon monoxide (CO) | ≤0.2 ppm (ISO 14687:2019) | Poisonous to fuel cell catalysts | Important for PEM fuel cell operation |

**Table 6.1** (Continued)

| Parameter | Specification/ Standard | Description | Relevance/Application |
|---|---|---|---|
| Carbon dioxide ($CO_2$) | ≤2 ppm (ISO 14687:2019) | Can affect the overall combustion process or fuel quality | Necessary for sensitive combustion and electrochemical uses |
| Sulfur compounds | ≤0.004 ppm (ISO 14687:2019) | Sulfur severely poisons catalysts | Ensures catalyst longevity |
| Particulates | No detectable particles >5 µm | Avoid clogging or abrasion in fuel cells | Maintains system integrity |
| Ammonia ($NH_3$) | ≤0.1 ppm | Common in some production methods like electrolysis with nitrogen-based feedstocks | Impacts fuel cell durability |
| Hydrogen production source | Renewable (e.g., solar and wind)—verified via LCA or certification (e.g., CertifHy and ISO 14064) | Production must ensure zero- or low-carbon emissions | Ensures green hydrogen classification |
| Carbon intensity | ≤1 kg $CO_2$e/kg $H_2$ (EU Standard), or as per regional regulations | Defines threshold to qualify as "green" hydrogen | Regulatory compliance and sustainability claims |
| Energy source certification | Renewable Energy Guarantees of Origin (REGO), I-REC, or similar | Certifies that electricity used comes from renewable sources | Transparency and traceability of green hydrogen production |
| Production efficiency | Electrolyzer efficiency ≥65–70% | Ratio of energy input to hydrogen energy output | Operational performance metric |
| Water quality for electrolysis | Deionized or distilled water meeting ASTM D1193 Type II or better | Prevents scale and damage to electrolyzers | Essential for safe and efficient electrolysis |
| Storage pressure | 350 bar (Type III), 700 bar (Type IV) | Standard pressures for hydrogen storage tanks | Depends on end-use (vehicles, industry, etc.) |
| Standard reference | ISO 14687:2019, SAE J2719, EU Certify, ASTM D7651, IEC 62282 | International standards for hydrogen fuel quality | Widely accepted across global hydrogen value chains |

Green hydrogen production must meet stringent quality standards and specifications to ensure its safe, efficient, and effective use across applications such as fuel cells, industrial processes, and energy storage. One of the most critical parameters is hydrogen purity. For fuel cell-grade applications, a purity of ≥99.999% (5.0 grade) is essential, while general industrial uses may accept purities between 99% and 99.99%. High purity ensures optimal energy conversion and prevents degradation of sensitive equipment. Moisture content is another vital factor—hydrogen must be nearly moisture-free, with a typical maximum of ≤1 ppm for high-purity applications, to prevent corrosion and maintain the integrity of storage and fuel cell systems. Similarly, oxygen content must be tightly controlled (≤0.1%) to avoid performance reduction and safety risks. Trace impurities such as carbon monoxide ($CO \leq 0.1$ ppm) and carbon dioxide ($CO_2 \leq 2$ ppm) are harmful to fuel cell membranes, even in minimal concentrations, while hydrocarbons (≤0.1 ppm) can contaminate or interfere with hydrogen-based systems.

Sulfur compounds, known for their corrosive and catalyst-poisoning effects, must be minimized to levels at or below 0.1 ppm. Particle contamination can lead to system blockages and reduced efficiency; hence, hydrogen must be filtered to remove particles larger than 1 μm and maintain particulate matter below 0.1 mg/m$^3$. Additionally, the thermodynamic properties of hydrogen—such as temperature and pressure—must align with specific application requirements. Typically, hydrogen is stored at high pressures (350–700 bar), and temperatures range around 15–30 °C, depending on use cases. Since hydrogen is odorless and colorless, some systems may include a trace scenting agent like mercaptan under controlled conditions to aid in leak detection, enhancing safety. Table 6.2

**Table 6.2** Certification and Compliance Requirements for Green Hydrogen Quality Control

| Standard/ Body | Standard Name | Overview | Key Requirements |
|---|---|---|---|
| ISO | ISO 14687:2019—Hydrogen Fuel Quality | Specifies hydrogen fuel quality for fuel cells in transport, stationary, and portable uses. | • Purity ≥99.999%<br>• Impurity limits ($CO$, $CH_4$, $NH_3$, sulfur compounds, moisture)<br>• Regular testing procedures |
| | ISO 9001:2015—QMS | Outlines requirements for Quality Management Systems | • Process control Documentation and traceability<br>• Continuous improvement |

**Table 6.2** (Continued)

| Standard/ Body | Standard Name | Overview | Key Requirements |
|---|---|---|---|
| ASTM | ASTM D7501-21 | Defines hydrogen purity and testing methods for fuel cell use | • Purity >99.999%<br>• Impurity limits (CO, $CO_2$, $N_2$, $O_2$)<br>• GC and MS testing methods |
| European Union | EU Hydrogen Strategy & RED II | Focus on sustainability and green hydrogen certification | • Renewable origin certification<br>• Carbon footprint <2 kg $CO_2$/kg $H_2$<br>• LCA and carbon intensity compliance |
| | REPowerEU & European Clean Hydrogen Alliance | Initiatives to scale green hydrogen production in the EU | • Certification of origin<br>• Supply chain transparency<br>• Carbon footprint reporting |
| U.S. DOE & NIST | DOE Hydrogen Program & NIST Standards | Promote hydrogen tech development and quality control | • Hydrogen purity standards<br>• Gas quality parameters (pressure, temperature)<br>• Safety protocols for production, storage, transportation |
| EPA & OSHA (USA) | EPA Environmental Standards | Regulates emissions and waste in hydrogen production | • $CO_2$, $NO_x$ emissions limits<br>• Water quality in electrolysis |
| | OSHA Safety Standards | Workplace safety in hydrogen handling | • Leak detection systems<br>• Handling and safety protocols |
| IEC | IEC 62282—Fuel Cell Technologies | Standards for fuel cell system design and safety | • Safe system design and operation<br>• Storage, distribution under specified P-T conditions |
| ISO | ISO 22734—Electrolysis of Water | Guidelines for electrolyzer system operation | • Hydrogen purity and impurity thresholds<br>• Electrical, thermal, pressure safety standards |
| Japan | National Hydrogen & Fuel Cell Standards | Japan's hydrogen production and fuel cell regulations | • Hydrogen purity compliance<br>• Renewable-based production must meet environmental sustainability norms |

illustrates certification and compliance requirements for green hydrogen quality control.

Furthermore, green hydrogen must be produced using 100% renewable energy sources such as solar, wind, or hydropower. Electrolysis systems should target an energy consumption of 50–55 kWh per kg of hydrogen to ensure sustainable and efficient operations. Storage and handling of hydrogen require compliance with strict safety protocols, including the use of pressure-tested storage tanks and adherence to international standards like ISO 14687-2. Regulatory compliance is also crucial—green hydrogen producers must be certified by relevant bodies and conform to ISO and national standards to ensure product quality, safety, and environmental responsibility. Finally, the carbon intensity of production must remain minimal, ideally close to zero or below 2 kg $CO_2$-equivalent per kg of hydrogen, to qualify as truly "green."

Meeting these standards is essential for the successful integration of green hydrogen into energy, transport, and industrial systems. Adherence guarantees not only performance and safety but also environmental sustainability supporting global efforts toward carbon neutrality and a cleaner energy future. Figure 6.4 illustrates a schematic diagram of gas purity testing. Figure 6.5 illustrates oxygen, carbon dioxide, and carbon monoxide detection of green hydrogen.

Testing and inspection protocols are crucial for ensuring the purity and safety of green hydrogen throughout its production, storage, and distribution phases. These protocols help verify that the hydrogen meets required standards for its intended applications, particularly for fuel cells, industrial use, and transportation. The key testing and inspection protocols for ensuring the purity and safety of green hydrogen are given in Tables 6.3–6.10.

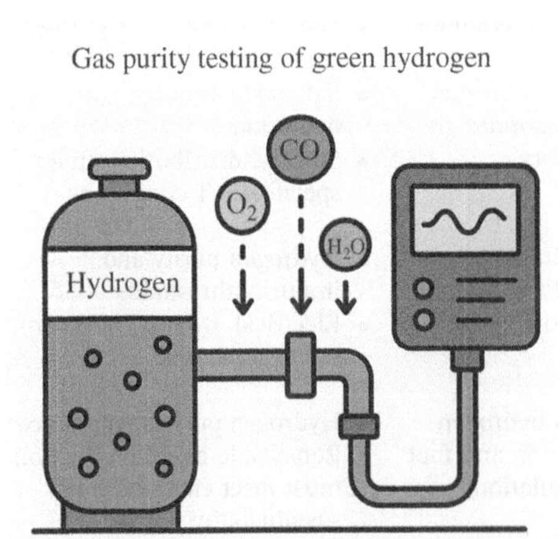


**Figure 6.4** Schematic Diagram of Gas Purity Testing

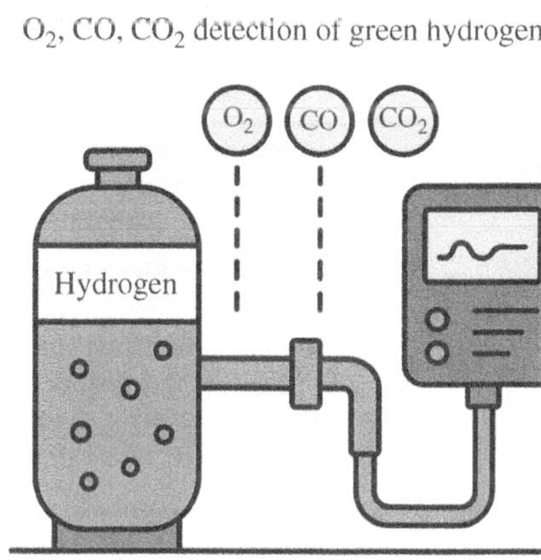


**Figure 6.5** Oxygen, Carbon Dioxide, and Carbon Monoxide Detection of Green Hydrogen

**Table 6.3** Gas Purity Testing

| **Objective** | **To determine the overall purity of hydrogen and detect impurities like $O_2$, $N_2$, CO, $CO_2$, $CH_4$, etc.** |
|---|---|
| Apparatus/ equipment | • Gas chromatograph (GC) with thermal conductivity detector (TCD) or flame ionization detector (FID)<br>• Calibration gas standards<br>• Sampling cylinder |
| Procedure | • Collect hydrogen gas sample using a clean, evacuated gas sampling cylinder.<br>• Inject the sample into the GC.<br>• Run the chromatogram using calibrated gas standards.<br>• Record and analyze the concentration of hydrogen and other gases. |
| Precautions | • Ensure no contamination during sampling.<br>• Calibration of GC is crucial before each test. |

Implementing robust testing and inspection protocols is essential for ensuring the purity, safety, and efficiency of green hydrogen production, storage, and distribution systems. These protocols help maintain hydrogen quality and safeguard against potential hazards that could affect both production and end-use applications, particularly in fuel cells and other sensitive systems. Regular monitoring and compliance with industry standards ensure that green hydrogen is safe, sustainable, and reliable for various industrial, transportation, and energy applications.

**Table 6.4** Water Content (Moisture) Testing

| **Objective** | **To measure the moisture ($H_2O$) content in hydrogen gas.** |
|---|---|
| Apparatus/equipment | • Hygrometer or dew point analyzer (chilled mirror or aluminum oxide sensor)<br>• Moisture analyzer (e.g., $P_2O_5$ sensor) |
| Procedure | • Connect the hydrogen gas line to the dew point analyzer.<br>• Allow the system to stabilize and reach steady flow.<br>• Record the dew point and convert to ppm using calibration charts or built-in conversion.<br>• Alternatively, use the hygrometer to measure moisture content directly. |
| Precautions | • Avoid exposure to ambient air.<br>• System should be moisture-free before testing. |

**Table 6.5** Oxygen Content Testing

| **Objective** | **To determine trace oxygen concentration in hydrogen gas.** |
|---|---|
| Apparatus/equipment | • Electrochemical oxygen analyzer<br>• Paramagnetic oxygen sensor<br>• Gas sampling line |
| Procedure | • Purge the sampling line to eliminate residual gas.<br>• Introduce hydrogen sample to the oxygen analyzer.<br>• Allow sensor to stabilize and record the oxygen concentration (in ppm). |
| Precautions | • Calibration is essential using known $O_2$ gas mixtures.<br>• Avoid backflow to prevent contamination. |

**Table 6.6** Carbon Monoxide and Carbon Dioxide Detection

| **Objective** | **To detect and quantify CO and $CO_2$ levels in hydrogen gas.** |
|---|---|
| Apparatus/equipment | • Nondispersive Infrared (NDIR) analyzer for CO and $CO_2$<br>• Gas chromatograph (alternative method) |
| Procedure | • Introduce hydrogen sample to the NDIR analyzer.<br>• Allow instrument to stabilize and analyze gas flow.<br>• Record the CO and $CO_2$ concentrations.<br>• Confirm results with GC if high precision is required. |
| Precautions | • NDIR analyzers are specific; check for cross-sensitivity.<br>• Avoid moisture in sample as it affects NDIR sensors. |

**Table 6.7** Sulphur and Nitrogen Compound Testing

| **Objective** | **To measure trace levels of sulfur (e.g., $H_2S$ and $SO_2$) and nitrogen compounds (e.g., $NH_3$ and $NO_x$).** |
|---|---|
| Apparatus/ equipment | • Gas chromatograph with sulfur chemiluminescence detector (GC-SCD)<br>• Chemiluminescence analyzer for $NO_X$<br>• Electrochemical sensors |
| Procedure | • Pass hydrogen through filters and traps to remove moisture and particulates.<br>• Inject sample into GC-SCD or specific analyzers.<br>• Identify and quantify the presence of sulfur or nitrogen compounds. |
| Precautions | • Avoid contamination from tubing and valves.<br>• Calibration with trace-level gas mixtures is required. |

**Table 6.8** Particle Contamination Testing

| **Objective** | **To detect solid particulate matter in hydrogen gas.** |
|---|---|
| Apparatus/ equipment | • Particle counter or optical particle analyzer<br>• Filter membrane unit (0.2–1 μm)<br>• Microscope or scanning electron microscope (SEM) |
| Procedure | • Pass a known volume of hydrogen gas through a filter membrane.<br>• Remove and analyze the filter using a microscope.<br>• Count and categorize particles by size and concentration.<br>• For advanced analysis, use SEM for morphology and composition. |
| Precautions | • System must be particle-free before testing.<br>• Ensure consistent gas flow rate. |

**Table 6.9** Temperature and Pressure Monitoring

| **Objective** | **To measure and monitor storage or line temperature and pressure for safety and performance.** |
|---|---|
| Apparatus/ equipment | • Pressure transducers (rated up to 700 bar)<br>• Digital thermocouples or RTD sensors<br>• Data logger system |
| Procedure | • Install sensors at designated points (inlet, outlet, storage).<br>• Calibrate sensors as per manufacturer recommendations.<br>• Continuously monitor and record values.<br>• Use alarms for deviation from specified ranges (e.g., 15–30 °C and 350–700 bar). |
| Precautions | • Sensors should be hydrogen-compatible and explosion-proof.<br>• Check for thermal lag in readings. |

**Table 6.10** Hydrogen Leak Detection

| **Objective** | **To identify hydrogen leaks for safety compliance and accident prevention.** |
|---|---|
| Apparatus/equipment | • Portable hydrogen gas detector (electrochemical or TCD-based)<br>• Soap solution (basic method)<br>• Thermal imaging for pressurized leaks |
| Procedure | • Survey valves, joints, pipelines, and fittings using a hydrogen leak detector.<br>• Move slowly and check for audible/visual alarms.<br>• For precise areas, apply soap solution and observe for bubbles.<br>• Document any detected leaks and tag the location for repair. |
| Precautions | • Use in well-ventilated areas.<br>• Always test detectors in known hydrogen environments before deployment. |

> ***Key: Risk management in inventory**: Identifying, assessing, and mitigating potential hazards in hydrogen storage, handling, and delivery to ensure operational safety and compliance.*

### 6.3.1 Quality Assurance Measures Throughout the Hydrogen Production Process

QA measures throughout the hydrogen production process are essential to ensure that the produced hydrogen meets industry standards, is free from contaminants, and is safe for use in fuel cells, transportation, and other applications. The QA process spans the entire production lifecycle, from feedstock preparation to hydrogen storage and delivery. Following are the key QA measures that should be implemented throughout the hydrogen production process:

**Feedstock Quality Control:** The quality of feedstock is the foundation of reliable hydrogen production. Water used for electrolysis must be highly purified, free from salts, minerals, and organic compounds that could reduce electrolyzer efficiency or compromise hydrogen purity. In processes such as SMR, the natural gas feedstock requires strict inspection to ensure it contains minimal sulfur and other harmful contaminants that could impair catalyst performance. To meet these requirements, pretreatment protocols including filtration, purification, and conditioning systems are implemented, thereby eliminating undesirable substances before they enter the production system. This ensures consistency in hydrogen generation and minimizes the risks associated with impurity-induced inefficiencies.

**Process Control and Monitoring:** Once feedstock quality is secured, effective process control and monitoring guarantee operational stability. Hydrogen production requires precise management of parameters such as temperature, pressure, flow rate, and electrical current. Continuous real-time monitoring of these factors ensures that the system remains within defined operating conditions. Advanced automation and control systems further enhance accuracy by adjusting variables dynamically, minimizing the potential for human error. Standardized operating procedures (SOPs) provide an additional layer of consistency, ensuring batch-to-batch uniformity in hydrogen production and preventing fluctuations that could impact final purity levels.

**Purity Testing During Production:** Maintaining high hydrogen purity is one of the most critical aspects of QA, especially since applications such as fuel cells demand hydrogen that exceeds 99.99% purity. Continuous monitoring tools such as gas chromatographs and mass spectrometers are employed to deliver real-time data on hydrogen composition, enabling immediate detection of impurities. Complementing these technologies, regular sampling and laboratory analyses provide detailed insights into contaminants such as carbon monoxide, carbon dioxide, methane, nitrogen, sulfur compounds, and moisture. To uphold stringent purity standards, advanced purification systems—including PSA, membrane separation, and cryogenic distillation—are integrated into the production line, effectively removing contaminants and ensuring compliance with application-specific requirements.

**System Integrity and Safety:** The integrity and safety of production systems are central to reliable hydrogen generation. Routine inspections of equipment such as electrolyzers, reactors, and pipelines help detect early signs of wear and minimize the risks of failure. Hydrogen-specific leak detection systems and gas sensors are employed to identify leaks at the earliest stage, preventing hazards associated with hydrogen's flammability. In addition, safety mechanisms such as pressure relief valves undergo frequent testing to ensure they function properly under high-pressure conditions, thus safeguarding both personnel and infrastructure.

**Energy Efficiency and Performance Testing:** Given the high energy demand of hydrogen production, energy efficiency is a vital component of QA. Monitoring systems track energy consumption during production, benchmarking performance against expected standards to identify inefficiencies. Optimization algorithms adjust parameters such as current, voltage, and flow rates in electrolyzers or temperature and pressure in reformers to achieve maximum efficiency without compromising hydrogen quality. Regular performance testing of equipment further ensures that degradation or system inefficiencies are quickly identified and corrected, reducing operational costs and enhancing sustainability.

**Hydrogen Storage and Transportation Safety:** Hydrogen storage and transportation present additional challenges that require rigorous QA practices. Storage tanks are inspected regularly for corrosion, leaks, and material degradation and are often made of robust materials such as advanced composites or steel designed to withstand high pressures. Temperature and pressure monitoring systems are employed to maintain safe and stable storage conditions. During transportation by pipelines, tankers, or cylinders, QA protocols ensure that hydrogen containers are designed to prevent contamination and maintain purity levels, guaranteeing that quality is preserved from the production site to the end-user.

**End-Product Quality Control and Certification:** Before distribution, hydrogen undergoes final quality checks to ensure compliance with end-use specifications. Applications such as automotive fuel cells require ultra-high-purity hydrogen (greater than 99.999%), which is verified through final certification tests. In addition, QA frameworks ensure regulatory compliance with international standards such as ISO 14687 for fuel cells and ASTM D7501 for industrial use. Comprehensive documentation and traceability systems record testing results, certifications, and production histories, providing transparency, accountability, and regulatory alignment.

**QA for By-products:** Hydrogen production often yields by-products such as oxygen from electrolysis or carbon dioxide from SMR. QA protocols extend to these outputs to ensure they meet environmental or industrial quality standards for reuse or safe disposal. By-product purity testing, combined with continuous environmental monitoring, ensures that emissions remain within regulatory limits and that the environmental footprint of production is minimized.

**Continuous Improvement and Feedback Loop:** A robust QA system is not static but evolves through continuous improvement mechanisms. Root cause analysis (RCA) of detected quality issues enables producers to implement corrective measures that prevent recurrence. Regular employee training ensures that production teams remain updated on the latest QA methodologies, safety practices, and regulatory requirements. Feedback systems involving downstream users, such as fuel cell manufacturers, provide real-world performance insights, guiding further refinements to quality and efficiency.

**Third-Party Audits and Inspections:** Finally, independent verification through third-party audits strengthens credibility and trust in hydrogen production. External auditors assess production systems for compliance with international standards and highlight areas for improvement. Obtaining certifications from globally recognized organizations such as ISO and Technischer Überwachungs-Verein (TÜV) not only confirms adherence to quality requirements but also enhances customer confidence, supporting the wider adoption of hydrogen in clean energy markets.

Implementing comprehensive QA measures throughout the hydrogen production process ensures that the produced hydrogen meets the highest standards of purity, safety, and efficiency. From feedstock quality control to final product testing, each step must be carefully monitored to avoid contaminants, minimize waste, and optimize performance. Continuous improvement and adherence to international standards are crucial to maintaining quality and meeting the growing demand for green hydrogen across various industries.

## 6.4 Risk Management in Inventory Control of Green Hydrogen

As the global energy landscape shifts toward sustainable alternatives, green hydrogen has emerged as a key player in the transition to a low-carbon economy. Produced through electrolysis powered by renewable energy sources, green hydrogen offers a clean and efficient solution for various industries, including transportation, power generation, and industrial manufacturing. However, its widespread adoption comes with significant challenges, particularly in inventory control and risk management.

Risk management in inventory control is crucial for ensuring the safe, cost-effective, and reliable supply of green hydrogen. Unlike conventional fuels, hydrogen presents unique risks related to storage, transportation, demand fluctuations, and regulatory compliance. Key risks include high-pressure storage hazards, potential leakage, perishability concerns, and supply chain disruptions due to the reliance on intermittent renewable energy sources.

To mitigate these risks, companies must adopt advanced inventory control strategies that incorporate real-time monitoring, predictive analytics, blockchain-based tracking, and AI-driven demand forecasting. Additionally, adherence to global safety standards such as ISO 19880-1 for hydrogen fueling stations and stringent regulatory frameworks is essential to minimize operational risks and ensure supply chain efficiency.

### 6.4.1 Identification and Mitigation of Risks Associated with Green Hydrogen Inventory Management

In green hydrogen inventory management, identifying and assessing risks involves recognizing potential hazards and uncertainties that could affect the supply, storage, and distribution of hydrogen. Key risks include:

- **Production Risks**: Disruptions in renewable energy sources or electrolyzer failures could impact hydrogen production.

- **Storage Risks**: Green hydrogen is typically stored at high pressures or low temperatures, which poses risks such as leakage, explosions, or equipment failures.
- **Transportation Risks**: The transportation of hydrogen via pipelines, tankers, or cylinders can face risks like accidents, leaks, or contamination.
- **Supply Chain Disruptions**: Fluctuations in demand, unexpected supply chain delays, or geopolitical factors could disrupt inventory availability.
- **Quality Control Risks**: Impurities in hydrogen or deviations in production quality may affect its use in fuel cells or industrial processes.
- **Environmental Risks**: The production and storage of hydrogen have environmental impacts, such as carbon footprint concerns in non-green hydrogen production or hazardous material handling in production and storage. Figure 6.6 illustrates risks associated with green hydrogen inventory management.

Assessing these risks involves evaluating their probability, potential impact, and the vulnerability of the system to such disruptions. Tools like risk matrices, fault tree analysis, and failure mode effects analysis (FMEA) can help identify and assess risks. Mitigating risks in green hydrogen storage and handling involves implementing strategies that reduce the likelihood of hazards and minimize their impact. Key strategies include:

**Safety Systems**: Installation of advanced leak detection and monitoring systems in storage facilities to promptly detect leaks and prevent accidents.

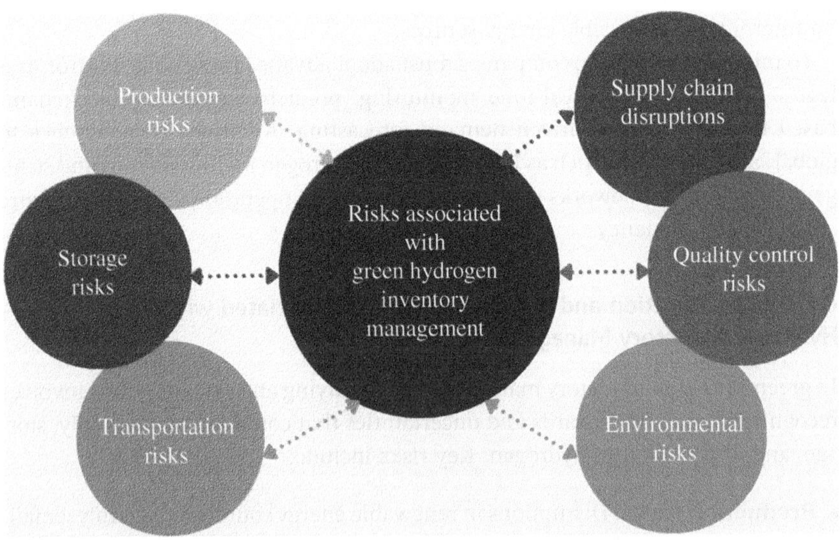


**Figure 6.6** Risks Associated with Green Hydrogen Inventory Management

Automated shutdown systems can stop the flow of hydrogen in case of detected anomalies.

**Robust Infrastructure**: Designing storage tanks, pipelines, and valves to withstand high pressures or cryogenic temperatures. Utilizing materials resistant to hydrogen embrittlement and corrosion.

**Controlled Environment**: Maintaining storage facilities at optimal temperatures and pressures. For example, hydrogen storage in cryogenic conditions ($LH_2$) or at high pressure in gas form requires precise temperature and pressure controls.

**Training and Safety Protocols**: Ensuring all personnel are trained on proper hydrogen handling, emergency response procedures, and the use of safety equipment (e.g., fire suppression systems and personal protective equipment).

**Regular Maintenance and Inspection**: Scheduling regular inspections of storage equipment, valves, and pipelines to ensure that they meet safety standards and are functioning correctly. Preventive maintenance can identify vulnerabilities before they lead to failures.

**Redundancy**: Implementing redundant systems and backup power sources to ensure operations continue smoothly if primary systems fail.

### 6.4.2 Contingency Planning for Addressing Unforeseen Events or Emergencies in Green Hydrogen Inventory Control

Contingency planning is essential for addressing emergencies or unforeseen events that may disrupt green hydrogen inventory control. This includes preparing for accidents, natural disasters, or unexpected demand surges. Key components of contingency planning are:

**Emergency Response Plans**: Detailed plans outlining immediate actions in case of emergencies like leaks, explosions, or equipment malfunctions. These plans should include procedures for evacuation, firefighting, and containment of hazards.

**Spill and Leak Containment**: Creating containment strategies for hydrogen spills or leaks, including secondary containment systems, pressure relief valves, and vapor recovery systems.

**Alternative Supply Chain Routes**: Identifying alternative routes and methods for transporting hydrogen in case of disruptions in primary transportation channels (e.g., road, rail, and pipeline).

**Spare Capacity**: Maintaining backup storage capacity to handle surges in hydrogen demand or to store excess hydrogen during production outages.

**Collaboration with Local Authorities**: Developing communication and coordination protocols with local emergency services, fire departments, and environmental agencies for coordinated responses during emergencies.

**Testing and Drills**: Regularly testing and conducting drills to ensure that emergency procedures are effective and that employees are prepared to act swiftly in case of an emergency.

### 6.4.3 Regulatory Compliance, Safety Standards and Monitoring System for Risk Management in Green Hydrogen Operations

Green hydrogen operations must comply with various safety standards and regulatory requirements to ensure safe production, storage, and distribution. Key compliance standards include:

**ISO 14687 (Hydrogen Fuel Quality)**: This standard specifies the purity levels and impurities allowed in hydrogen, ensuring that the hydrogen meets quality standards for safety and performance.

**ISO 9001 (Quality Management Systems [QMS])**: A certification that ensures hydrogen production facilities maintain effective QMS to ensure consistency and safety in operations.

**ISO 22734 (Hydrogen Generation by Electrolysis)**: A standard for the safe operation of electrolyzers used in hydrogen production, ensuring safety in hydrogen generation.

**IEC 62282 (Fuel Cell and Hydrogen Technologies)**: Provides technical requirements for hydrogen storage, transport, and use, focusing on minimizing risks associated with hydrogen systems.

**Occupational Safety and Health Administration (OSHA)**: In the United States, OSHA regulations govern worker safety in hydrogen production and handling facilities, requiring proper training, safety protocols, and personal protective equipment.

**Environmental Protection Agency (EPA)**: Ensures that hydrogen production, especially from nonrenewable sources, meets environmental safety standards, particularly regarding emissions and waste disposal.

**Department of Transportation (DOT)**: Regulates the safe transport of hydrogen, ensuring that pipelines, vehicles, and storage systems comply with safety standards.

## 6.5 Total Quality Management (TQM) Principles of Green Hydrogen

As the world shifts toward clean and sustainable energy, green hydrogen has emerged as a promising solution to reduce carbon emissions and support global decarbonization efforts. Produced through electrolysis powered by renewable energy sources, green hydrogen offers a zero-emission alternative for industries such as transportation, power generation, and manufacturing. However, ensuring the efficiency, safety, and reliability of green hydrogen requires the implementation of TQM principles across its entire value chain.

TQM is a systematic approach focused on continuous improvement, customer satisfaction, and operational excellence. In the context of green hydrogen, TQM principles are essential for maintaining high production standards, minimizing inefficiencies, ensuring safety compliance, and optimizing supply chain processes. Given the technical complexities of hydrogen production—such as purity control, energy efficiency, and storage safety—adopting a quality-centric approach is critical for industry-wide success.

Key TQM principles, including customer focus, process optimization, employee involvement, data-driven decision-making, and continuous improvement (Kaizen), play a vital role in ensuring the reliability and sustainability of green hydrogen. Additionally, compliance with international quality and safety standards, such as ISO 9001 for quality management and ISO 14687 for hydrogen purity, ensures that green hydrogen meets global benchmarks for efficiency and performance.

TQM is a comprehensive approach focused on continuous improvement in all organizational processes, with the goal of meeting or exceeding customer expectations. TQM principles emphasize quality at every level of an organization and involve all employees. In the context of green hydrogen, these principles can be applied to improve product quality, ensure sustainability, and enhance customer satisfaction.

**Application of TQM in Green Hydrogen:**

**Process Improvement**: Continuously refining the production, storage, and distribution processes of green hydrogen to ensure efficiency and reduce environmental impact.

**Sustainability**: TQM principles can guide efforts to produce green hydrogen in an environmentally responsible manner, ensuring that processes are sustainable and aligned with renewable energy goals.

**Quality Control**: Establishing robust QMS to ensure the purity and safety of green hydrogen, addressing issues such as contaminants or impurities during production.

### 6.5.1 Leadership Commitment and Employee Involvement to Quality and Continuous Improvement

Leadership plays a crucial role in the successful implementation of TQM. Effective leadership ensures that quality standards are maintained and that continuous improvement becomes a part of the organization's culture.

**Leadership in Green Hydrogen:**

**Vision for Sustainability**: Leaders in green hydrogen companies must have a clear vision for sustainable energy and quality management to foster innovation and competitiveness.

**Commitment to Quality**: Leadership must set clear expectations for quality and allocate resources for ongoing training, technology upgrades, and R&D to drive continuous improvement.

**Encouraging Innovation**: Leaders should foster an environment where innovative solutions for improving green hydrogen production, storage, and utilization are continuously explored.

Employee involvement is central to TQM. Engaged and empowered employees contribute to problem-solving, process improvements, and the development of solutions to meet quality standards.

**Employee Engagement in Green Hydrogen:**

- **Training and Development**: Employees should be trained in the latest green hydrogen technologies, sustainable practices, and quality standards to empower them to contribute effectively.
- **Problem-Solving**: Encouraging employee participation in identifying inefficiencies or areas for improvement within the hydrogen production process.
- **Collaboration and Teamwork**: Cross-functional teams within organizations can help solve complex problems related to green hydrogen production, where employees bring diverse perspectives and expertise.

In TQM, customer satisfaction is the ultimate measure of quality. Meeting customer expectations ensures long-term success and a competitive edge.

**Customer Focus in Green Hydrogen:**

- **Understanding Customer Needs**: Continuous research and feedback collection to understand customer needs, ensuring the production of high-quality green hydrogen that meets or exceeds expectations.
- **Transparency and Accountability**: Green hydrogen producers must communicate clearly with customers about the benefits, quality, and environmental advantages of their products.
- **Customer Education**: Promoting awareness of the role of green hydrogen in sustainable energy solutions and addressing any concerns related to safety, pricing, or availability.

Data is a crucial element in driving improvement and ensuring quality in TQM. Data-driven decision-making helps organizations monitor performance, identify trends, and make informed decisions for continuous improvement.

**Data-Driven Decisions in Green Hydrogen:**

- **Performance Metrics**: Establishing KPIs to assess the quality of the green hydrogen production process, including yield, energy efficiency, emissions, and cost.
- **Continuous Monitoring**: Using real-time data from sensors and automation systems to monitor production processes, identify potential issues, and make immediate corrective actions.
- **Predictive Analytics**: Leveraging advanced analytics and AI to predict and mitigate potential issues in the production process, ensuring higher-quality outputs and reducing waste.

TQM in green hydrogen not only ensures high-quality products but also contributes to the broader goal of creating a sustainable and efficient hydrogen economy. By focusing on quality, leadership commitment, employee empowerment, customer satisfaction, and data-driven decision-making, organizations can establish a strong foundation for success in this rapidly growing sector.

## 6.6 Quality Tools and Techniques of Green Hydrogen

As the global demand for sustainable energy solutions rises, green hydrogen has emerged as a clean and renewable alternative to fossil fuels. Produced through electrolysis powered by renewable energy sources, green hydrogen is a key enabler of the energy transition, supporting sectors such as transportation, power generation, and industrial processes. However, ensuring its purity, efficiency, and safety requires the implementation of advanced quality tools and techniques across its production, storage, and distribution processes. Quality management in green hydrogen is essential to address challenges such as hydrogen purity control, leakage prevention, efficient storage, and process optimization. Various quality tools and techniques, including Six Sigma, FMEA, Statistical Process Control (SPC), and TQM, play a crucial role in maintaining high standards. Additionally, real-time monitoring, AI-driven predictive maintenance, and blockchain-based traceability are being integrated to enhance process reliability and regulatory compliance. Adherence to global standards such as ISO 14687 for hydrogen purity, ISO 9001 for quality management, and ISO 19880 for hydrogen fueling infrastructure ensures that green hydrogen meets industry benchmarks for performance and safety. Implementing these quality tools and techniques not only improves operational efficiency but also builds trust in green hydrogen as a viable and scalable clean energy source.

SPC is a powerful quality tool used to monitor and control production processes. It involves using statistical methods to track and analyze process variations, ensuring that the process remains stable and within set quality limits.

**SPC in Green Hydrogen:**

**Process Monitoring**: SPC can be used to monitor key parameters in green hydrogen production such as pressure, temperature, and purity levels of the hydrogen produced. By collecting data on these variables, organizations can ensure that the process stays within acceptable limits and deviations are detected early.

**Control Charts**: Implementing control charts to track the variability of hydrogen production over time, identifying out-of-control points that may indicate underlying problems in the system.

**Real-Time Data**: SPC techniques, such as real-time data collection and analysis, allow operators to make immediate adjustments to the process, reducing defects and ensuring consistent product quality.

### 6.6.1 Failure Mode and Effects Analysis (FMEA) and Root Cause Analysis for Identifying and Addressing Potential Quality Issues

FMEA is a systematic approach to identifying and addressing potential failures in a process, product, or system. It involves evaluating the impact of different failure modes and prioritizing them based on their potential severity.

**FMEA in Green Hydrogen:**

**Identifying Risks**: FMEA helps in identifying potential failure modes in green hydrogen production, such as equipment malfunctions, contamination, or improper storage conditions that could affect product quality or safety.

**Impact Assessment**: By analyzing the potential effects of failures, FMEA allows organizations to prioritize risks and address the most critical ones, ensuring that corrective actions are taken before they lead to significant quality problems.

**Mitigation Strategies**: Once risks are identified, FMEA helps develop mitigation strategies, such as routine maintenance, quality control checks, or the implementation of more robust systems to reduce the likelihood of failures.

RCA is a technique used to identify the underlying causes of quality-related problems, rather than just addressing symptoms. It aims to resolve issues permanently by eliminating their root causes.

**RCA in Green Hydrogen:**

**Investigating Process Failures**: When quality problems arise in the production of green hydrogen, RCA techniques can be applied to investigate the process and determine the root cause of issues such as impurities, efficiency losses, or failures in storage and transportation.

**5 Whys Technique:** A common RCA tool, the 5 Whys method, can be used to ask "why" multiple times to drill down to the underlying causes of a problem. For example, if hydrogen production is not meeting purity standards, a series of "why" questions can help uncover process or equipment issues.

**Corrective Actions**: Once the root cause is identified, the organization can implement corrective actions to eliminate the problem and prevent recurrence, improving the overall quality of the green hydrogen product.

### 6.6.2 Lean and Six Sigma Methodologies for Improving Efficiency and Reducing Waste

Lean and Six Sigma are methodologies that focus on eliminating waste, improving efficiency, and reducing defects in processes. Together, these approaches help optimize operations and ensure high-quality products.

**Lean and Six Sigma in Green Hydrogen:**

**Lean Principles**: Lean techniques, such as value stream mapping and continuous flow, can be applied to identify and eliminate waste in the green hydrogen production process. This could involve streamlining energy consumption, reducing idle times, or minimizing unnecessary steps in the production cycle.

**Six Sigma**: Six Sigma focuses on reducing process variation and improving consistency. In green hydrogen production, Six Sigma can be applied to minimize variations in production rates, energy efficiency, and hydrogen purity, ensuring a higher level of consistency and quality in the final product.

**DMAIC (Define, Measure, Analyze, Improve, Control) Process**: The DMAIC methodology can be used to drive continuous improvement in green hydrogen production, focusing on improving process efficiency, reducing waste, and ensuring high-quality standards.

## 6.7 Integration of Quality Management Systems of Green Hydrogen

With the rising global demand for clean energy solutions, green hydrogen has emerged as a key driver of the energy transition. As a zero-emission fuel produced through electrolysis powered by renewable energy, green hydrogen has vast applications in transportation, power generation, and industrial sectors. However, ensuring its efficiency, safety, and sustainability requires a structured and integrated approach to QMS. The integration of QMS in green hydrogen production and supply chains is essential to maintain high purity levels, optimize efficiency, reduce operational risks, and ensure regulatory compliance. By aligning with international quality and safety standards such as ISO 9001 for quality management, ISO 14687 for hydrogen purity, and ISO 14001 for environmental management, industries can enhance the reliability and scalability of green hydrogen technologies.

A QMS is a formalized system that documents processes, procedures, and responsibilities for achieving quality objectives. Integration of a QMS in green hydrogen production involves embedding quality management principles into all aspects of the hydrogen value chain, from production to delivery. This integration is essential for maintaining product consistency, minimizing waste, and meeting environmental sustainability goals.

**Key Elements of QMS Integration:**

**Process Standardization**: One of the first steps in integrating a QMS is standardizing all production processes, which includes the design and development of systems for the production, purification, and storage of hydrogen. By developing standardized procedures, the green hydrogen production process becomes predictable, ensuring consistent product quality.

**Documentation of Procedures**: Proper documentation forms the backbone of any QMS. In green hydrogen production, this involves detailing every step in the process, including safety protocols, quality control measures, and maintenance schedules. By ensuring that procedures are well-documented, organizations can maintain transparency and ensure that all team members follow consistent processes.

**Continuous Improvement Framework**: A QMS should support a continuous improvement mindset. For green hydrogen, this means regularly revisiting and revising processes, introducing new technologies, and incorporating feedback from stakeholders, such as customers and regulators, to enhance the overall quality of the product.

**Risk Management**: In green hydrogen production, several risks—such as equipment failure, contamination, and process inefficiencies—can impact the quality of the hydrogen. A well-integrated QMS involves identifying potential risks at each stage of production and implementing mitigation strategies, such as regular maintenance, contingency planning, and quality control checks.

**Resource Allocation**: Integration of QMS also involves ensuring that the necessary resources—such as skilled personnel, equipment, and technology—are in place to support the QMS. This includes the allocation of resources for regular training and investing in new technologies that can improve the production process.

**Supplier QA**: Suppliers of raw materials or equipment play an essential role in the quality of green hydrogen. Integrating QMS involves establishing quality expectations and monitoring supplier performance to ensure that materials and equipment meet the required standards.

One of the fundamental aspects of a QMS is its alignment with external regulatory requirements and industry standards. For green hydrogen, this includes compliance with environmental, safety, and product quality regulations, as well as adherence to industry-specific standards.

**Regulatory Requirements:**

Green hydrogen production is subject to various national and international regulations, especially regarding safety, environmental impact, and emissions reduction. In addition to adhering to general industrial standards, green hydrogen producers must comply with specific regulations concerning energy use, carbon footprints, and the environmental impact of hydrogen production processes.

Examples of key regulations and standards include:

1) **ISO 9001**: ISO 9001 is a widely recognized standard for quality management. Adhering to ISO 9001 ensures that green hydrogen producers establish effective quality systems, meet customer expectations, and adhere to legal requirements.
2) **ISO 14001**: This standard focuses on environmental management systems, which are crucial for green hydrogen producers aiming to reduce environmental impact and ensure sustainability in hydrogen production.
3) **Hydrogen Safety Standards (e.g., ISO 14687-2)**: These standards define the specifications for hydrogen purity and safety. In line with these standards, green hydrogen producers must implement strict controls over hydrogen storage and transportation to avoid contamination or hazardous situations.
4) **Local Regulations on Renewable Energy**: Many countries have specific regulations for renewable energy projects, including hydrogen production. These regulations often include specific energy efficiency targets, carbon reduction goals, and sustainability measures, which green hydrogen producers must integrate into their QMS.

**Industry Standards**

Industry standards provide a framework for producing green hydrogen at the desired quality level. For example, the **IRENA** and **Hydrogen Council** set global standards for hydrogen production, emphasizing the use of renewable energy, technological innovation, and sustainability. By aligning their QMS with these standards, green hydrogen producers can ensure that their products meet the expectations of both consumers and regulators.

Successful implementation of a QMS in green hydrogen production requires collaboration across various functions within the organization. Cross-functional collaboration and communication are critical to ensure that quality management principles are consistently applied at every stage of production, from R&D to marketing and customer service.

**Key Aspects of Cross-Functional Collaboration:**

- **Research and Development (R&D)**: R&D teams play an essential role in improving production processes, introducing new technologies, and refining hydrogen quality. Cross-functional collaboration between R&D, production, and QA teams ensures that innovations are implemented effectively and that quality standards are maintained.
- **Production and Operations**: The operations team is responsible for day-to-day production activities, and they must be in constant communication with QA and maintenance teams to identify any potential issues early. This collaboration helps ensure that processes stay within the established parameters, preventing quality deviations.
- **QA and Control**: The QA team must work closely with every other department to verify that the processes and products meet quality standards. Collaboration with production, R&D, and supply chain teams helps the QA team identify potential areas for improvement and enforce corrective actions when necessary.
- **Supply Chain Management**: Effective communication between the supply chain and production teams is necessary to ensure that all materials and components meet quality standards. Suppliers must be part of the QMS integration, with regular audits and performance assessments.
- **Customer Service and Feedback**: Quality is not only about internal processes; it's also about meeting customer expectations. Customer service teams must provide feedback to the production and quality control teams regarding product performance, reliability, and customer satisfaction, creating a feedback loop for continuous improvement.
- **Leadership and Management**: Cross-functional collaboration must also include leadership. Management must foster a culture of quality, encourage interdepartmental communication, and allocate resources to support QMS implementation. Leadership commitment to quality ensures alignment of the entire organization toward achieving quality objectives.

### 6.7.1 Training and Development Programs to Build a Culture of Quality and Continuous Improvement

Training and development are vital components of any effective QMS. In the green hydrogen sector, where technology and regulatory requirements evolve rapidly, it is essential to ensure that employees are well-equipped with the necessary skills and knowledge to maintain high-quality standards.

**Key Elements of Training and Development:**

1) **Ongoing Training**: Employees at all levels, from operators to senior managers, should receive regular training on quality management principles, regulatory compliance, new technologies in hydrogen production, and the company's quality policies.
2) **Skills Development**: Training should not be limited to generic quality management practices; it must also focus on specific skills required for green hydrogen production. This includes training on hydrogen production processes, safety standards, environmental regulations, and data analysis for monitoring performance.
3) **Knowledge Sharing**: Encouraging knowledge sharing across departments can foster a culture of continuous improvement. Workshops, seminars, and cross-departmental meetings can be used to facilitate the exchange of ideas and best practices.
4) **Quality Circles**: Establishing quality circles—small groups of employees from different departments working together to solve quality-related problems—can help create a sense of ownership and encourage employee involvement in continuous improvement.
5) **Leadership Development**: Management must be trained to lead quality initiatives, make data-driven decisions, and foster a culture of quality. They should be equipped with the tools to guide teams through the implementation of QMS, resolving challenges, and ensuring ongoing improvement.
6) **Certifications**: Offering employees opportunities to obtain industry-recognized certifications in QMS, such as Six Sigma or ISO 9001, can further enhance their knowledge and commitment to quality.

Continuous monitoring and periodic review are essential for ensuring that the QMS remains effective and continues to meet organizational objectives. By monitoring QMS performance, green hydrogen producers can identify inefficiencies, address emerging issues, and make data-driven decisions to improve overall performance.

**Key Aspects of Monitoring and Review:**

1) **Performance Metrics**: Establishing KPIs is critical for measuring the success of the QMS. These could include production efficiency, product quality, customer satisfaction, compliance with regulatory standards, and environmental impact.
2) **Audits and Inspections**: Regular internal and external audits ensure that the QMS is functioning as intended and identify areas for improvement. External

audits are especially important for verifying compliance with regulatory requirements and industry standards.

3) **Feedback Loops**: Collecting feedback from customers, suppliers, and employees is essential for reviewing the effectiveness of the QMS. This feedback should be analyzed and used to make necessary adjustments.
4) **Management Reviews**: Senior management must conduct periodic reviews of the QMS to evaluate its effectiveness, ensure it aligns with organizational goals, and identify areas for improvement. These reviews should involve cross-functional teams to provide a comprehensive evaluation of the system.
5) **Corrective and Preventive Actions (CAPAs)**: A key feature of QMS is the implementation of CAPAs. If a quality issue is identified, it is essential to investigate its root cause, implement corrective actions, and take preventive measures to avoid recurrence.

## 6.8 Conclusion

The chapter on Inventory Control and TQM presents a comprehensive overview of modern practices essential for the efficient handling and QA of Green Hydrogen, a cornerstone in the global transition toward clean energy.

We began by exploring the fundamentals of inventory management in the context of green hydrogen, emphasizing its critical role in balancing production with demand while ensuring safety and sustainability. As green hydrogen infrastructure continues to evolve, inventory planning and forecasting emerge as vital tools, helping organizations to anticipate supply chain fluctuations, optimize production schedules, and reduce operational costs.

Effective inventory control strategies were then discussed, focusing on minimizing waste and optimizing storage capacity. These strategies are particularly important given the volatile and highly flammable nature of hydrogen, demanding strict safety and efficiency protocols. We also examined JIT inventory approaches, which reduce holding costs and improve responsiveness in the distribution of green hydrogen, aligning supply with real-time demand while supporting lean operational models.

Advanced inventory tracking and monitoring systems for hydrogen storage facilities were reviewed, highlighting how digital technologies—such as IoT, sensors, and AI—enhance visibility, accuracy, and safety across the supply chain. These systems ensure traceability and support compliance with regulatory standards.

To uphold product integrity, robust quality control measures, including testing and inspection, are indispensable. These processes not only guarantee the purity and usability of hydrogen but also foster trust and reliability in end-use applications such as fuel cells and industrial processes. Furthermore, the chapter addressed risk management in inventory control, identifying potential hazards and proposing mitigation strategies. This includes contingency planning, safety stock optimization, and emergency response protocols critical for managing uncertainties in supply and demand.

Finally, the chapter integrated the principles of TQM and their practical application across the green hydrogen inventory lifecycle. Emphasizing customer-centricity, continuous improvement, and cross-functional collaboration, TQM frameworks support the establishment of a quality-first culture, ensuring that both inventory practices and product outputs meet or exceed industry benchmarks.

In summary, the convergence of inventory control techniques and TQM principles not only enhances the operational efficiency and safety of green hydrogen systems but also contributes significantly to their long-term sustainability and success. As green hydrogen emerges as a transformative energy solution, embracing innovative inventory and quality management practices will be key to building a resilient and future-ready hydrogen economy.

## 6.9 Case Study

**Case Study: Total Quality Management of the Green Hydrogen Production at HyGreen Energy Pvt. Ltd.**

**Background**

HyGreen Energy Pvt. Ltd., an Indian startup focused on renewable energy innovation, began green hydrogen production in 2022 with a vision to support the country's transition to clean energy. With increasing pressure to deliver high-purity hydrogen while minimizing environmental and operational risks, the company adopted a **TQM** approach across its production chain to ensure continuous improvement, customer satisfaction, and regulatory compliance.

**Challenges Faced**

Before implementing TQM, the company faced several operational challenges:

- **Inconsistent hydrogen purity levels**, leading to client dissatisfaction.
- **Frequent downtime in electrolyzers**, impacting production efficiency.
- **Inefficient inventory management**, resulting in waste and high storage costs.
- **Lack of standardized procedures**, causing variability in production outputs.
- **Safety concerns**, particularly during storage and transportation.

**TQM Implementation Strategy**

HyGreen Energy Pvt. Ltd. deployed TQM principles with a clear roadmap focusing on quality across people, processes, and performance metrics.

1) **Customer Focus**
   - Regular feedback loops were established with industrial clients to understand specific purity and delivery requirements.
   - Customer satisfaction surveys were integrated into the feedback system to align process improvements with client expectations.
2) **Continuous Improvement (Kaizen)**
   - Small, incremental changes were implemented in the production process.
   - Operators and engineers participated in regular brainstorming sessions to identify bottlenecks and suggest improvements.
3) **Employee Involvement and Training**
   - Employees were trained in Six Sigma methodologies and hydrogen-specific safety and quality standards (like ISO 14687).
   - Quality circles were formed to monitor daily tasks and report issues in real-time.
4) **Process Approach**
   - SOPs were documented for each stage—from water purification, electrolysis, compression, to storage.
   - Real-time process monitoring through SCADA systems was introduced to ensure consistency.
5) **Fact-Based Decision-Making**
   - A data-driven approach was used to track KPIs such as hydrogen purity, defect rate, energy consumption per kg of $H_2$, and on-time delivery rate.
   - RCA tools like Fishbone Diagrams and 5 Whys were employed to resolve recurring quality issues.
6) **Risk Management Integration**
   - A dedicated team conducted risk assessments (FMEA) for critical equipment like electrolyzers, compressors, and storage tanks.
   - Emergency protocols were updated, and drills were conducted monthly.
7) **Supplier Quality Management**
   - Supplier audits were performed for key inputs like deionized water systems, renewable electricity providers, and storage cylinders.
   - Supplier scorecards helped track compliance with quality and sustainability metrics.

**Results and Outcomes**

Within a year of implementing TQM, the company observed:

- **Purity levels improved** from 96.5% to 99.999% (fuel cell grade).
- **Operational efficiency** increased by 22% due to reduction in downtimes.

- **Customer satisfaction** scores rose by 35% in annual surveys.
- **Inventory holding costs** decreased by 18% through lean storage strategies.
- **Waste generation** was reduced by 30% through better control and reuse mechanisms.
- **Zero safety incidents** were reported for six consecutive quarters.

The case of HyGreen Energy Pvt. Ltd. highlights how TQM can be a powerful tool in establishing green hydrogen production as a reliable, sustainable, and scalable industry. By embedding quality in every step—from raw material to customer delivery—the company not only achieved operational excellence but also gained competitive advantage in an emerging and high-stakes sector. This case illustrates that in the context of green energy, quality is not just about compliance—it is a commitment to innovation, safety, and sustainability.

## 6.10 Exercise

1. Explain the importance of inventory planning and forecasting in green hydrogen production. How does it impact supply chain efficiency?

2. What are the key inventory control strategies to minimize waste and optimize storage capacity in green hydrogen facilities?

3. Describe the just-in-time (JIT) inventory management approach and its benefits in green hydrogen distribution.

4. How can inventory tracking and monitoring systems improve the safety and efficiency of green hydrogen storage?

5. Discuss the role of integrated inventory management systems in optimizing supply chain operations for green hydrogen.

6. What are the key quality standards and specifications that must be met in green hydrogen production?

7. Describe common testing and inspection protocols used to ensure the purity and safety of green hydrogen.

8. What quality assurance measures should be implemented throughout the hydrogen production process?

9 Explain the importance of certification and compliance requirements in green hydrogen quality control.

10 How can continuous improvement initiatives enhance the quality of green hydrogen production and distribution?

11 Identify and assess three major risks associated with green hydrogen inventory management.

12 What are some risk mitigation strategies to minimize potential hazards in green hydrogen storage and handling?

13 Explain the role of contingency planning in addressing unforeseen emergencies in green hydrogen inventory control.

14 Discuss key regulatory compliance and safety standards for risk management in green hydrogen operations.

15 How can monitoring and reporting mechanisms help track and manage risks in green hydrogen inventory control?

## 6.11 Objective Type Questions

1 **Which of the following is a primary goal of inventory planning in green hydrogen production?**
 a) Increase fossil fuel reserves
 b) Reduce storage capacity
 c) Balance supply with production demand
 d) Increase carbon emissions

2 **What is the key benefit of just-in-time (JIT) inventory management for green hydrogen?**
 a) Overproduction
 b) Reduced handling costs and storage waste
 c) High inventory holding costs
 d) Greater lead times

3 **Which tool is most suitable for tracking green hydrogen storage levels in real time?**
 a) Manual inventory logs
 b) Digital inventory management system

c) Paper-based ledgers
d) None of the above

4 **An integrated inventory system ensures:**
a) Disconnection from supply chains
b) Fragmented logistics
c) Seamless coordination between production, storage, and distribution
d) Isolation of operational units

5 **Which of the following is critical for ensuring green hydrogen safety and usability?**
a) Purity testing
b) Color labeling
c) Cost estimation
d) Packaging design

6 **What do certification and compliance standards for green hydrogen ensure?**
a) Marketing strategy
b) International trade tariffs
c) Quality, safety, and reliability
d) Low production costs

7 **Inspection protocols in green hydrogen production are primarily used to:**
a) Enhance packaging
b) Ensure consistent purity and safety
c) Maximize profits
d) Reduce water usage

8 **Continuous improvement initiatives help in:**
a) Reducing product quality
b) Increasing defects
c) Enhancing production processes over time
d) Avoiding compliance

9 **Which of the following is a key risk in storing green hydrogen?**
a) High taxation
b) Flammability and leakage
c) Excess solar energy
d) Underemployment

**10 Contingency planning in inventory control focuses on:**
- **a)** Increasing order frequency
- **b)** Responding effectively to unexpected disruptions
- **c)** Removing automation systems
- **d)** Maximizing downtime

**11 Which of the following is a component of effective risk mitigation?**
- **a)** Ignoring safety rules
- **b)** Regular safety training and system checks
- **c)** Manual error tracking
- **d)** Disorganized storage

**12 What is the purpose of monitoring and reporting mechanisms in inventory risk control?**
- **a)** Increase legal risk
- **b)** Track production profit
- **c)** Identify and address hazards promptly
- **d)** Delay decision-making

**13 Which principle is central to total quality management?**
- **a)** Top-down communication only
- **b)** Employee exclusion
- **c)** Continuous improvement and customer focus
- **d)** Isolated departmental efforts

**14 In TQM, leadership commitment primarily means:**
- **a)** Cutting training budgets
- **b)** Passive support
- **c)** Driving and modeling quality initiatives
- **d)** Ignoring feedback

**15 Why is customer satisfaction important in green hydrogen quality management?**
- **a)** It determines fuel pricing.
- **b)** It ensures quality meets user expectations.
- **c)** It increases carbon content.
- **d)** It supports marketing campaigns.

**16 Failure mode and effects analysis (FMEA) is used to:**
- **a)** Increase variation
- **b)** Eliminate root causes

c) Predict and prioritize potential failures
d) Reduce inspection costs

**17 Root cause analysis (RCA) focuses on:**
a) Temporary fixes
b) Identifying symptoms
c) Identifying the fundamental source of problems
d) Avoiding accountability

**18 Which of the following is a lean methodology aim?**
a) Increase waste
b) Delay improvements
c) Enhance efficiency by eliminating waste
d) Remove automation

**19 A Quality Management System (QMS) is effective when it is:**
a) Isolated from operations
b) Focused solely on marketing
c) Integrated across functions and aligned with industry standards
d) Disconnected from employee training

**20 Why is training essential in quality management systems?**
a) To increase production time
b) To build a culture of quality and empower employees
c) To avoid accountability
d) To reduce communication

## 6.12 Glossary

1) **Inventory Control**: A systematic approach to managing the ordering, storage, and utilization of materials to minimize waste, reduce costs, and ensure availability for hydrogen production and distribution.
2) **Inventory Management**: The overall process of monitoring and controlling stock levels, storage conditions, and movement of green hydrogen and related resources to balance demand with supply efficiently.
3) **Inventory Planning**: The strategic process of determining the right quantities of green hydrogen and related inputs to maintain optimal supply without excessive storage or shortages.

4) **Forecasting Models**: Analytical methods used to predict demand and supply variations in green hydrogen based on production cycles, energy usage patterns, and market fluctuations.
5) **Advanced Inventory Control Strategies**: Innovative methods applied to manage hydrogen inventory efficiently, focusing on waste reduction, energy-saving storage techniques, and sustainability goals.
6) **Waste Minimization**: Processes aimed at reducing excess hydrogen loss, leakage, or spoilage through better inventory practices and sustainable management.
7) **Energy-Efficient Storage**: Storage methods that reduce energy consumption while maintaining the required conditions for safe and effective green hydrogen storage.
8) **Environmental Impact**: The effect of inventory practices on the environment, including emissions, resource consumption, and ecological footprint in green hydrogen operations.
9) **Just-in-Time (JIT) Inventory Management**: An inventory approach where hydrogen and resources are procured and delivered only when needed, reducing excess stock, minimizing costs, and optimizing resources.
10) **Inventory Tracking Systems**: Technologies used to monitor hydrogen levels, movement, and storage conditions in real-time, ensuring accuracy, efficiency, and visibility across facilities.
11) **Monitoring Systems**: Digital or automated systems that track hydrogen quality, safety, and compliance, supporting proactive decision-making in inventory control.
12) **Traceability**: The ability to track hydrogen from production to distribution, ensuring accountability, safety, and compliance with industry standards.
13) **Compliance**: Adhering to regulatory, environmental, and safety standards in handling, storing, and distributing green hydrogen.
14) **Quality Control Measures**: Systematic processes such as inspections and audits designed to ensure hydrogen meets required purity, safety, and performance standards.
15) **Testing and Inspection**: Rigorous evaluation of hydrogen quality, pressure levels, and storage safety to ensure integrity and prevent risks in distribution.
16) **Risk Management**: Frameworks and practices for identifying, assessing, and mitigating uncertainties such as supply disruptions, storage failures, or demand fluctuations in hydrogen inventory systems.
17) **Failure Mode**: A way in which a process or system could fail, used in quality management to identify weaknesses and prevent potential risks in hydrogen storage and supply.
18) **Six Sigma**: A data-driven methodology for improving processes, reducing defects, and enhancing quality control in hydrogen inventory and distribution.

19) **Total Quality Management (TQM)**: A holistic management approach focused on continuous improvement, customer/stakeholder satisfaction, and achieving operational excellence across hydrogen supply chains.
20) **Continuous Improvement**: An ongoing effort to improve processes, quality, and efficiency in hydrogen inventory systems to achieve long-term sustainability.
21) **Stakeholder Satisfaction**: Meeting the expectations and needs of all parties involved in the hydrogen supply chain, including producers, distributors, regulators, and consumers.
22) **Operational Excellence**: Achieving superior performance in hydrogen production, storage, and distribution through efficiency, safety, and quality-focused practices.
23) **Sustainable Hydrogen Supply Chains**: Inventory and quality management systems designed to balance environmental responsibility, economic viability, and energy efficiency in green hydrogen distribution.

## References


1 Singla, M. K., Gupta, J., Safaraliev, M., Nijhawan, P., & Oberoi, A. S. (2024). Hydrogen storage in activated carbon for fuel cell-powered vehicles: a cost-effective and sustainable approach. *International Journal of Hydrogen Energy*, 58, 446–458 https://doi.org/10.1016/J.IJHYDENE.2024.01.117.

2 Song, B., Wang, X., Kang, Y., & Li, H. (2024). Research on hydrogen leakage diffusion and safety analysis in hydrogen fuel cell vehicles with regard to leakage location and ventilation ports. *International Journal of Hydrogen Energy*, 83, 173–187 https://doi.org/10.1016/J.IJHYDENE.2024.08.105.

3 Sun, L., Jia, J., Wang, Q. L., & Zhang, Y. (2024). A novel multiphase DC/DC boost converter for interaction of solar energy and hydrogen fuel cell in hybrid electric vehicles. *Renewable Energy*, 229, 120779 https://doi.org/10.1016/J.RENENE.2024.120779.

4 Tuncer, D., & Yilmaz Ulu, E. (2024). Contribution of regenerative suspension module to charge efficiency and range in hydrogen fuel cell electric vehicles. *International Journal of Hydrogen Energy*, 75, 547–556 https://doi.org/10.1016/J.IJHYDENE.2024.03.219.

5 Vilberger, M. E., Popov, N. S., Domakhin, E. A., Anibroev, V. I., & Mosin, M. E. (2024). Increasing the energy efficiency of an electric vehicle powered by hydrogen fuel cells. *International Journal of Hydrogen Energy*, 85, 406–415 https://doi.org/10.1016/J.IJHYDENE.2024.08.099.

6 Wang, G., Wang, J., Hu, C., Lv, W., & Mao, H. (2024). Simulation of hydrogen leakage and diffusion in an enclosed garage of mixed parking of fuel-cell vehicles. *Next Research*, 1, 100073 https://doi.org/10.1016/J.NEXRES.2024.100073.

7 Wang, Q., Wang, Y., & Chen, Z. (2025). Stackelberg-equilibrium-based collaborative charging management strategy for renewable fuel vehicles in regional integrated electricity-hydrogen system. *Applied Energy*, 377, 124617 https://doi.org/10.1016/J.APENERGY.2024.124617.

8 Wu, Y., Gu, W., Huang, S., Wei, X., & Safaraliev, M. (2025). A four-layer business model for integration of electric vehicle charging stations and hydrogen fuelling stations into modern power systems. *Applied Energy*, 377, 124630 https://doi.org/10.1016/J.APENERGY.2024.124630.

9 Xia, W., Ren, Z., Li, H., & Pan, Z. (2024). A data-driven probabilistic evaluation method of hydrogen fuel cell vehicles hosting capacity for integrated hydrogen-electricity network. *Applied Energy*, 376, 123895 https://doi.org/10.1016/J.APENERGY.2024.123895.

10 Ye, T., Zhao, S., Lau, C. K. M., & Chau, F. (2024). Social media sentiment of hydrogen fuel cell vehicles in China: evidence from artificial intelligence algorithms. *Energy Economics*, 133, 107564 https://doi.org/10.1016/J.ENECO.2024.107564.

11 Zhang, Q., Chen, J., & Ihara, T. (2024). Assessing regional variations in hydrogen fuel cell vehicle adoption: an integrative approach using real-world data and analytic hierarchy process in Tokyo. *Applied Energy*, 363, 123014 https://doi.org/10.1016/J.APENERGY.2024.123014.

12 V. Khare, C.J. Khare, M.A. Bhuiyan, "Prediction of tidal energy resources at Gulf of Kutch and Gulf of Khambhat by data mining," *Cleaner Energy Systems*, 2023, 6, 100090.

13 V. Khare, C.J. Khare, M.A. Bhuiyan, "Design, optimization, and data analysis of solar-tidal hybrid renewable energy system for Hurawalhi, Maldives," *Cleaner Energy Systems*, 6, 100088, 2023.

14 V. Khare, A. Jain, M.A. Bhuiyan, "Assessment of hydro energy potential from rain fall data set in India through data analysis," *e-Prime – Advances in Electrical Engineering, Electronics and Energy*, 6, 100290, September 2023.

15 Khare, V., & Chaturvedi, P. (2023). Design, control, reliability, economic and energy management of microgrid: a review. In *Advances in Electrical Engineering, Electronics and Energy* (Vol. 5, 100239).

16 V. Khare, M.A. Bhuiyan, "Perspective of renewable energy in the BRICS country," *Advances in Electrical Engineering, Electronics and Energy*, 5, 2023, 100250.

17 Khare, V., & Chaturvedi, P. (2023). Solar energy system concept change from trending technology: A comprehensive review. In *Advances in Electrical Engineering, Electronics and Energy* (Vol. 4, 100183).

18 V. Khare, P. Chaturvedi, Krishnkant, M. Mishra, "Design and operational management of sustainable multi-networks green energy system for study area at India," *Materials Today Proceedings*, 2023. https://doi.org/10.1016/j.matpr.2023.05.575

19 Zhou, J., Weng, S., Phuang, Z. X., Tan, J. P., Farooque, A. A., Wong, K. Y., & Woon, K. S. (2024). Exploring intra-regional hydrogen production alternatives for fuel cell vehicles via greenhouse gas-based life cycle supply chain (GHG-LCSC) optimization. *International Journal of Hydrogen Energy*, 81, 1322–1337 https://doi.org/10.1016/J.IJHYDENE.2024.07.304.

# Index



*Hydrogen Technologies in Decarbonizing Industries and Energy Systems*, First Edition.
Vikas Khare, Sanjeet K. Dwivedi, Monica Bhatia, and K. K. Pant.

Published 2026 by John Wiley & Sons, Inc.

## *f*

## t



## u



## v

Printed and bound by CPI Group (UK) Ltd, Croydon, CR0 4YY

17/09/2026

14964234-0007